Sustaining Seas

Sustaining Seas

Oceanic Space
and the Politics of Care

Edited by Elspeth Probyn,
Kate Johnston, and Nancy Lee

ROWMAN &
LITTLEFIELD
——— INTERNATIONAL ———
London • New York

Published by Rowman & Littlefield International Ltd.
6 Tinworth Street, London, SE11 5AL, United Kingdom
www.rowmaninternational.com

Rowman & Littlefield International Ltd. is an affiliate of Rowman & Littlefield
4501 Forbes Boulevard, Suite 200, Lanham, Maryland 20706, USA

With additional offices in Boulder, New York, Toronto (Canada), and Plymouth (UK)
www.rowman.com

British Library Cataloguing in Publication Data

A catalogue record for this book is available from the British Library
ISBN: HB 978-1-7866-1283-0
PB 978-1-7866-1387-5

Library of Congress Cataloging-in-Publication Data Is Available

ISBN 978-1-78661-283-0 (cloth: alk. paper)
ISBN 978-1-78661-387-5 (pbk: alk. paper)
ISBN 978-1-78661-284-7 (electronic)

∞™ The paper used in this publication meets the minimum requirements of
American National Standard for Information Sciences—Permanence of Paper
for Printed Library Materials, ANSI/NISO Z39.48-1992.

Contents

Acknowledgments

The editors acknowledge the assistance of School of the Philosophical and Historical Inquiry and the Sydney Environment Institute at the University of Sydney in supporting the conference that led to this edited book. They also thank Gurdeep Mattu at Rowman & Littlefield International for taking on this project.

Elspeth acknowledges Australian Research Council Discovery Project funding for the Sustainable Fish Lab. Elspeth would also like to thank her co-editors, Kate and Nancy, for their enthusiasm and hard work in putting this volume together.

Kate and Nancy would like to thank their co-editor Elspeth for her support and guidance throughout the editorial process.

Sustaining Seas

An Introduction

The conference from which this book rises was titled *Sustaining the Seas*. It was a wonderful, generous, and generative gathering of scholars, activists, artists, poets, practitioners, and Indigenous custodians of sea country. But, as we came to think about this book, we stumbled over the "the" in the title. Increasingly, we felt that the definite article that bridges *Sustaining* and *Seas* seems to act to foreground human agency. Who or what can sustain the seas? We wondered: surely it is hubris to even obliquely posit that as a species we could sustain the seas when all about is the shocking evidence that we simply have not. The image of the cover eloquently attests to this. The plastic in the sea overwhelms our title. Plastic may be the most obvious point of public debate at the moment about the lack of care for the ocean, but, as our authors reveal, there are a multitude of other issues that crowd within oceanic spaces. They also suggest ways of seeing, of probing, of using "fingeryeyes" (Hayward 2010) to understand different dimensions of the oceanic and the aquatic.

In a recent article, Paul Gilroy critiques "the high-altitude theorising" that has become a calling card of cultural thinking in the Anthropocene. Against this, he turns our attention to a "lowly watery orientation" (Gilroy 2018, 10). From this position we are able to grasp the oceanic in its entangled roles as "the ancient epiphany of human being" as it forms "intersecting planetary webs of trade, information, and accumulation" (2018, 9). This then conjoins the sublime of the seas that brings forth poetry, art, and philosophy, and the pragmatics (and the effects) of humans living with and on the seas. As Stacy Alaimo puts it, "the sea surges through the bodies of all terrestrial animals, including humans—in our blood, skeletons, and cellular protoplasm" (Alaimo 2016, 118). But the seas also carry and contain the historical, cultural, and economic materiality at the very base of human life.

In a modest way, *Sustaining Seas* engages with the different pulls of the oceanic. Our collection foregrounds conversations across, and within, several disciplines including practitioners of different specialties (artists, writers, planners, policy makers) about how to sustain the seas, as they sustain us. As a whole, our authors aim to build a better understanding of what it means to care for aquatic places and their biocultural communities. We hope to provide readers with new conceptual framings, as well as grounded case studies with a wide geographical and cultural breadth that might inspire more research and practice. The chapters take the reader in different directions, and sometimes into whirlpools of historical fact, fictions, and legal frictions. This book assumes that understanding complexity, including social, cultural, colonial, ecological, and economic interconnections, is crucial to any interrogation of the marine human.

These are complex issues, in a complex time when the effects of anthropogenic change are abundantly clear, even as they are unevenly shared. *Sustaining Seas* is truly interdisciplinary with all the difficulties that entails. We have witnessed an oceanic turn in several disciplines including cultural studies, geography, literature, and environmental humanities, consolidated in neologisms such as "Blue Humanities" and "Blue Cultural Studies." Increasingly, academics, activists, and industry are working against the limitations of working in silos, and new and exciting interdisciplinary collaborations are flourishing. Collaborations such as those documented here have brought forward unique ways to think about problems and solutions. Innovative and creative collaboration is sorely needed in a time which, for many interconnected reasons (political, military, cultural, and the race to feed an overflowing human population), is being hailed as "the new ocean age" (Han-il cited in Sanchez-Velasco et al. in this volume).

Sustaining Seas is divided into six sections. However, they overlap and mix across these dividers just as the ocean sweeps over man-made barriers.

PRACTICES OF CARE

Multiple ways of caring for the seas are foregrounded in this section, which provide readers with frames through which to think about what caring means, and how we might practice care of and with oceans. Here care is nuanced; it is contextualized and historicized. Who can care, where and how, for what and whom, are questions that must always underlie any discussion of care. This is to productively acknowledge the limits of care and caring.

Green opens chapter 1 with a sobering account of the alarms that have been ringing for decades in regard to oceanic distress in Cape Town. These accounts foreground the exigency of formulating a new oceanic regime shift to manage diverse ocean problems. Green's argument makes it clear that this new regime can only come through truly interdisciplinary ocean scholarship. In chapter 2, Probyn brings together clandestine marine migrants and fishers from the global south, and lines in the sea set out by the United Nations Convention on the Law of the Seas to argue for what she calls, after Gilroy (2018), a *watery humanism*. Here we see how colonialism persists in regulatory designations of who fishes where, and the unequal effects of global overfishing in the global south. Johnston and Pratt present in chapter 3 a case study of a workshop/walkshop set on Sydney Harbour. Through the creation of contemporary field guides, they explore the relationship between noticing and caring. The concerns of care turn their attention to the ground as they and their participants in the guides walk, talk, and reflect on the mixed histories of Blackwattle Bay in Sydney. Through a political ecology lens and an analytic comparison of tuna fisheries in the Indian Ocean, Andriamahefazafy and colleagues in chapter 4 detail significant social interactions that are often absent in tuna fishery studies but that have consequences in the regulation of tuna in the Indian Ocean. Their analysis brings to the fore how care can inform more than human governance in humane and meaningful ways.

FISH AS FOOD: CONSUMING AND SUSTAINING

Section 2 travels into the visceral realms of fish as food. Garcia Garcia and colleagues in chapter 5 consider the discursive relationship between fish as represented in fishery policy and fish as represented in consumer cultures and postharvest governance models to reveal a disconnection, which they analyze in relation to future policy making. In chapter 6, Truninger and colleagues draw on ethnographic field work in Portugal and the United Kingdom from their collaborative project to compare ways in which consumers mobilize knowledge to determine freshness and how freshness relates to other consumer ethical concerns. Using the case study of an iconic Sydney Chinese restaurant known for its live fish tanks and seafood cuisine, in chapter 7 Lee explores the restaurant as an "affective habitus," demonstrating the ways in which different groups create meaning, knowledge, and develop attachments to place and in turn how this comes to figure in how people care for the seafood they consume. Drawing on fieldwork in Indonesia, in chapter 8 Budiastuti draws on a more than human assemblage and on globalization theories to consider the local changes to the ways in which Islamic beliefs and categorization of catfish as nonhalal relates to class, ethnicity, and boundary formation in a globalized context. In chapter 9, Sharp develops the concept

of the *ruin economy* alongside a concept of *salvage economy*, to better understand the case study of fish-waste salvage practices in New Zealand.

GOVERNANCE AND REGULATING THE OCEANS

As Rayfuse puts it in chapter 10, "Out of Sight, Out of Mind," the sheer
distance of the high seas from human habitation makes regulating fishing
activities a considerable challenge. There are also all sorts of challenges that
arise from governing coastal and inland fisheries. In short, governing aquatic
spaces and resources is a deeply vexed and vexing problem. Section 3 offers
diverse disciplinary and geographical perspectives into the challenges of caring better for the seas and instituting and practicing that care through systems
of governance. Rayfuse, in chapter 10, brings to the fore the particular legal
challenge of regulating fisheries on the high seas. With her expertise in international marine law, she considers the role of regional fisheries management
organizations in the conservation and management of high-seas fisheries and
assesses their efficacy, particularly in light of the new challenge of climate
change. Using a case study from New Zealand, Le Heron and colleagues in
chapter 11 focus on the genesis, emergence, and achievements of marine participatory initiatives that have appeared in the twenty-first century, to consider
the evolving analytical and political spaces in participatory initiatives and their
potentialities for forging sustainable processes based on wider engagement
of diverse actors. In chapter 12, in the South African postapartheid context,
Norton considers the relationship between policy and policing. Drawing on
her deeply immersed ethnographic fieldwork amongst Marine Compliance
Inspector units, she argues that the penalization of harm has superseded the
prevention of harm as the *de facto* goal. This brings to light the uncomfortable
correlation between care and harm in fishery governance. Mann's chapter 13
orients the reader to the small-scale fishers of the Global South where more
than half a billion people depend on fisheries for their livelihoods. She argues
that fisheries are a classic example of the exploitation of the commons, where
property rights are incomplete, and migration patterns make rights to fish difficult to enforce. Her chapter explores how the concept of food sovereignty has
been employed to assert these rights, and define responsibilities.

EMBODYING THE MARINE

What is it to be a body immersed in marine environments and what is it to
be an aquatic body? Section 4 presents a range of short chapters, some of

which are accompanied by images and artworks, which explore such onto-logical questions. Neimanis's interconnected pieces in chapter 14 are deeply informed by feminist, queer, and decolonial theories of embodied ethics to offer a meditation on breathing and the ocean. She does this alongside the work of renowned Australian artist Janet Laurence's 2015 work *Deep Breathing: Resuscitation for the Reef.* Through a poetic chapter 15, Hamilton explores how drains have helped to maintain the fantasy of individuality despite what we know about interrelationality. In both her artwork and accompanying chapter, artist Nicholson in chapter 16 recontextualizes maternal–fetal toxic-ity and environments. Highlighting the absurdity of government advice that promotes individual responsibility for one's environment, Nicholson draws attention to the embodied toxification inherited through aquatic environments and across disparate cultures caused by corporate and government actions. In a similar vein, in chapter 17, Bossell creates artworks to elicit new imaginar-ies of belonging within a network of marine human life forms.

LIVING HUMAN AND MARINE ECOSYSTEMS

Seascapes are produced through natural and cultural conditions. In section 5, authors, scholars, and practitioners explore this socioecological relationship by presenting cutting-edge design and interdisciplinary practice. Collectively, these case studies encourage readers to situate themselves in more-than-human ecosystems made up of biological, cultural, artistic, and technological elements. Some of the case studies (Reef Urbanism, Oystertecture, Buoyant Ecologie's Float Lab, Grandeza) foreground design innovation and ask ques-tions, such as: What new biocultural systems do these technologies enable? How are they situated within a geopolitical context and what are some of the environmental crises they may prevent? In the wake of Superstorm Sandy's devastation of the eastern coastline of the United States, SCAPE describes in chapter 21 how their project, Living Breakwaters, along with other projects instigated by Governor's Office of Storm Recovery were conceived to con-nect physical, social, and ecological resilience. In conversation with SCAPE, Marcus in chapter 19 describes how his team at Buoyant Ecologies is build-ing the Float Lab as a model for interspecies collaboration that is rooted in mutual cooperation and benefit rather than purely human self-interest. In a similar vein, Sánchez-Velasco and colleagues use chapter 20 to explore how South Korea's extraordinary archipelago of artificial reefs transforms the country into a terrestrial-marine, nature-culture continuum. In chapters 18 and 22, Vergés and colleagues and Williams and colleagues, respectively, continue this conversation that challenges the practice of presenting environmental problems within disciplinary boundaries. Set on Sydney's coastline, Vergés and colleagues use chapter 18 to showcase Operation Crayweed, a project that combines science, community engagement, and art to raise awareness about the importance of seaweeds through the restoration of crayweed (*Phyllospora comosa*), a dominant seaweed that disappeared from the coastline over thirty years ago. Coauthored by a scientist, two artists, and a human geographer, chapter 22 explores how, why, and in what ways interdisciplinary collaborations can provoke new approaches to thinking about environmental problems.

THINKING WITH SEAS

What might be offered (tools for thinking, practicing, doing politics) in aquatic imaginings and modes of representations as they are recorded in literary works and other cultural forms? These are overarching questions of this short but inspiring final section, section 6. In chapter 23, "The Sea Is Time," Ajumeze explores the cultural representation of the Niger Delta's watery ethos in relation to time-space and capitalist modernity. He does so by drawing on J. P. Clark-Bekederemo's dramatic text, *The Raft*, which depicts how the seas and rivers always already embody the apparatus of time. Lavery's chapter 24 considers representations of the Southern Ocean and explores what a theory from the south might look like in cartography and thought. Lavery coins the term the *oceanicity* of the Southern Ocean, and exhorts us to think "from the ocean." Through a detailed reading of fiction representing the Southern Ocean, Lavery charts the nonhuman depth and scale of the seas and conjures a lyrically political tableau of seas at the bottom of the world, away from the Northern Hemisphere's parochialism—where another version of Gilroy's offshore humanism may come to flourish.

MESSAGES IN A BOTTLE

Long ago, in seemingly simpler times, the conceit of sending a message in a bottle and throwing it into the seas signified a hopeful connection between humans and the seas: a human wish carried by the sea. Now, of course, plastic bottles that swirl in the immense Pacific garbage patch send an unambiguous message.

What messages do we hope that readers of our beautiful sprawling collection might take from the considerations of *Sustaining Seas*? As editors, we hope that several key points emerge. First, as researchers and practitioners we must think carefully about care. Practices of care need to be situated contextually and in the specificities of their conjunctures. We need to ask why should people care, before we insist on care as a first principle. This entails asking modest yet far-reaching questions about the conditions of possibility for care. Race, gender, class, and colonialism are entangled elements of the seas.[1] To reflect on our subtitle, this is to say that oceanic space provides the necessary conditions of possibility for a politics of care. Second, as so many of the mind-expanding experiments documented in this collection demonstrate, true interdisciplinarity starts at the very beginning—or even before—and requires all involved to practice care, to listen carefully to what is being said about where and how care for the oceanic is to be practiced. Third, we hope that this book helps to further what we might call *marine human sciences*, a space where research and action on the more than human and the more than oceanic meet and prosper. Finally, it is to remember that the seas sustain all life, and that homage to this needs to be reflected in not just research but poetry, art, living, and laughter.

NOTE

1. Elspeth thanks Alifa Bandali for this phrasing and for this reminder.

REFERENCES

Alaimo, Stacy. 2016. *Exposed: Environmental Politics and Pleasures in Posthuman Times*. Minneapolis: University of Minnesota Press.

Gilroy, Paul. 2018. "'Where Every Breeze Speaks of Courage and Liberty': Offshore Humanism and Marine Xenology, or, Racism and the Problem of Critique at Sea Level: The 2015 Antipode RGS-IBG Lecture." *Antipode* 50, no. 1: 3–22.

Hayward, Eva. 2010. "Fingeryeyes: Impressions of Cup Corals." *Cultural Anthropology* 25, no. 4: 577–99.

Section 1

PRACTICES OF CARE

Chapter One

Ocean Regime Shift

Lesley Green

The first alarm bell was red.

At the time, the apartheid state feared the "Red Tide" of Communism the most, but it was a different kind of "Red Tide" that landed on the shores of the Marine Nature Reserve that was named for apartheid's chief architect, H. F. Verwoerd. In that 1989 red tide event in False Bay, an estimated forty tons of abalone had died along the southern Cape coast east of False Bay (Horstman et al. 1991), along with thousands upon thousands of creatures in the kelp forest: crayfish, mussels, sea urchins. It was not the first red tide event, but it was the first that was so extreme.

The second alarm bell was also red and began in the early 1990s: red beaches were caused by mass strandings of the rock lobster *Jasus lalandii*, also known in Afrikaans as *kreef* and in the fisheries management literature as West Coast Rock Lobster (WCRL). They simply walked up, out of the ocean, out of the official "take zones" that had been set up by fisheries quota managers to manage their extraction and prevent their extinction. They beached themselves on the shore by the hundreds of thousands—sometimes it seemed millions. When a lobster stranding was particularly bad, the army would be called in, and having traveled from water to sand, military helicopters then ferried them by air to a landfill.

The third alarm bell was rung by the words "ecological regime shift" in the mid-1990s. The phrase derives from invasion biology, and perhaps also from the Gulf wars of the time, in which regime change was the objective. Marine ecologists deployed the words in South Africa to describe a mass migration of lobsters from Cape Point that crossed the sea bed of False Bay to a rocky shore about a hundred kilometers away called Cape Hangklip. This was not just a case of more eggs and better larval recruitment on site—it was an actual mass migration of adults. Studies of their stomach contents showed they had changed their diet, eating different species and setting off different trophic cascades: one of which added to the already extreme extinction pressures on *Haliotis midae*, the local abalone. *J. lalandii* was changing the ecology of the kelp forest.

The fourth alarm bell started ringing in 1997 and rang for sixteen years. Local fishing communities who lived off the sea were classified as subsistence fishers under the marine environmental laws developed during the presidency of Nelson Mandela. The category, and its legal definition as fishers who would eat but not vend marine harvests, denied them the right to sell their catch. Over the years, they chained themselves to Parliament; won battles in the Equity Court; and mobilized nationwide to develop a Small-Scale Fishers' Policy under which they could be recognized as small-scale fishers, could reject individual quotas for specific species, and would not be criminalized as poachers. Research showed that fishers were so distrustful of the controlling hierarchy of marine science that many took it as a point of pride to not comply with fisheries regulations (Schultz 2010).

The "Small-Scale Fisheries Sector for South Africa" (Republic of South Africa 2012) policy was gazetted in 2012, allowing for the catch of a "basket" of species, not limited to the species that were being scientifically managed and to which "subsistence" fishers had generally been allocated the right to harvest only one. Although the new policy said very little about ecology, it was expressed with an emerging discourse of care for the sea and for community members to live well, without hunger. But its implementation was handed over to the Department of Trade and Industry, whose first step was to require every fishing community to form a proprietary company. The extraction of financializable marine biomass, calculated with reference to applied mathematicians' total allowable catch, remained the defining relationship.

The fifth alarm was raised in 2014, when the army was called in to fight against the extinction of *Haliotis midae*, better known as *perlemoen* or *abalone*, as part of a newly thriving global illicit trade in wildlife products believed to be aphrodisiacs.

The situation was surreal. Fishers may well ask, *Why and how in our democracy can a perlemoen have better representation in Parliament than a fisher?*

Is it not surreal to mobilize a war machine to protect a snail?

Can the army protect the perlemoen from the lobsters that are invading the kelp forests?

Would the army restore practices of care for the ocean?

The sixth alarm came in 2014 and 2015 from kayakers and divers who reported occasional floating ponds of human excrement offshore from central Cape Town. I paddled regularly with some of the kayakers, and when several acquired serious gastroenteritis after practicing their kayak rolls, they started questioning the city's claim that the marine sewer outfalls were adequately diluted by the sea. A microlight pilot, Jean Tresfon, provided photographs of the plume emanating from the outfalls, and a local filmmaker, Mark Jackson, made a short film on the city's sewage outfalls, which pump forty to fifty million liters of effluent to sea every day. The city's response was to issue a "cease and desist" letter to the filmmaker and the aerial photographer and to argue that the two scientists who were interviewed for the film—epidemiologist Jo Barnes and environmental chemist Leslie Petrik—were unscientific. They claimed that the occasionally high *Escherichia coli* counts were not from the sewage pipe but from stormwater runoff. According to that argument, then, the sea was unable to dilute stormwater run-off but did not have the same problem with untreated sewage; it also suggested a complete failure of municipal street-cleaning. Yet there were no equivalent cases of illness caused by central-city street-side puddles, nor was there any public outcry about the lack of street sanitation.

The seventh alarm rang in 2017: changing winds and rising air and sea temperatures put Cape Town into an unprecedented drought after three years of low rainfall. The city's dams were about 35 percent full at the end of the rainy season. Rainfall levels were the lowest in recorded history, far lower than even the most highly regarded climate science laboratory had foreseen. Cape Town's solution was to set up emergency desalination plants, situated next to the same marine sewer outfall that the kayakers, filmmaker, and aerial photographer had flagged. The announcement of the pending desalination plant came at around the same time that Leslie Petrik, Jo Barnes, and I were working with graduate researchers Cecilia Sanuso, Adeola Abegunda, and Melissa Zackon to collect water and marine samples to test them for molecular compounds that could only have come from sewerage and not from storm water. The test results found that 100 percent of the indicator compounds we had tested for were present in the water and in sea creatures. These included high-schedule drugs and multiple household chemicals, many of which were endocrine disruptors and carcinogens, all of which were bioaccumulating in the filter feeders at the shoreline: starfish, sea urchins, seaweeds, limpets, mussels. Our team rushed a paper to the *South African Journal of Science* (*SAJS*), making the case that molecular compounds in the seawater from the sewage outfalls represented a public health hazard, and that the desalination plant was being sited inappropriately.

Two years later, the desalination provider sued the city of Cape Town, and the multimillion-rand plant ceased to function. The provider's claim: the seawater was up to four hundred times more polluted than the seawater data provided by the city in its invitation to tender (Gosling 2019).

The eighth alarm rang two days prior to the release of our *SAJS* article in November 2017, when Xanthea Limberg, a member of the city's Mayoral Committee for Water and Sanitation, released a press statement on the scientific study commissioned in 2015 and completed six months earlier, and which, until then, the City had refused to release. The press release stated: "A study by the Council for Scientific and Industrial Research (CSIR) into the City's sea marine outfalls has confirmed that they pose no significant risk to human health and do not measurably affect inshore water quality or the wider environment."

But the actual CSIR study (2017, 224) had concluded, "It is . . . illogical and indeed irresponsible to imply that effluent discharged through the outfalls is not impacting on the marine receiving environments or posing a potential human health risk. Indeed, the notion of no impact to a marine receiving environment in the context of effluent discharge is unfounded."

None of the CSIR scientists, however, challenged the City's press release.

These alarms paint a picture of an ocean ecology in crisis. But they also speak to a crisis in urban ocean governance, and the inability of South African marine sciences to address this. This is not because South African marine scientists are not good at what they do: on the contrary, the passion of southern African marine scientists for the species that they represent—scientifically and politically—is infectious. Certainly, the major regional marine science journal, the *African Journal of Marine Sciences*, struggles to include the social sciences. In its focus on natural sciences, however, it offers excellence, study after study.

But therein lies the trouble: I could find not a single study that considers the risks that I have listed as a whole.

So although alarm bells have been going off about local marine ecology with respect to harmful algal blooms, falling lobster counts, the extinction risks to abalone, the changing kelp forest ecology in the wider region, the over-harvesting of fish stocks, and the regional implications of ocean warming and ocean acidification and ocean plastics, it is fair to say that a science has not yet emerged that is able to work integratively. The result is that the field has been unable to address the anthropogenic aspects of the marine crises as more than issues that have "social dimensions" that require "regulation," "compliance," and "enforcement."

The terms themselves reflect the ascendance of the paradigm of behavioral economics, which assumes that social behavior can be gamed by environmental managers using game theory. Behavioral economics itself is a product of neoliberalism that has financialized nature via the language of "ecosystem services"—and in that paradigm, social relationships with marine species are reduced to a calculus of specific demographic rates of predation: the "take" of recreational fishers, commercial fishers, and small-scale fishers. Relations of care for a bay, or value placed on the sea, have no existence in this approach.

Sciences that are split along the nature-culture divide and believe neoliberal thought and practices to be neutral and objective, and which focus only on the demonstration of points of data without attention to the big picture, are wholly unable to confront the terraforming neoliberalism that shapes imagination, action, and policy.

The influence of urban sanitation on the surrounding urban oceans has barely begun to attract local marine researchers' attention, whose focus has been on species range, balance, and numbers in relation to fishing—not on urban ecology. Yet since the 1960s, the effect of urban expansion has been among the most significant changes in human presence on the planet.

In what follows, I want to explore the multiple crises in Cape Town's urban ocean ecology to make an argument that, responding to the ecological regime shifts of our time, requires a regime shift in ocean scholarship—an ocean regime shift. To sustain the seas, scholars who care for them need to reclaim the big picture.

The sciences need not work in solely reductionist ways. The claim that scientists ought not do big-picture work that includes the politics of addressing environmental crises is a trap created and sustained by those who want to continue neoliberal extractivist approaches to nature unchallenged.

That trap operates in a number of ways.

In some contexts, the argument is that scientists only have authority in their own fields of study and may not comment on politics. That argument is often used to silence climate scientists. But the argument that nature is separate from society is neither scientific nor empirically verifiable. On the contrary, it is unscientific. Cutting-edge scholarship in contemporary sciences in fields as diverse as climate change, epigenetics, and neurobiology have all demonstrated that nature is not separate from society. Epigenetics points out that genes can be switched on and off by social stressors. Neuroscience has found that neurological pathways are not only biologically laid down but are formed experientially. And climate science demonstrates that planetary climate is being altered by policies and administrations that make specific social, political, and economic choices about energy and food and consumption.

In other contexts, the argument is that a scientist can only take a view on society if they have quantitative social science data that can be worked up into a mathematical model of a social system. The problem here is that "social systems" are not empirically observable: they too are operations of the very fictions that are decried by their proponents. Moreover, the world of relationships is not empirically measurable in the same way that, say, sand

and water might be. In any scalable model of the cosmos from nanoparticles and quarks to the expanse of the known universe, nowhere is there a point at which relationships appear; nor is there any observable "social system." In other words, representing human collective choices requires something other than a research method designed to enumerate objects in space.

Reducing human life to numbers by turning people into measurable objects with predictable pathways offers the illusion of control and neutral governability, but its effects are anything but value-free. Human life and relations are not what the theory of *Homo oeconomicus*—economic human—claims them to be. Economics is not the definitive social science, because behavioral economics is not universal. The figure of *H. oeconomicus* is, ultimately, brutal in its reduction of relational complexity to sets of single on-off switches. Translated into systems of techno-governance, it strips lives and choices down to variables in algorithms in the name of equality and justice and expressibility in science, technology, engineering, and mathematics (the STEM sciences). It is a catastrophic claim to "social science" that does not have traction in social science journals, because it strips lived relations out of the picture, brutalizing people with incomprehensible decisions about their futures and causing the sort of mass resentment that drives the New Right to reclaim relations, care, and value under the banner of nationalism. In insisting that the figure of *H. oeconomicus* is the only way to do environmental science for environmental governance, free-market fundamentalists have captured the authority of science and natural scientists' capacity for imagining alternatives.

In this context, the humanities and social sciences have a critically important role to play in reclaiming the "big picture" currently captured by neoliberal technocrats.

What form of scholarship will enable Capetonians to think, act, and desire "living together well with the ocean"?

As a white South African scholar thinking in, with, and alongside the Global South, "living together well with the ocean" entails identifying and addressing the entanglement of modernist thought with coloniality. As such, it requires that we consider the capital and the commons and the undercommons; extinctions as well as exclusions; race and gender objectification; and living beings along with nonliving rocks, water, and life.

With those broad terms of reference, I want to try out a few approaches: with Trinh Minh Ha, to think not about but alongside (Chen 1992). With Elspeth Probyn (2016), to think "athwart."[1] With Vinciane Despret (2016), to ask what animals would say if we asked them the right questions. With Aime Cesaire (1972), to think about the necessity of recognizing the surreal in the real. Each of these approaches offer a line of flight from the prevailing gods of reason (Latour 2007).

Why might that matter to environmental sciences and environmental governance? Why should colleagues in marine sciences consider the social sciences and humanities? My argument is this: our time is one in which neoliberal governance and market logics have actively sought to capture the authority of science. Reductionist science (and reductionist social science) offers only points of data that can be narrated by others. In doing so, they have surrendered the power of story.

This chapter is an invitation to the environmental sciences to think with and alongside various actors in an urban ocean ecology in this "anthropocene" moment, where natural and social sciences can no longer be thought of as separate systems. The goal is not to replace, but to recompose into a whole the species concerns that have already been framed in regional marine sciences. The method I propose is to go as deeply as possible into a range of entangled situations to begin to understand their linkages in terms other than "natural systems" and "social systems."

THINKING WITH AND ATHWART

I. Blue Flag Science

The waters of Table Bay sparkle when you drive into the city bowl, looking toward Robben Island, where Nelson Mandela, Robert Sobukwe, and others were imprisoned. Follow the highway into Strand ("Beach") Street, which runs alongside the Castle. Near here, in 1654, the Dutch colonist Jan Van Riebeeck took, in the name of the Flag of Orange, fourteen wheelbarrow loads of a shoal of fish caught by the sloop anchored offshore and filled its nets with one cast.

Today the beaches of Cape Town are under a different flag: the Blue Flag, which describes itself on its website as "a world-renowned eco-label trusted by millions around the globe." The brand is held up by the City Council as international proof that its beaches are safe enough for tourists and are ecologically healthy.

Are they?

When kayakers with whom I had been paddling in 2014 began to get ill, they slowly joined the dots: when the northwest wind blows—as it does in winter from April to September each year—it brings the Cape's rains; the seas are colder and the contents of the warmer sewage outfall are more likely to rise to the surface. Around the same time that the kayakers became ill, social media went viral with the message that the city of Cape Town was sending fifty million liters of filtered but untreated sewage out to sea every single day.

The rich irony was that the city was simultaneously fighting activists protesting the lack of waterborne sewerage in the shacklands where homes are made of scraps of wood and iron. Sanitation activists under the leadership of Andile Lili had begun dumping the contents of domestic toilets in public places, including the steps of the Western Cape Provincial Government. Known politely as "poo flingers," their actions also inspired Chumani Maxwele, a University of Cape Town student activist who threw fecal waste at the campus statue of Cecil John Rhodes, starting the decolonization movement (Green 2015). In every instance, a hazardous materials company was called in to clean up—but the city itself was sending filtered but untreated human waste into the ocean every day, both at Granger Bay near the planned desalination plant site, and at the nearby Camps Bay on the Atlantic coast.

At issue here is the form that scientific question takes. Blue Flag science asks whether water can be demonstrated to be safe for a tourist to swim in: the research question is simplified into a requirement for data on whether the seawater contains fecal and streptococcal bacteria above specified limits. That question is further reduced by the mandate that those tests be conducted only twice a month. The assessment of the data pays no attention to the methodological differences that would come into play if the findings were to be peer-reviewed, such as the time of day the water was sampled, how deep, how often, what winds had blown, how long it took, and under what conditions the sample got to the laboratory bench.

Blue Flag data is a textbook case of a science that asks a question that addresses the concerns of capital—in this case to keep the tourist economy booming—rather than seeking to inquire of the effects of the sewer outfalls on the commons: that is, on sea water, marine species, or public health. It is science in the service of economic growth that provides officials and elected officials with performance indicators verified not under peer review but under the sign of a trademark: Truth with a capital T and a tiny TM. Such a science is designed to shut down questions to allow the flow of capital, rather than to facilitate consensus-based governance on health and well-being.

It is inconceivable that there are no geological traces of more than fifty million liters of sludge and effluent poured daily into Table Bay, without regard to changing winds, currents, changes in the thermocline, or the difference between the flow rates on the surface waters and the deeper waters of the bay, which moves five times slower than the top water that sparkles in the sunshine. The disposal of toilet paper into the ocean, albeit disintegrated, constitutes a geological shift in the regional ocean as plant cellulose—which would previously have fed the soils as leaves or rotting wood—now enters the marine ecosystem. Then there are the microplastics from the maceration of sanitary towels, plastic bags, and earbuds in the sewage pump stations and from the soaps in the city's millions of washing machines. And there are the soaps and detergents and drain cleaning agents sold by every supermarket and corner shop in the city; every bottle of disinfectant and shampoo and sunscreen; all the medicines from every pharmacy and hospital; and the detergents and paint thinners and oil-cleaning products from every home, abattoir, and factory floor.

Sewer outfalls change ocean chemistry in ways that in turn change the ocean ecology—in ways that are both so micro that they cannot be seen without the help of a microscope and so macro that they can at times be seen from space. In this regard, the archetypal example is the harmful algal bloom.

II. Harmful Algal Blooms

Officially called *dinoflagellates*, some harmful algal blooms (HABs) are neurotoxins. Others are individually harmless but can bloom so massively that they take up all the oxygen in the water. Scientific literature on HABs notes their association with disruptions of the nitrogen cycle from commercial agriculture and urban sewage discharge (in which urine provides vast quantities of ammonia, and vast supplies of bacteria). Add to that ocean warming and wind changes associated with climate change, and the future is looking bright for HABs.

As fascinating as they are destructive, dinoflagellates come in many more shades than red—some are green, some are black, and some bring a breath-taking "ocean fire" phosphorescence to the waves, which local Western Cape fishers used to rely on to fish at night "before the lights went out" around the 1960s (Van Sittert 2015). Viewed under electron microscopes, the variation in dinoflagellate bodies is astonishing.[2]

Prompted by the unexpected and catastrophic red tides that bloomed in False Bay in 1988 and 1989, and again in 1995 and 1996, doctoral researcher Lizeth Botes set out to understand what species were blooming, and why.

In the care taken over every name and every claim, Botes's (2003) species sampling is surely an example of the best of scientific research. The study looked at natural phenomena to try to explain the algal blooms: specifically, sea surface temperatures and winds in relation to the deep-water upwelling that occurs when winds drive warmer surface waters away and the deeper, colder, and nutrient-rich bottom waters come to the surface. Indeed, in a world of nature-without-people, this process of wind-driven upwelling would be the sole basis of the regional marine ecology. But in the mid-1980s, a great many other factors were at work on the shores of False Bay, including the rise of commercial agriculture with the extensive use of fertilizers; the rise of chemical industries; and the rise of new shack settlements on the Cape

Flats—specifically Khayelitsha—without sanitation services, so raw sewage flowed through storm-water channels.

Did these three factors contribute to a perfect storm that spawned the unprecedented algal blooms of False Bay? It is difficult to argue the contrary. A dissertation by Carl Rundgren in 1992 shows the major sewage, industrial, and storm water discharges into False Bay in 1986, along with chemical factories. The Mitchells Plain and Monwabisi outfalls, and the Sand Rivers, Seekoei, Kuils, and Eerste River systems, were serving a rapidly growing population with associated increases in fertilizer and pesticide use, along with effluent discharge from new wastewater treatment facilities.

Thus, into False Bay in the 1980s flowed a vast soup of new nitrates, including nitrogens associated with urine, fertilizers, and other nitrate users such as meat-processing factories.

Effluent-rich waters are perfect for some algae, which can be autotrophic (producing their own food like a plant, by photosynthesis) and heterotrophic (consuming other species for their food). Put them in a situation in which they have nitrogens and phosphates that stimulate photosynthesis, along with a delicious soup of bacteria, and the conditions are right for a perfect storm. Dinoflagellates—marine algae—are highly responsive to changes in ocean chemistry, climate, and bacteria. Even changes in atmospheric dust from stronger winds far away have an effect. When unseasonally strong winds blow in the Sahara desert, where sands contain iron particles, they can foster algal blooms in the Caribbean and subtropical Atlantic (Westrich et al. 2016).

And yet, with the nature-culture divide in mind, and determined to believe in the possibility of being apolitical, the science of algal blooms in South Africa could not see the urban ecology of the Apartheid state in its willful disregard of the sanitation needs of black South Africans. Nor could it name as anything but neutral the political decisions of the current City Council to prioritize data favorable to tourism over data that addresses the concerns of city residents. Nor could it change the course of management decisions over extinction risks to abalone and crayfish, which are certainly aggravated by poachers but are also driven by the toxicity that arises when urban oceans are used as sewers.

Viewed in "anthropocene" terms, the situation is less a matter of natural and social systems and more a matter of a political biogeochemistry. With that concept in mind, let us return to the issue of lobster walkouts and lobster-led ecological regime shift.

III. Lobster Walkouts[3]

J. lalandii offers much to think with. The best ecological science can model the new lobster-led ecological regime but has yet to fully understand either why or how the lobsters moved from the West Coast to the southern shores of the Western Cape.

Lobster motivation is difficult to assess: there is no Sigmund Freud for lobsters—although there used to be: it was none other than a young Sigmund Freud who demonstrated that lobster neural structures are of the same order of complexity as mammals'. That thread of biological thought was cut short by the rise of radical behaviorism (Sacks 2014).

"The nature of *J. lalandii*" comes to us through the legacy of a behaviorist biology. In the South African scientific literature, there are at least two quite different scripts for understanding *J. lalandii*.

One is the *J. lalandii* already described previously: "*J. lalandii* the regime shifter" of kelp forest ecology, enrolled in the ecological imaginaries of invasion biology and conservation policy. Brendon Larson's 2005 critique of combat metaphors in invasion biology is important: he argues them to be problematic in that they lead to an inaccurate perception of invasive species, contribute to public misunderstanding of what is at issue, and naturalize militaristic thinking that is often counterproductive to conservation.

The other script of *J. lalandii* appears in the extractive science of fisheries management, where the single-species community plays the part of a capital asset in the national bank, whose growth or decline is projected via the agency of a regularly revised algorithm that in turn dictates how much of the capital asset may be withdrawn. This *J. lalandii*—"*J. lalandii* the algorithm"—enrolls the lobster in the machine economy: the cold chain of its supply and demand.

Although the two accounts of *J. lalandii* are very different, they share an insistence that they are not "political" accounts of nature. Both fisheries activists and government officials routinely argue the contrary. To maintain the fiction of the separation of nature from politics, both are constrained to maintain another fiction: that the lobster is a creature capable only of a mechanistic response to stimuli, and not a sentient, responding being.

Because the case of *J. lalandii* offers a situation in which the lobsters have chosen to do what marine and coastal managers did not expect—their choices were off-script—it is useful to consider whether *J. lalandii*'s representation is part of the problem. To say in Amerindian contexts of thought that the scripts of extraction and predation that define the *J. lalandii* fishery have been refused by lobsters is not likely to be met with riposte. Similarly, to think like a lobster, among South African fishers, is not an outrageous anthropomorphism but a reasonable recognition that all creatures search out places and spaces that offer well-being. Lobster fishers along the West Coast are insistent that crayfish are highly responsive creatures that dislike lobster discards being thrown in the water. The observation would not surprise an evolutionary ecologist whose question is, "What new choices did creatures have to make in order to survive?" Only in the framework in which nonhumans are commodified objects for extraction is it possible to assert that lobsters are incapable of response to the conditions of the Anthropocene.

Reading more widely about lobsters than the *J. lalandii* literature in South African debates, I was astonished to find research in ethology that suggests both interspecies and cross-species communications between fish and crustacea about the availability of food and the likelihood of predation (Laland, Atton, and Webster 2011; Webster, Ward, and Hart 2008; Webster et al. 2013). The complexity of lobsters' neurological systems, which so intrigued the young Freud, is evidenced in studies of lobster behavior ("neuroethology") (Kravitz 2000): their complex social lives include rituals of aggression and dominance (Sato and Nagayama 2012); they have a capacity for magnetic navigation (Boles and Lohmann 2003) and an ability to migrate long distances and adapt to new environs (Bertelsen et al. 2010); they are highly sensitive to smell, using chemical signaling to communicate; they have a strong fear response, as evidenced in their rapid, long-distance movement away from tagging sites in an acoustic tracking study (Atkinson, Mayfield, and Cockroft 2005); and they can, on rare occasions, choose to move in a mass migration. Lobsters, this literature suggests, are far more sentient, and far more collectively responsive, than behaviorist fisheries sciences have imagined.

Vinciane Despret (2016) asks, "What would animals say if we asked them the right questions?" What are the quanta of well-being that the lobsters are looking for, in a slowly warming ocean so close to a city's sewerage flows?

IV. Quotas and the Army

The current means of calculating the total allowable catch and enforcing quotas produces an unworkable moral economy of extraction of life as objects: it establishes an ontology of unbecomings in the name of conservation. Where governance turned the small-scale fishers' emerging language of concern into proprietary companies, the rekindling of care for the ocean was erased. The closure of the commons and the rise of a new class of extractive owners was mainstreamed, which inevitably will require policing and militarization. Is this a sustainable approach to conservation? Reclaiming care for the ocean requires a different kind of relation than ownership and extraction; extinction risks not manageable via business administration but via care for the commons.

Those who have invented and taught the master's degree in business administration in the last thirty years have done enormous damage insofar as that mind-set has led to the capture of the commons (see Brown 2015). I would like to see the rise of a new degree series: the MCA—master's of commons administration—or perhaps the DCC—the doctorate of commons care.

Why? The situations I have described have something important in common. Fishers have struggled against single-species quotas conceptualized by and for a capitalist accounting of the sea. Lobsters appear to be attempting to escape the high-take zones, the sludge and chemistry of sewer outfalls, and the low oxygen of harmful algal blooms that arise more frequently in a context where officials argue that dumping sewage in the sea is better political economics than establishing wastewater treatment facilities. Perlemoen are struggling to survive the poachers, who are struggling to survive the predatory economy and the new lobster regime. Ocean users are fighting the brand-management approach to water quality, which is hopelessly inadequate on a day-to-day basis. What is in common here?

Listening together to the dilemmas that each are facing, it seems that they have in common a battle against a particular kind of reasoning, a particular kind of logic that has become dominant in the era of neoliberalism. "The gods of reason," Bruno Latour calls them; they are forms of argument that are so tightly interwoven, and so mutually affirming, that they are all but intractable. They are rock-paper-scissors types of closed systems, in which the only decision is which one of them will win: technical efficiency, economic profitability, or a specific kind of scientific objectivity. In my view, the most important, fundamental error made by many climate scientists is to assume that neoliberal values and relations are capable of unmaking the geological effects of the Anthropocene. A regime shift in scholarship is required. But for this to happen, researchers and scholars in the social sciences and humanities must become active in building alliances and partnerships with communities, scientists, and thinkers who can see the madness of the gods of reason, and be willing to conceptualize forms of ecological science that are captive neither to the financialization of the environment nor the ideology of *Homo oeconomicus*.

In seeking to think through the "Big Picture" in the ongoing Capetonian marine crises, a number of lines of approach have emerged toward a regime shift in ocean scholarship.

First: Environmental management sciences have become reductionist and focused on the territorial lines that can be drawn on a map with reference to jurisdictions and governmental mandates. Regime-changing scholarship, by contrast, is integrative, following flows of microbes and molecules that are not bound by lines of property. Regime-changing scholarship must address the commons and challenge the claims, methods, and authority of science made under the flag of economic growth.

Second: Marine science that researches nature without society cannot see the molecular flows of the human and is unable to explain the effects of political ideas or urban ecology on biogeochemistry.

Third: Regime-changing ocean management sciences will not render marine creatures as extractables but will prioritize the desires of creatures, including humans, to survive and live well. A partnership of the social sciences and humanities with the sciences can challenge and change the representation of ocean species as objects.

Fourth and finally: Rather than partnering with the army and the economy to govern marine ecology, an ocean regime shift requires recentering the forms of care articulated by small-scale fishers in the drafting of the small-scale fishers' policy, to overcome the destructive "thingification" that undergirds the conditions leading to an Anthropocene. After all, managing three thousand kilometers of South African coastline through the police and military will never fully succeed and will continue to alienate fishers from fisheries managers. Working with fishers to reclaim the ethics of care that they themselves have articulated will be a far more sustainable approach to sustainability. Extending that ethics of care to urban coastal management could profoundly shape the terms of current city ocean management.

NOTES

1. "Thinking athwart" is Elspeth Probyn's conceptualization of a humanities-framed fisheries and ocean scholarship, propounded in *Eating the Ocean*.

2. See the images and notes in Lizeth Botes's 2003 PhD dissertation, pages 60–61.

3. An earlier version of this section was published as Lesley Green, "Calculemos *Jasus lalandii*: Accounting for South African Lobster," in Bruno Latour with Christophe Leclerq, *Reset Modernity!* (Cambridge, MA: Massachusetts Institute of Technology Press, 2016).

REFERENCES

Atkinson, L. J., S. Mayfield, and A. C. Cockroft. 2005. "The Potential for Using Acoustic Tracking to Monitor the Movement of the West Coast Rock Lobster *Jasus lalandii*." *African Journal of Marine Science* 27 (2): 401–8.

Bertelsen, R., K. Maxwell, T. Matthews, and J. Hunt. 2010. "Caribbean Spiny Lobster Movement and Population Metrics at the Western Sambo Ecological Reserve." Paper presented at "Linking Science to Management: A Conference and Workshop on the Florida Keys Marine Ecosystem." Duck Key, Florida, October 19–22, 2010.

Blue Flag. Home page. Accessed January 2, 2017. http://www.blueflag.global.

Boles, Larry C. and Kenneth J. Lohmann. 2003. "True Navigation and Magnetic Maps in Spiny Lobsters." *Nature* 421 (January 2): 60–63.

Botes, Lizeth. 2003. "Taxonomy, Distribution and Toxicity of Dinoflagellate Species in the Southern Benguela Current, South Africa." PhD diss., Department of Zoology, University of Cape Town.

Brown, Wendy. 2015. *Undoing the Demos*. Cambridge, MA: Massachusetts Institute of Technology Press.

Cesaire, Aime. 2000 [1972]. *Discourse on Colonialism*. Translated by Joan Pinkham. New York: Monthly Review Press.

Chen, Nancy. 1992. "'Thinking Nearby': A Conversation with Trinh Minh Ha." *Visual Anthropology Review* 8 (1): 82–91.

City of Cape Town Media Department. 2017. "City of Cape Town Statement: CSIR Confirms Sewage Outfalls Pose No Significant Risks." November 21. Accessed June 2, 2019. http://pr.africannewsagency.com/general/City-of-Cape-Town-statement-CSIR-confirms-sewage-outfalls-pose-no-significant-risks-2958308.

Council for Scientific and Industrial Research (CSIR). 2017. *Cape Town Outfalls Monitoring Programme*: Chapter 7. Accessed June 2, 2019. http://resource.cape-town.gov.za/cityassets/Media%20Centre%20Assets/CT-Outfalls-Report.zip.

Despret, Vinciane. 2016. *What Would Animals Say If We Asked the Right Questions?* Minneapolis: University of Minnesota Press.

Gosling, Melanie. 2019. "Cape Town Council's Desalination Debacle: Seawater 400% More Polluted than City of Cape Town's Tender Data Indicated." News24, May 27. Accessed June 2, 2019. https://www.news24.com/SouthAfrica/News/va-desalination-debacle-seawater-400-more-polluted-than-citys-tender-data-indicated-20190527.

Green, Lesley. 2015. "The Changing of the Gods of Reason." *e-FLux* Supercommunity: Apocalypsis, June 9. Accessed June 2, 2019. http://supercommunity.e-flux.com/texts/the-changing-of-the-gods-of-reason/.

Horstman D. A., S. McGibbon, G. C. Pitcher, D. Calder, L. Hutchings, and P. Williams. 1991. "Red Tides in False Bay, 1959–1989, with Particular Reference to Recent Blooms of *Gymnodinium sp.*" *Transactions of the Royal Society of South Africa* 47 (4 and 5): 611–27.

Kravitz, E. A. 2000. "Serotonin and Aggression: Insights Gained from a Lobster Model System." *Journal of Comparative Physiology* A 186.

Laland, K. N., N. Atton, and M. M. Webster. 2011. "From Fish to Fashion." *Philosophical Transactions of the Royal Society B: Biological Sciences* 366: 958–68.

Larson, Brendon M. H. 2005. "The War of the Roses: Demilitarizing Invasion Biology." *Frontiers in Ecology and Environment* 3 (9): 495–500.

Latour, Bruno. 2007. "The Recall of Modernity: Anthropological Approaches." Translated by Stephen Muecke. *Cultural Studies Review* 13 (1): 11–30.

Probyn, Elspeth. 2016. *Eating the Ocean*. Durham, NC: Duke University Press.

Republic of South Africa. 2012. "Small-Scale Fisheries Sector for South Africa." *Government Gazette* 564, no. 35455 (June 20). Pretoria, South Africa.

Rundgren, Carl David. 1992. "Aspects of Pollution in False Bay, South Africa (with Special Reference to Subtidal Pollution)." MSc diss., Department of Zoology, University of Cape Town.

Sacks, Oliver. 2014. "The Mental Life of Plants and Worms." *New York Review of Books*, April 24. Accessed April 19, 2019. http://www.nybooks.com/articles/archives/2014/apr/24/mental-life-plants-and-worms-among-others/.

Sato, Daisuke and Toshiki Nagayama. 2012. "Development of Agonistic Encounters in Dominance Hierarchy Formation in Juvenile Crayfish." *Journal of Experimental Biology* 215: 1210–17.

Schultz, O. 2010. "Belonging to the West Coast: An Ethnography of St Helena Bay in the Context of Marine Resource Scarcity." Masters diss., Department of Social Anthropology, University of Cape Town.

Van Sittert, Lance. 2015. "The Fire and the Eye: Fishers' Knowledge, Echo-Sounding and the Invention of the Skipper." *Marine Policy* 60 (C): 300–308.

Webster, M. M., N. Atton, W. J. Hoppitt, and K. N. Laland. 2013. "Environmental Complexity Influences Association Network Structure in Fish Shoals." *The American Naturalist* 181 (2) February: 235–44.

Webster, M. M., A. J. W. Ward, and P. J. B. Hart. 2008. "Shoal and Prey Patch Choice by Co-occurring Fishes and Prawns." *Proceedings of the Royal Society B: Biological Sciences* 275.

Westrich, J. R., A. M. Ebling, W. M. Landing, J. L. Joyner, K. M. Kemp, D. W. Griffin, and E. K. Lipp. 2016. "Saharan Dust Nutrients Promote *Vibrio* Bloom Formation in Marine Surface Waters." *PNAS* 113 (21): 5964–69.

Chapter Two

"The Sea Is Empty"

Fishers, Migrants, and a Watery Humanism

Elspeth Probyn

"The sea is empty" (cited in Jacobs 2017): I remember thinking this some years ago snorkeling in the Mediterranean off Sicily. Tourists prodded at a poor lone octopus, seemingly the only life in what looked like a barren moonscape of an ocean.

Of course, the sea wasn't empty. It was filled with memories and material shards of history—of shipwrecks and bodies. On October 3, 2013, one of the worst migrant shipwrecks occurred on the coast of Lampedusa off Sicily when a boat carrying 500 passengers capsized leaving only 155 survivors.

Dead bodies wash up on shores. Years prior to this tragedy, in 2006, thirty thousand migrants managed to reach the Canary Islands—with some seven thousand people who drowned trying to make the crossing (International Organization for Migration [IOM] 2018). The lucky ones arrived on Fuerteventura, described in travel brochures as "a veritable oasis on the Atlantic. White and luminous, the island boasts huge stretches of golden shiny beaches washed by crystal-clear blue waters that will make you feel as if you were in paradise."[1] They came to the Canaries in *cayucos* or *pateras* or *pirogues*, one thousand miles across open ocean in small wooden boats held together with rusty nails (McAllister and Prentice 2018), migrants fleeing poverty. That poverty has different roots but a major one is the lack of, or more properly, the theft of fish. West Africa has become a global hub of illegal fishing, losing an estimated $1.3 billion a year to the trade, according to a report from the Africa Progress Panel, with Senegal alone accounting for $300 million—around 2 percent of its gross domestic product (McAllister and Prentice 2018). China's "distant-water fishing" fleet (e.g., those that fish outside of their two-hundred-nautical-mile exclusive economic zone [EEZ]) is a prime culprit, which targets West African nations; their vessels report only an estimated 8 percent of their catch (Jacobs 2017).

A perfect storm: the phrase is overused but perhaps appropriate here. Climate change and overfishing are creating the conditions that send mostly young men from the Global South on rickety boats to try to reach Europe. These are fishers made into migrants by global unsustainable fishing policies. These are all matters that solicit us, call to us to care. They tend, however, to be used to unilaterally interpellate, and constitute, communities of care. The physical plight of the ocean has, of course, risen to the surface in relatively recent times, propelled in part by lush documentaries that combine beautiful photography of marinescapes and marine beings mangled in plastic. Overfishing has become framed in a cruel scenario that sees commercial fishers painted as the rapists of the sea, and lauds a so-called ethics of refusing to eat fish. As I have graphically argued elsewhere (Probyn 2016), there is a dumbing down of caring about the ocean and her inhabitants. There I argued against a simplifying of the ocean, and urged more attention to the complexity of relations within and of the more-than-human marine. Here I change gears to consider care through the prism of the unequal positioning of the Global North and South in terms of fishing and oceans. Drawing on Paul Gilroy's (2018) figuration of "offshore humanism," I consider how the visceral proximities of (over)fishing, clandestine migration, and the United Nations Convention on the Law of the Seas inadvertently set in motion a spatialization of the seas that places coastal nations of the Global South at risk of losing their precious more-than-human marine resources.

"The sea is empty": the words are from Zhu Delong, a seventy-five-year-old fisherman in eastern China. In the economic opening of China from the 1970s, much of China's development was on its coastline. As Christopher Carolin puts it, "[T]he local marine environment paid a staggering price in terms of pollution, overharvesting, and critical habitat loss" (2015, 136). The United Nations Food and Agriculture Organization (FAO) estimates that "some 50 percent of China's total offshore ocean area is considered severely polluted or unfishable" (Carolin 2015, 136). This has irretrievably damaged China's domestic fishing industries with 40 to 50 percent unemployment in regional fishing centers in China. In 2013, Tabitha Grace Mallory wrote that the expansion of China's distant water fishing was "currently driven more by employment concerns than by food needs" (2013, 100). In the years since, the rise of the middle-class desire for seafood in China has further added to pressures on the world's fish stock.

Of the many phenomena that unite fishers across the world, the bleak present and even bleaker future with fewer and fewer fish is the most salient. Climate change drives fish stock away from their usual habitats as the waters grow warmer, and they move either up north or down south toward the poles. Overexploitation is the other common factor but it is one that needs to be differentiated. It is an abstraction that masks the vast differences between small-scale fishers and the industrial trawlers that comb the seas, often engaged in the illegal, unregulated, and unreported theft of fish. As fish grow ever-scarcer and smaller, many of these gargantuan ships are sieving the waters for small fish that are now known as "industrial grade forage fish" (in the Fishmeal and Fish Oil Organization's[2] terminology) capturing the fish that are the staple diet of the poor to be ground into fishmeal. These ships are part of the 3,400-strong Chinese distant-water fishing fleet, out for fish to feed China's massive aquaculture industry. It is estimated that the annual demand for fish meal in China is 2.5 million tons. As Zhang Chun puts it, "China is catching more fish to feed other fish than the Japanese catch to eat themselves" (2017). To put that into perspective, Japan is one of the major global fisheries.

In this chapter, I want to connect the conditions of possibility that render the loss of fish and the loss of human life as intertwined outcomes. The old Chinese fisherman on the shore haunts me as he contemplates the near-complete devastation of China's domestic fisheries and coast. Does he know that China is killing off the fisheries of developing nations far away? Does he care? News images of "clandestine marine Senegalese migrants" (Ifekwunigwe 2013) who are un- or under-employed fishers forced to try to enter Europe traveling by fishing boat are seared into my mind. Their livelihoods shattered, their possible deaths will destroy families and communities who had hoped for remittances. The Chinese fisherman and the Senegalese migrants are united by the lack of fish, the lack of their livelihoods and ways of life.

Moustapha Balde, a young Senegalese man, puts it succinctly: "Foreigners complain about African migrants coming to their countries, but they have no problem coming to our waters and stealing all our fish" (in Jacobs 2017). As Matteo Raffaelli, the director of the recent documentary *Mareyeurs* (2017) about the plight of young Senegalese fishers, states bluntly, "it is wrong to separate the issue of migration from that of sustainable resource use."[3] Many are the victims of what Paul Zeleza calls "diasporas of structural adjustment" forced on developing nations by the World Bank and other international bodies (cited in Ifekwunigwe 2013, 219). As Jeanne Koopman recounts:

> In 1985, Senegal signed a structural adjustment agreement with the World Bank that compelled the government to cut back on services to small farmers. In 1986, the government stopped subsidizing fertilizer and fuel for irrigated rice production. Farmers' input costs jumped from a quarter to a third of the harvest. In 1987, the government stopped providing seasonal credit. This forced plot owners to send more male dependents out of the villages so that non-farm earnings could help pay the rising cost of inputs. (2009, 265)

Many of these men headed for the coast to try their luck at fishing, adding to the already crowded inshore and the depletion of fish. Faced with the empty seas, they dream of heading for El Dorado over the seas.

OFFSHORE

In Wolof, the local Senegalese language, *Barça or Barsaax* is simple rhyming slang that illustrates the far from simple journey men are willing to embark on to reach Europe. It means "Barcelona or die" and is an apt description of the sacrifice thousands of men make for their families.[4]

This cry echoes with what Paul Gilroy refers to as "offshore humanity"—those vulnerable, infrahumans torn away from land and stranded offshore (2018, 18). This a gendered situation: it is overwhelmingly men who leave but when the young men leave, they leave behind mothers, sisters, daughters, girlfriends, and wives—who may have helped find the money for the journey. Yayi Bayam Diof, the mother of a son who perished at sea, has formed an association of grieving mothers. As she bluntly puts it, "[T]his is . . . to remind all of those who say 'Barça or Barsaax' (Barcelona or death) that we, the women will have to carry the burden of their untimely deaths, day by day" (cited in Ifekwunigwe 2013, 224). They have lost sons, and may be in debt from paying for their unsuccessful journeys. They may have been depending on their sons to send back money. Senegal is the fourth largest recipient of remittance monies in sub-Saharan Africa. Fatou Cisse (n.d.) writes that "the World Bank (2010) estimates that remittances sent through formal channels increased from US$344 million in 2002 to US$1,288 million in 2008." Families need the remittance from offshore migrants, and, as we will see, the country trades its fishing resources for remittance. Such is the pressure that migrants who have been deported back to Senegal will make the journey again and again, often without telling their families. In Luna Vives's study, they told her they were "'ashamed' to face their families, and in particular their mothers" who had taken loans to pay for their sons' trip (Vives 2017, 190).

In 2015, Gilroy gave the *Antipode* talk to the Royal Geographical Society and the Institute of British Geographers (RGS-IBG) annual conference, where he expanded on the figure of the offshore migrant. It was an incisive and tough call to geography, and cultural theory more generally. He punctured what he called "high-altitude theorising" that under the "enthusiasm for the Anthropocene . . . must be appreciated as part of the contemporary crisis of *radical* thought and imagination" (2018, 10).

Gilroy is, of course, one of the great forces behind British black cultural studies and ethnic and African-American studies, and a key analyst of raciology. Gilroy uses theory, literature, histories, and ideas with erudition and deadly grace.

I sat in the auditorium thrilled by the span of Gilroy's erudition, his political sensibility, and his insistence on a "lowly watery orientation" (2018, 10). Gilroy's grasp of history and politics is legendary, and the cut and thrust of his argument against much of Anthropocene thinking was electrifying, especially perhaps in a room full of scholars gathered under the conference's rubric of "Geographies of the Anthropocene" (Gilroy 2018).

Gilroy's argument in the published version of that talk (Gilroy 2018) extends many of the themes he has developed over the years. For instance, one can feel the pull of Herman Melville's *Moby Dick*, which has been an important source for Gilroy as well as for the venerated Trinidadian writer and activist C. L. R. James who has long influenced Gilroy's thinking. As Gilroy writes, "For James, Melville's relationship to the sea and its global commerce had enabled him to see the future of capitalism more clearly than any other writer of that time" (2018, 6). James was in prison on Ellis Island, awaiting deportation when he wrote *Mariners, Renegades and Castaways: The Story of Herman Melville and the World We Live In* (1953). He variously described himself as a Trotskyist, Leninist, and humanist-Marxist, but he saw in *Moby Dick* a cutting indictment of American capitalism. We can hear James's inspiration in his description of how Melville brought together disparate themes:

Nature, technology, the community of men, science and knowledge, literature and ideas are fused into a new humanism, opening a vast expansion of human capacity and human achievement. (James, cited in Gilroy 2010, 623–24)

It is clear that Gilroy's insistence on building a new vision of humanism is deeply inflected by James. Over the years, Gilroy has progressed this project through several books and articles. As he says elsewhere, "I've got very much drawn into thinking about the technological processes, objects and systems that mediate our relationship to the constitution of racialized bodies" (cited in Bell 1999, 28). In his *Antipode* lecture, the "planetary humanism" inspired by Fanon and Césaire becomes an "offshore humanism":

My hope is that it can be excavated from the unique conceptual space in which combative antiracist humanism has repeatedly confronted colonialism, racism and nationalism . . . with the performative, infrahumanizing potency of that troubling neofascist and racist rhetoric [on the issue of migrants] in mind, we should be prepared to ask what we might now imagine to distinguish ourselves, our vulnerability and our precarious relationship with one another as human beings? (Gilroy 2018, 16)

Gilroy's article recursively engages with the case of Antonis Deligiorgis, the Greek soldier who rose to fame on the front pages of British newspapers as the hero who saved twenty drowning Syrian and Eritrean migrants in one of the many boat arrivals to Greece in the summer of 2015. The UN's International Organization for Migration (IOM) states that 3,771 died in attempts to enter Europe via the Mediterranean Sea in 2015 (IOM 2018). The vast majority in 2015 were from Syria, Afghanistan, and Iraq, many of whom traveled to Greece from Turkey on inflatable boats. In an astonishing feat of logistics, the nongovernmental organization United for Intercultural Action compiles lists of deaths since 1993. From 1993 to May 5, 2018, 34,361 died trying to get to Europe. The *Guardian* recently republished their "List" along with an infographic that represents those 34,361 deaths as a sea of red and blue dots, which include the 27,000 drowned. The "List" counts fifty-six pages of tightly framed tables that include columns for names, destination, cause of death, and source of information. For example:

> 29/04/1819 N. N. (1 man) Africa 16 drowned in shipwreck off Cap Falcon, Oran (DZ) on way to Spain; 3 missing, 19 rescuedObsAlgerie/Caminando/EFE/Réf/QUOTI/IOM (www.unitedagainstracism.org)

Gilroy carefully describes Deligiorgis's action to save all those he could from drowning that day in Crete. He offers Deligiorgis's example as a springboard for a wider thinking about how "his bravery could have something to teach the rest of the EU about primal, humanitarian responsibility to and for others less fortunate than oneself" (2018, 18). Acknowledging the pitfalls of focusing on one man, Gilroy (2018, 18) argues that "there are other things going on in this shoreline drama":

> Deligiorgis and the people he saved were all soaking wet. The rescuers had to battle against relentless waves that "kept coming and coming." Their salty saturation communicates something of the way that being human is transformed when the solidity of territory is left behind. We are afforded a glimpse of vulnerable, offshore humanity that might, in turn, yield an offshore humanism. (18)

Of course this was but one situation, and the pages and pages of names of those who have died trying to flee catastrophic circumstances at home can dwarf this lone heroic figure. But Gilroy's tactic is to draw attention to "small acts" (Gilroy 1993) of humaneness which might spark a "*creative* re-enchantment of the human" (Gilroy 2018, 19).

Perhaps because of his less than laudatory comments about the state of Anthropocentric thinking, Gilroy was roundly attacked. Andrew Baldwin didn't wait until the talk was published before he went on the attack:

> That Gilroy locates his new humanism in the spontaneous act of rescue is laudable; recalibrating what it means to be human amid the tragedies of contemporary migration in the Eastern Mediterranean is a pressing task. I am, however, less convinced that this humanism can be applied uncritically to the plight of climate refugees, our yet-to-come Other par excellence, without at the same time giving full heft to the colonial and racial histories that mark its body. (Baldwin 2017, 301)

It seems astonishing that this critic—a white, Canadian geographer based in the United Kingdom—could not be aware of Gilroy's continuous contributions to new ways of reading "colonial and racial histories," and gleaning from them the seeds to rework the politics of the present. As Sindre Bangstad puts it:

> Gilroy's work speaks to practically anyone interested in the vexed questions relating to "race," racism and nationalism in our time, and offers a trenchant multi-pronged critique of various forms of essentialism from the point of view of an engaged postcolonial scholar long committed to anti-racism and anti-colonialism and the furthering of multicultural cultures of conviviality. (Bangstad 2019)

Perhaps what got to this academic was Gilroy's commanding call to replace the high theory wrangling about human exceptionalism with a "determination to destroy the bitter stratification of that species along the lines specified by race" (2018, 11).

I admit that I squirmed at times during Gilroy's talk. The manuscript of my book, *Eating the Ocean*, was trundling its way through publication. And it was much too late to rectify my superficial engagements with race. I could but blush inwardly at my obsession with "more-than-human fish" (Probyn 2016) ignoring that the entire history and present of "eating the ocean" is riven by role of "the infra-human": "the slaves from many parts of Africa who were exchanged for rum, cloth, salt-cod and other commodities" (Gilroy 2018, 5). How had I avoided the fact that the "belligerent, oceanic operations [of the economic infrastructure of Atlantic modernity] were possible only because those vulnerable people were, and often still are, judged to belong to nature rather than to history, society, or culture"? (Gilroy 2018, 5).

In what follows I want to consider how the degradation of fishing results in "clandestine migration." This is to consider how some boat migrants are bound up in the local manifestation of global politics and problems around fishing rights. Climate change exacerbates, of course, the precarious state of oceans and fish. But less nebulous forces are also stripping the ocean bare.

GUEDJ AMOUL BANKASS ("THE SEA HAS NO BORDER") (BEATLEY AND EDWARDS 2018)

Jayne O. Ifekwunigwe's account of the Africans "dying to go to Europe" places them as "the latest geopolitical manifestations along the interconnected historical continuum of transatlantic slavery and mercantile imperialism" (2013, 218). She connects Africans "crammed like sardines into vessels and transported across hostile waters" within a history that spans "more than five centuries and [that is] still unfolding" (2013, 220, 229). This has produced and continues to reproduce a "Living Africa [as] vast, heterogeneous, complex, and contradictory and thus can be partially . . . reduced to the lowest common denominators of AIDS, corruption, tribalism, wars, ethnic cleansing, poverty, famine, and political violence" (2013, 229). Ifekwunigwe places today's young Senegalese men, these "clandestine boat migrants," as part of this historical present, but not as "synonymous with the inhumanity of transatlantic slavery" (2013, 229). As she puts it, African slaves "were undesired but desirable commodities" whereas present-day clandestine African migrants are "unwanted and 'officially' unnecessary" (229). In Spain, as that country's unemployment rate soared there were fewer and fewer jobs that the Spanish wouldn't do—the previously unwanted, badly paid, and insalubrious ones that Senegalese migrants had been happy to take.

In 2005 and 2006, the preferred route of Senegalese men trying to get to Europe was by sea:

> The distance between Dakar, Senegal's capital, and the Canary Islands, is 938 miles. In the flat-bottomed wooden boats of 14 to 18 meters this trip takes between five to eight days, depending on weather conditions, the navigation skills of the captain and the quality of the boat. The boats in general have no roof and hold a capacity of 50 to 80 persons, depending on the smuggler and the size of the boat. (Poeze 2013, 25)

The boats would have traveled along the Canary Current, a wind-driven surface current that unites Senegal and the Canary Islands. This current causes upwellings where the nutrient-dense cooler waters are brought to the surface. This is what makes the coast of western African so fecund with marine life, and provided the once plentiful fish for Senegalese fishers. From Poeze's research, Ifekwunigwe quotes a man who attempted this route:

We were on the open sea for eight days. The fourth day, water started to enter the boat. At that time, food and water was also finished . . . the fifth day, we encountered a big European ship that was fishing for tuna. They gave us bread and milk and a little bit of water, but they refused to take us with them. (cited in Ifekwunigwe 2013, 229)

What tragic irony: a large industrial tuna boat passes by the noses of former fishermen in their small *pirogues*, men who could have only dreamed of catching such a valuable fish. The fish they would have previously caught was the small sardinella, the staple of Senegalese diets. But now foreign-based industrial fishing take this once plentiful fish meal and grind it into fishmeal in the many Chinese-owned factories in neighboring Mauritania (Tacon and Metian 2009). This small fish was especially important for the forty thousand women processors who used to cure the fish for sale across the entire country. This labor-intensive production and trade of low-quality fish products is crucial to the food security of not only Senegal but much of western Africa (Béné, Lawton, and Allison 2010).

LINES IN THE SEA

The United Nations Convention on the Law of the Sea (UNCLOS) is a key player in the changing fortunes of fishing nations. It is hard to believe it was only in 1982 that this global convention set out the jurisdiction of the oceans, effectively nationalizing what Hugo Grotius referred to in 1608 as *Mare Liberum*, "the Freedom of the Seas" (Grotius 1916). It took another twelve years to ratify it. In an article published in 1981, Parzifal Copes set out the terms that were to be agreed on in the 1982 Convention:

The fisheries provisions . . . are defined with reference to an exclusive economic zone (EEZ) extending beyond a 12-mile territorial sea to a distance of 200 nautical miles from a coastal baseline. Article 56 of the Draft Convention accords to each coastal state in its EEZ "sovereign rights for the purpose of exploring and exploiting, conserving and managing the natural resources." . . . Article 61 makes the coastal state responsible for conservation of the living resources (i.e., fisheries) in its EEZ, and Article 62 requires it to promote the objective of optimum utilization of these resources. (1981, 218)

Copes states that by "the beginning of 1981 nearly all the world's 200-mile coastal zones were attached by national claims to fisheries jurisdiction" (1981, 218). With the hindsight of the decades since then, it becomes clear what an entangled mess these Articles set in motion. As Elizabeth Havice puts it, UNCLOS "at once demarcated and blurred the territorial bounds of state" (Havice 2018, 1283). While Article 56 sets out that all coastal nations receive their entitlement to their EEZs, Article 61 puts the onus of protecting those resources to nation states, no matter the extent of their economic or coastal security development. Then Article 62 provides a kicker—coastal states have to authorize other states to catch the "surplus" resources they do not have the means to exploit. As Copes states, "essentially, the rules constitute a commandment that 'thou shalt not waste fish,' imposing a moral obligation on the coastal state to be reasonable in sharing resources in excess of its own capacity to utilize them" (1981, 218).

At the time, many commentators in the Global North considered that this would be to the benefit of developing states. As Copes stated in 1981:

> For the time being a good case can be made for extensive co-operative arrangements between distant-water countries and developing coastal states. The latter are often not in a position, technologically or economically, to utilize fully their fish resources. . . . [T]hey may obtain much needed foreign exchange from access fees they charge. (1981, 224)

In 1981, Tony Loftas also saw the potential for benefit: "as part of its mandate to assist in fisheries development and management, the UN [FAO] focused at the time on helping the countries of the Third World to secure the maximum benefits from the new ocean regime" (Loftas 1981, 229).

However, distant-water fishing nations are generally countries with large domestic market requirements for fish that cannot be met from their own coastal fish resources (Loftas 1981, 223). At the time, those distant-water fishing nations consisted of the countries that now constitute the European Union, Japan, South Korea, and the Soviet Union. No one would have envisioned China's present might. Copes was hopeful that the arrangements between developing coastal nations and these powerful players would result in the distant-water country supplying "fish [to the nations who owned the EEZs] for their domestic market at a reasonable cost, and to help the coastal country to develop its own fishing industry in the longer term. The requirements may include the construction of port, handling, and processing facilities, the supply of equipment and the training of local labour" (1981, 224).

Unfortunately, according to Pierre Failler and Thomas Binet, even the relatively well-regulated European Union has "failed to meet their stated aim, improved management of fish stocks—indeed, they have contributed to fisheries' degradation" (2011, 168). It seems hard to believe that at the time of the Convention it would not have been obvious what might happen with the provision that states unable to fully exploit their resources should turn to distant-water fishers. As Failler and Binet point out, "the technologically advanced EU vessels compete with partner-country fleets, particularly small-scale artisanal fishing boats for increasingly scarce fish stocks" (2011, 168).

There are no prizes for guessing which side wins.[5] In addition, countries in the Global South deal with an uneven playing ground whereby the European and Chinese fishing fleets are massively subsidized. The Chinese government spent $22 billion on subsidies to the fishing industry between 2011 and 2015 (Jacobs 2017). But all countries have to play within the World Trade Organization (WTO) guidelines on subsidies. In a twist "on the new green" (Green 2019), the EU subsidies that go to restructuring the European fishing industry and reduced fishing intensity are dubbed "green" by the WTO. However, once ships arrive in a partner country's or subregion's waters, they increase fishing intensity and participate de facto in overfishing. From the European point of view, they are "green subsidies," but for the partnering developing nation, they are regarded by the WTO as "red" and bad subsidies (Failler and Binet 2011, 168).

As Failler and Binet conclude, "the surplus issue disappears, despite its intended centrality in states' decision-making" (2011, 168). The provisions of the 1982 UNCLOS, however well-intentioned they were—and it has to be remembered the immensity of formulating and ratifying UNCLOS—have resulted in the current situation. As Failler and Binet put it, this "perfectly illustrates a short-term view of the situation: financial contributions from fisheries licenses and sales contribute greatly to public revenue in partner countries, even as they harm national fleets and marine ecosystems" (Failler and Binet 2011, 169).

Senegal, like many African governments, has tended to prefer short-term financial compensation from fisheries access agreements over developing its own informal domestic sector in part because, as a result of its somewhat chaotic nature, it is harder to extract revenue from local fishermen than from state governments and multinational fishing concerns (Brown 2005, 2). But as Mario Alberto Da Silva, a West African artisanal fisherman states, "[F]or us, it has no sense or benefit because the industrial fishing boats don't leave us any chance of survival. They fish right up to the coast without being stopped and the government doesn't have the means to control their activities. If the government would listen to us, we wouldn't sign an agreement with people who catch everything, even the small fish" (in Brown 2005).

As I mentioned before, the biggest threat to global fishing stocks, China, was not even envisioned in the 1982 Convention. As a 2012 European Parliament report states, "[T]he beginning of Chinese distant-water fishing is relatively recent, starting in 1985, in an attempt to restructure the capture fisheries sector by moving fishing capacity from Chinese coastal waters to the open oceans. This process started when one of China's state-owned fishing enterprises expanded outward, notably to Africa" (Blowmeyer et al. 2012). Since then, Chinese distant-water fishing fleets have become the elephant in the sea when it comes to sustainability. But many other nations benefit from its tremendous fishing power: "[T]he EU imported €1.5 billion of fishery products from China in 2010, about 6% of all fish consumed" (Blowmeyer et al. 2012).

This plays out in diverse and destructive ways: for instance, along the Senegal-Mauritania border competition over dwindling fish pits Saint-Louis fishermen against the coast guard of their northern neighbor. In Mauritania, the Chinese have built twenty fishmeal factories. These are the last straws for overfished seas. A 2018 report states, "The hungry machines of [the Chinese owned] Africa Protéine are producing fishmeal—a nutrient-laden powder that fuels the $160 billion aquaculture industry" (Green 2018). As Abdou Karim Sall, president of an association of small-scale fishermen in Senegal, states, "in four or five years, there won't be any fish stocks left; the factories will close, and the foreigners will leave. We'll be left here without any fish" (Green 2018). The sea really will be empty.

If you don't have any fish, where do you go? On small fishing boats to Europe. Thus, the figures with which I started this chapter meet: the Chinese fisherman contemplating empty oceans, the desperate Senegalese former fishers aboard their small *pirogues* heading for the distant coasts of Europe. And underlying them is the scaffolding of the sea that UNCLOS released. Article 62.2's dictate about ensuring the exploitation of a national surplus of fish unleashed a tsunami of overexploitation that leaves everyone high and dry.

Sustainable fishing makes the difference between staying and leaving in Senegal and in so many coastal nations in the Global South—and North. Are those leaving to be labeled greedy economic migrants and treated with disdain? Or are they climate change refugees—what Baldwin calls "our yet-to-come Other par excellence"? (2017, 301). Both labels are abstractions, taking us away from the racialized historical conditions that produce them. And, as I argued at the outset, the language of care can distance us from the visceral intersections of climate change, migration, and unequal North-South relations. However, beyond the terms we use, what matters is rendering those conditions visible so that we recognize that unsustainable fishing practices and regulations will continue to force people offshore—caught adrift in a rapidly warming ocean. It is connecting the dots—be they the ones that mark migrant deaths at sea, or the flaunting of national EEZs, and the precise ways that climate change affects differentially—that needs to be the basis of a watery humanism.

NOTES

1. H10 Hotels, "H10 Ocean Dunas." https://www.h10hotels.com/en/fuerteventura-hotels/h10-ocean-dunas.

2. In 2012, IFFO changed its name from IFFO The Fishmeal and Fish Oil Organization to IFFO The Marine Ingredients Organization. "The Fishmeal and Fish Oil Industry: IFFO's History." http://www.iffo.net/fishmeal-and-fish-oil-industry-iffos-history.

3. "Senegal's Migrants: The Lure of 'Somewhere Else,'" *DW* (May 12, 2016). https://www.dw.com/en/senegals-migrants-the-lure-of-somewhere-else/a-36590613.

4. "Senegal's Migrants."

5. See Havice (2018, 1290) for a compelling analysis of how in certain cases such as tuna fishing in the western and central Pacific Ocean, negotiations have "resulted in an unprecedented expression of power of the smallest states—states generally understood as vulnerable to and marginalized by global economic relationships—over the largest and most powerful."

REFERENCES

Baldwin, Andrew. 2017. "Climate Change, Migration, and the Crisis of Humanism." *Wiley Interdisciplinary Reviews: Climate Change* 8, no. 3: e460.

Bangstad, Singre. 2019. "A Planetary Humanism Made to the Measure of the World." *Africa Is a Country*. https://africasacountry.com/2019/02/a-planetary-humanism-made-to-the-measure-of-the-world.

Beatley, Meaghan and Sam Edwards. 2018. "Overfished: In Senegal, Empty Nets Lead to Hunger and Violence." GlobalPost/Public Radio International (PRI). https://gpinvestigations.pri.org/overfished-in-senegal-empty-nets-lead-to-hunger-and-violence-e3b5d0c9a686.

Béné, C., R. Lawton, and E. H. Allison. 2010. "Trade Matters in the Fight against Poverty": Narratives, Perceptions, and (Lack of) Evidence in the Case of Fish Trade in Africa. *World Development* 38, no. 7: 933–54.

Bell, Vikki. 1999. "Historical Memory, Global Movements and Violence: Paul Gilroy and Arjun Appadurai in Conversation." *Theory, Culture & Society* 16, no. 2: 21–40.

Blowmeyer Roland, Ian Goulding, Daniel Pauly, Antonio Sanz, and Kim Stobberup. 2012. *The Role of China in World Fisheries*. European Parliament Directorate-General for Internal Policies. http://www.europarl.europa.eu/meetdocs/2009_2014/documents/pech/dv/chi/china.pdf.

Brown, Oli. 2005. "Policy Incoherence: EU Fisheries Policy in Senegal." United Nations Development Programme. http://hdr.undp.org/sites/default/files/hdr2005_oli_brown_29.pdf.

Carolin, Christopher. 2015. "The Dragon as a Fisherman: China's Distant Water Fishing Fleet and the Export of Environmental Insecurity." *SAIS Review of International Affairs* 35, no. 1: 133–44.

Chun, Zhang. 2017. "Chinese Aquaculture Is Driving Fisheries to the Brink." China Dialogue August 17. https://chinadialogueocean.net/455-chinese-aquaculture-is-driving-fisheries-to-the-brink/.

Cisse, Fatou. n.d. "Senegal." World Bank. http://siteresources.worldbank.org/EXT-DECPROSPECTS/Resources/476882-1157133580628/RMA_Ch8.pdf.

Copes, Parzival. 1981. "The Impact of UNCLOS III on Management of the World's Fisheries." *Marine Policy* 5, no. 3: 217–28.

Failler, Pierre, and Thomas Binet. 2011. "A Critical Review of the European Union West African Fisheries Agreements." *Oceans the New Frontier. AFD, IDDRI, TERI*: 166–70.

Gilroy, Paul. 2018. "'Where Every Breeze Speaks of Courage and Liberty': Offshore Humanism and Marine Xenology, or, Racism and the Problem of Critique at Sea Level: The 2015 Antipode RGS-IBG Lecture." *Antipode* 50, no. 1: 3–22.

———. 2010. "Planetarity and Cosmopolitics." *The British Journal of Sociology* 61, no. 3: 620–26.

———. 1993. *Small Acts: Thoughts on the Politics of Black Cultures*. London: Serpent's Tail.

Green, Lesley. 2019. "Is White the New Green?" Environmental Humanities Lecture Series. The Australian Museum, Sydney.

Green, Matthew. 2018. "Ocean Shock: Plundering Africa." Reuters. https://www.reuters.com/investigates/special-report/ocean-shock-sardinella/.

Grotius, Hugo. (1609) 1916. *Mare Liberum*. New York: Oxford University Press.

Havice, Elizabeth. 2018. "Unsettled Sovereignty and the Sea: Mobilities and More-Than-Territorial Configurations of State Power." *Annals of the American Association of Geographers* 108, no. 5: 1280–97.

Ifekwunigwe, Jayne O. 2013. "'Voting with Their Feet': Senegalese Youth, Clandestine Boat Migration, and the Gendered Politics of Protest." *African and Black Diaspora: An International Journal* 6, no. 2: 218–35.

International Organization for Migration. 2018. "World Migration Report." https://www.iom.int/news/iom-counts-3771-migrant-fatalities-mediterranean-2015.

Jacobs, Andrew. 2017. "China's Appetite Pushes Fisheries to the Brink." *New York Times* (April 30, 2017). https://www.nytimes.com/2017/04/30/world/asia/chinas-appetite-pushes-fisheries-to-the-brink.html?_r=0.

James, Cyril Lionel Robert. (1953) 2001. *Mariners, Renegades, and Castaways: The Story of Herman Melville and the World We Live In*. Hanover, NH: University Press of New England.

Koopman, Jeanne E. 2009. "Globalization, Gender, and Poverty in the Senegal River Valley." *Feminist Economics* 15, no. 3: 253–85.

Loftas, Tony. 1981. "FAO's EEZ Programme: Assisting a New Era in Fisheries." *Marine Policy* 5, no. 3: 229–39.

Mallory, Tabitha Grace. 2013. "China's Distant Water Fishing Industry: Evolving Policies and Implications." *Marine Policy* 38: 99–108.

McAlister, Edward, and Alessandra Prentice. 2018. "African Migrants Turn to Deadly Ocean Route as Options Narrow." *World News*. https://www.reuters.com/

article/us-europe-migrants-africa/african-migrants-turn-to-deadly-ocean-route-as-options-narrow-idUSKBN1O210J.

Poeze, Miranda. 2013. "High-Risk Migration: From Senegal to the Canary Islands by Sea." In *Long Journeys: African Migrants on the Road*, vol. 8, edited by Alessandro Triulzi and Robert Lawrence McKenzie, 45–66. Leiden: Brill.

Probyn, Elspeth. 2016. *Eating the Ocean*. Durham, NC: Duke University Press.

Raffelli, Mateo. 2017. Dir. *Mareyeurs*, Ocean Film.

Tacon, Albert G. J., and Marc Metian. 2009. "Fishing for Feed or Fishing for Food: Increasing Global Competition for Small Pelagic Forage Fish." *Ambio* 38, no. 6: 294–302.

Vives, Luna. 2017. "Unwanted Sea Migrants across the EU Border: The Canary Islands." *Political Geography* 61: 181–92.

Chapter Three

Speculative Harbouring at Blackwattle Bay

Interdisciplinary Pedagogies and the Politics of Care

Kate Johnston and Susanne Pratt

A shared pathway wraps around Blackwattle Bay and makes for a smooth and pleasurable walk. Beyond the Bay is the vast estuary of Sydney Harbour, a drowned river valley known as a *ria*. In many ways this is an idyllic scene—runners, picnickers, kayakers, and dog walkers move against the backdrop of the Sydney Harbour. Yet urban harbors are troubled harbors as a consequence of intense urbanization and population growth over the last two centuries (Banks et al. 2016, 217). Degradation of water and sediment quality, overexploitation of resources, hardening of shorelines, habitat loss, climate change, competing economic interests, and conflicts among the many users are just some global issues facing harbors (218). Sydney Harbour is no exception. Although it is often cited for its beauty, areas are still glutted by pollution from industrial and domestic waste (Dafforn et al. 2016; Davies and Wright 2014, 458), conservationists suggest its ecosystems are disrupted by invasive species (see Banks et al. 2016, 356; Bugnot et al. 2014), and its colonial history means that many Aboriginal people who use the Bay still experience injustices.

Blackwattle Bay's impending redevelopment by UrbanGrowth NSW Development Corporation (UrbanGrowth NSW), a New South Wales Government agency, raises a set of concerns around access and ownership. UrbanGrowth NSW calls this megaproject "The Bays Growth Centre." This "growth center" is set to "deliver innovation and attract jobs of the future, reinforcing Sydney's reputation as an internationally-competitive, resilient and prosperous global city to live, work and visit" (UrbanGrowth NSW Development Corporation 2017).

As locals to the area—either as a University neighbor (in the case of Susanne) or resident (in the case of Kate)—we have often walked the Bay and asked: what does walking tell us about the Bay, including its past, present, and future? How do we engage with the Bay, not only to experience pleasure but also to foster care? How might learning about its histories, ecosystems, cultural life, regulations, and future developments inform acts of care? What opportunities exist to critically engage with the ways these facets are framed? And who tells the story of the Bay, from what point of view, and why?

The local council, City of Sydney, has provided the English-speaking public with a series of signs. There's a sign about the industrial history, another about bees, and another explains the seawall flowerpot restoration project (see Strain, Morris, and Bishop 2017, 1). Signs also warn people not to fish west of Sydney Harbour Bridge (including in Blackwattle Bay) because of elevated levels of dioxins. Although these signs provide useful information, they offer only a partial story. They conform to a factual format familiar to nature trail signs or traditional nature field guides, rather than opening up questions or promoting reflection. Of course, there are other ways to read the Bay. Jakelin Troy[1] (professor of linguistics and director of Aboriginal and Torres Strait Islander Research at the University of Sydney) reminds us that even if there is no official signage about the Aboriginal history of the Bay, there are other signs in nature; you just need to learn how to read them. Nonconforming or disruptive actions can also draw attention to issues facing the Bay. For example, most evenings Kate encounters a man and his dog along the path from Rozelle Bay to Blackwattle Bay moving slowly, stopping to talk to passersby, always with a bag filled with rubbish, which he has picked up from the mudflats. This act is a reminder of the problem of rubbish in the harbor. Another example is the sight of people fishing in the Bay.

Blackwattle Bay became the site of a two-day interdisciplinary postgraduate walkshop—Speculative Harbouring—and is the focus of this chapter. The walkshop took place in 2017 and was centered around walking and the creation of a field guide—*Speculative Harbouring Field Guide to Blackwattle Bay*. Blackwattle Bay was chosen as a field site because, like many harbors, it is a site of material and cultural confluences and faces radical change through the impending redevelopment by UrbanGrowth NSW. Within plain sight, various compounding crises are occurring that raise questions of responsibility. For example, water, waste, and toxins from drains and engineered rivers mix with brackish sea water and tides, shaping aqueous flows between land, harbor, and ocean. Furthermore, histories of colonial violence continue to shape the present, and future, of the Bay. For example, although Aboriginal people may be consulted about ways to represent their heritage (New South Wales [NSW] Government 2017, 20) there seems to be little effort in the way of involving the local Aboriginal communities in the broader development of the Bay. Winkiel argues that a wide range of interdisciplinary approaches are needed to address "the systemic crises of the oceans," including responses that interrogate "human/nonhuman marine relations in order to rethink the ethics of human actions toward the nonhuman world so that they might be nondominating" (2019, 2). To generate a field guide that could carefully engage with rethinking ethics in the Bay we assembled an interdisciplinary team. The exchanges that ensued afforded the group an opportunity to notice a wider range of issues and use a variety of interpretative lenses than any one discipline alone could provide, it also helped us to move beyond a dominating separation of land and water that infrastructures in the Bay—fences, sea walls—and colonial legacies enact.

SPECULATIVE HARBOURING: A TWO-DAY WALKSHOP

Through this case study, this chapter suggests that *critical* field guides can be tools for learning in public—a form of public pedagogy with the potential to disrupt and reimagine dominant narratives. Our guide joins a growing number of modern field guides that share in this critical approach. As we shall elaborate, such guides can promote connection to, and care of, place by focusing attention to the often-overlooked aspects of the Bay and critically questioning multiple perspectives and knowledge frames. As Metzger writes, "to learn to pay attention is . . . fundamental to learning to care, as attention *sensitizes* us to that on which we focus our attention" (2014, 1004).

We do not claim that the field guide disrupted dominant narratives within a wide public (that is the field guide had a small audience); however, the coproduction of the guide and responses from the audience within which it circulated does demonstrate the potential for the genre. That is, there is potential when a field guide departs from its single disciplinary mode and historical focus on identification, to open up ways to critically engage with place, in particular, ways that might promote care and facilitate community consultation processes in order to co-create harbors for more humans and nonhumans to thrive.

Walking itself promotes a certain kind of noticing and engagement. We refer to Speculative Harbouring as a *walkshop* because of the emphasis on walking and creating. At a basic level, a walkshop can be conceived as a workshop undertaken through walking (Wickson, Strand, and Kjølberg 2015, 243). The kind of walking that we proposed through the walkshop was akin to what Stengers (2005) suggests is a slowing down, not in terms of speed but rather to enable the asking of critical questions. Springgay and Truman add to Stengers, saying this can "create openings where different kinds of awareness and practices can unfold" (2018, 15). Walking was also important for the kind of field guide that the group produced—a guide that was meant to be carried on a walk of the Bay, to help walkers slow down and critically engage with the Bay.

Day one was arranged around what Anna Tsing calls *Arts of Noticing*— "ways of teaching open yet focused attention" (2011, 19). For Tsing, there are many *arts* of noticing: taxonomy, drawing, music, and poetry, to name a few. Recalling taxonomy and a meeting with a curator of fungi at Copenhagen University's Botanical Museum, Tsing says, "Here, through naming, we notice the diversity of life" (2011, 192). Our interpretation of the arts of noticing is that any one art, while offering an opportunity to notice and connect, is partial. For this reason, we brought together an interdisciplinary group, including students from a variety of disciplines (architecture, government and international relations, fine arts, geography) as well as practicing artists, a professor of linguistics, a government employee, and experts with different experiences and knowledge of the Bay (the Aboriginal Elder Uncle Mark and the marine scientist Ross Coleman). We were interested in what new forms of engagement, perspectives, and reflections might come from interactions among this group.

Participants were invited to share a method from their respective discipline or practice that they then experimented with as we walked the Bay. In addition, participants attended to what the feminist philosopher Val Plumwood (2008, 147) calls *shadow places*—those places that are rarely considered but "whose degradation we as commodity consumers are indirectly responsible for." This included what we could call *shadow objects* and *issues*, such as seawalls, drains, plastics, the persistent yet invisible pollution, colonial history, contemporary Aboriginal connections to Country, and labor.

Our guiding questions were: "How might we better care for, and with, harbors? What modes of disciplinary and transdisciplinary practice, including art and activism, might best support this care work?" (Johnston and Pratt 2017, 5).

At the end of day one, we introduced participants to the idea of a *matter of care* and asked them to choose a particular matter that had arisen over the day and return to the Bay to engage with their matter. Maria Puig de la Bellacasa (2011) presents a feminist vision of care that goes beyond a moral disposition and implies more than concern.[2] Rather, care fosters "a strong sense of attachment and commitment to something" (89–90).

Day two moved from walking and arts of noticing to reflecting on representational practices and politics. We explored the role field guides could play in mobilizing different ways of noticing and caring in the Bay. Participants each developed a field guide entry to represent their matter of care and encourage readers to notice and engage with that matter as they too walk the

Bay. These pages combined to generate the *Speculative Harbouring Field Guide to Blackwattle Bay.*

INTERDISCIPLINARY CRITICAL PUBLIC PEDAGOGY

In reflecting on the use of field guides within the context of teaching and learning, we became interested in how field guides could be a form of public pedagogy that encourage a deeper engagement with and care of waterways and seas. Burdick, Sandlin, and O'Malley define public pedagogy as "various forms, processes, and sites of education and learning beyond or outside formal schooling" (2014, 2). Although our walkshop took place within the formal educational structure of a postgraduate event, in this chapter we are interested in positioning field guides as a tool for public pedagogy. The rationale for this emerges out of our desire to explore ways of facilitating interdisciplinary arts of noticing to engage with the core questions driving the walkshop. Complex problems, such as polluted harbors, require ways of knowing that move beyond disciplinary silos and operate in, outside, and across academia to engage multiple publics in shared matters of care.

In their broad survey of public pedagogy literature, Sandlin, O'Malley, and Burdick (2011) propose that public pedagogy consists of five key domains: citizenship within and beyond schools, pedagogical theory on popular culture and everyday life, informal institutions and public spaces as educative arenas, dominant cultural discourses, and public intellectualism and social activism. The Speculative Harbouring walkshop speaks to these domains by drawing on people's experience, fields of expertise, and particular matters of care throughout the walkshop and by using the field guide as a form of popular culture for teaching and learning. Furthermore, the walkshop took place outside of traditional educative space and it sought to question dominant cultural discourses about Blackwattle Bay. Finally, the field guide was circulated to passersby, including local residents, in what could be seen as a form of social activism.

However, not all forms of public pedagogy are activist in orientation. Salvio proposes that public pedagogy "at times imposes a hegemonic force, while at other times is used to enact cultural and political resistance and counter-hegemonic possibilities as well as generate critical engagements with knowledge that is difficult to recognize or to come to terms with" (2014, 101). We hence have preference for the term "*critical* public pedagogy" in this discussion, as proposed by Sandlin, O'Malley, and Burdick, to distinguish between forms of everyday culture that take a critical stance and can "decode and interrupt dominant ideologies of race, class, gender, sexuality, militarism, and neo-liberalism" (2011, 347). In this chapter, the "critical" places emphasis on the way the cocreation of field guides can be used to disrupt dominant discourses and to "generate critical engagements" around matters of care. Giroux (2004)—who is influential to the field of public pedagogy—elaborates on the critical potential of pedagogy: "As a critical practice, pedagogy's role lies not only in changing how people think about themselves and their relationship to others and the world, but also in energizing students and others to engage in those struggles that further possibilities for living in a more just society" (2004, 64).

Although scholarship on public pedagogy has been applied to a range of fields, including art (Zorrilla and Tisdell 2016) and civic participation (Hickey-Moody 2013), there is currently a paucity of research that explicitly links critical field guides to theory on public pedagogy. Considering the contemporary proliferation of such guides, it is important to consider their public function. It is outside the scope of this chapter to provide a full account of ways in which field guides are currently being used in formal and informal education or the long-term effects of field guides on creators, readers, and matters of care; however, we use our experience to open up possibilities for ways in which field guides can contribute to critical public pedagogy and to the care of waterways. In the next section—drawing on the legacy of field guides as an educational tool used both inside and outside formal educational settings—we propose that adaptions of the field guide genre can be created and used outside formal educational contexts as a form of critical public pedagogy. These guides offer opportunities to engage with contemporary discourses across, for example, popular culture, civic issues, and social activism.

FIELD GUIDES

Field guides cross boundaries, for example, between home and work, or academia and popular culture. They accompany walkers, traveling between homes, parks, beaches, bays, public libraries, botanical gardens, forests, and university bookshelves, to name a few sites. They are picked up and passed around by citizen scientists, artists, ornithologists, students, trained biologists, librarians, birders, and other publics. While researching *A Field Guide to Field Guides*, a bibliographic guide, Diane Schmidt (1999) encountered various responses from librarians she spoke to. Some suggested field guides were "lowbrow fluff" and others who worked with trained biologists and amateur scientists responded with enthusiasm (Schmidt 2006, 274). This boundary-spanning dimension of field guides makes them a useful and flexible format for interdisciplinary public pedagogy.

In this chapter, we distinguish between *traditional* field guides, which are "handy, inexpensive, portable identification guides" (Schmidt 2006, 274), and *critical* field guides, which loosely interpret and experiment with the field guide format to critique dominant discourse and "further possibilities for living in a more just society" (Giroux 2004, 64; see Floating Studio for Dark Ecologies [FSDE] on "experimental" field guides). In this section, we first briefly introduce the history of traditional field guides, then discuss critical field guides that informed our own guide.

Field guides with species names, keys, descriptive texts, alphabetical indices, and images to help with identification are now familiar to many (see Hawthorne and Lawrence 2006). For example, these different identification systems become handy tools when you want to distinguish between a sulphur crested cockatoo versus a white umbrella cockatoo. Early instances of field guides occurred before the fourteenth century in illustrated manuals, including herbals (Givens 2006; see also Farnsworth et al. 2013). Sara Scharf (2009) links the development of the field guide identification systems to eighteenth-century botanical identification manuals and the creation of different identification systems.

Rooted within this development are systems of ordering based on enlightenment ideals of man as master over nature, and on the ideal "that man, made rational in God's image, was uniquely positioned—primed, in fact—to understand His plan for the universe" (Scharf 2009, 89). Although there have been innovative developments in format, such as online versions and apps (Farnsworth et al. 2013), we are interested in those field guides that question enlightenment binaries such as man-nature, land-sea, and associated "mutually reinforcing dualisms" (Plumwood 1993, 443) such as mind-body, male-female, white-other. A critical questioning of enlightenment ideals and binaries is important as traditional field guides and their taxonomic structures can contain narratives that reinforce unjust distribution of power.

In contrast, contemporary critical field guides offer different forms of identification that move beyond natural sciences and explore different fields, such as drones (Pater 2013) and social ecologies (for example Sarah Kanouse 2010). Writing on the "resurgence of 'field guides' in a networked age," media studies scholar Shannon Mattern argues that "the field isn't a pre-existing territory that one enters; the researcher brings it into being by conceptually delimiting it, and then 'working' it" (2016, n.p.). Rather than proposing tools for identifying a supposedly passive nature from the position of an agential human, Mattern's point highlights that fields, and other matters, are brought into being through attention, which not only creates the field but also cocreates the actor bringing the field into being. This disrupts binaries of subject-object worldviews. Writing specifically on water and "aqueous politics of location," Neimanis (2013, 39; 2017) also highlights the coconstitution of humans and other "bodies of water," and the responsibilities that result from these entangled relations. She states, "our human responsibilities [are] within a more-than-human aqueous ecology—that is, an ecology in which humans and other bodies of water (animal, vegetable, meteorological, geophysical) are always already implicated, as lively agents, in one another's well-being" (Neimanis 2014, 6). We are curious how field guides as a form of public pedagogy can play a role in disrupting dominant worldviews that can have negative effects, such as when subject-object binaries engrained in the scientific method are used to perpetuate colonial taxonomies and ignore human relationships to different species and place.

A number of field guides experiment with the format, turning away from natural science and identification of different species to concerns that are simultaneously natural and cultural. For example, Ruben Pater's *Drone Survival Guide: Twenty-First-Century Birdwatching* (2013, n.p.) plays with the notion of birdwatching, and instead depicts silhouettes of "common drone species," ones often used in warfare or surveillance. He asks, "[O]ur ancestors could spot natural predators from afar by their silhouettes. Are we equally aware of the predators in the present-day?" The guides include information on hiding from drones and are printed on Chromolux ALU-E mirrored paper, a reflective material that enables users of the guide to potentially hide from drones by using the guide to reflect the sun at the drone's camera. The artist Sarah Kanouse's *Post-naturalist Field Kit for Saint-Henri* (2010) also plays with the tools of naturalists but seeks to challenge the division of nature and culture that many previous naturalist's field guides promote. As Kanouse states, the kit "updates the naturalist's toolbox for the exploration of the social ecologies of urban landscapes." The kit contains tools such as tweezers, air quality monitors, specimen jars, and field maps and cards that invite people to reflect on social and environmental concerns. Another example that challenges nature-culture distinctions in a water context is *A Field Guide to the Dark Ecologies of Newtown Creek* by the media art collective Floating Studio for Dark Ecologies (FSDE, 2017). Newtown Creek is a Superfund[3] site in New York that flows between Brooklyn and Queens. Describing their use of the field guide format, FSDE explains how it differs from traditional guides "couched in the language of 'Nature' and Natural History, [in which] entries point to things—whether species or stars—as objects within a rigid taxonomy" (FSDE 2017, introduction). Rather, the guide "draws on these roots, but departs radically in the way it organizes and presents you with information, giving more attention to place and ecology, embedding the system of entries in a framework of inquiry and investigation that offers you multiple ways to engage with Newtown Creek."

OUR FIELD GUIDE EXPERIMENT AT BLACKWATTLE BAY

As demonstrated previously, critical field guides can open questions, such as: can we decolonize the field guide? How might we tell environmental and social narratives differently? What tends to go unnoticed in traditional field guides? How might we present alternative modes of understanding that are sympathetic to others (human and nonhuman) and to water environments? We were particularly interested in how we could experiment with the field guide format to care better for the environments within which we walk, wade, work, and play around Blackwattle Bay. We set out to agitate any one disciplinary mode for the purpose of promoting what we call *careful noticing*—that is, "the kind of noticing which opens up possibilities of understanding, empathy and care, and different sensory engagements with place as well as attention to the less noticed aspects of the harbour" (Pratt and Johnston forthcoming).

After looking at a variety of field guides, the group gravitated toward the way in which the cards in the *Post-naturalist Field Kit for Saint-Henri* started with a question, a quote that elaborated on the question, and then different activities to engage with the question being posed.

Participants' entries critique singular dominant perspectives. For example, through a human geography lens, Kate Eager's (2017) contribution draws attention to the limits of mapping in representing human experience and land-sea confluences. Natalie Pearson, who is trained in heritage and museum studies asks, "What maritime histories can be told here? Whose stories and pasts are privileged in the telling of these maritime histories?" (2017, 15).

The collaborative work of Christine Winter, Louisa King, and Jakelin Troy allowed each participant to not only share their disciplines of linguistics, architecture, and government and international relations but to also move beyond these disciplines. Their entry includes Dharug names relevant to the Bay and also invites readers to think about the material relationship of human bodies and the Bay through minerals common to human bodies and bays (2017, 28).

Entries also draw on multiple senses and question subject-object relationships. For example, Jamaya Masters represents the embodied, physical movements of shuckers, experiencing the labor of another through repetition, and contrasts this to the cultural meanings associated with oysters, including leisure (2017). Kassandra Bossell (2017) employs the readers to imagine themselves as and empathize with the more-than-human inhabitants of the Bay. Joseph McDonald (2017) was inspired by Uncle Mark's provocations and in his entry encourages readers to listen to Country[4] and reflect on Indigenous-led place-making and the importance of "deep listening" to design. McDonald writes, "Listening to Country is one way that Aboriginal people connect to the land and water. . . . In the spirit of careful attentiveness, turn your attention to the full range of sounds you can hear" (2017, 25).

This entry also asks fundamental questions about development: who can participate in designing the future of Blackwattle Bay? Development was also the focus of Mark Elliot-Ranken (2017), who sets an activity for readers to draw a view of the Bay from a specific location dominated by private apartments, to consider the privileging of access to certain views. Caitlin Fargher (2017) encourages readers to consider the function of the oyster in building Sydney since colonization and asks what the role of the oyster in future development of the Bay might be.

The field guide offered the opportunity to draw attention to the *shadow places* (Plumwood 2008) and *objects*. For example, inspired by the work of feminist environmental humanities scholar Jennifer Mae Hamilton (see Hamilton 2017), we encouraged the group to notice drains and think about infrastructural limitations. Drains, as Hamilton writes, "are both metaphor and metonymy for this mess we're in" (2017, n.p.). Drains offer a way to reflect on how water moved through the area when the park was a tidal swamp, before European settlers modified the Bay, compared with the ways water (and waste and debris) move through today. Drains became the focus of Bernadette Smith's (2017) contribution "Beyond Drain," which encourages readers to look for and compare drains to speculate on ways that drains could be designed differently to reduce stormwater and waste in the Bay.

Through the case of the *Speculative Harbouring Field Guide to Blackwattle Bay* and a review of other critical field guides, we have argued that such guides, as well as the cocreative and interdisciplinary process itself, are important critical public pedagogy tools. These guides offer a pedagogical platform and process to question dominant narratives of place and nature, especially legacies that see an artificial separation of nature and culture, object and subject, sea and land. These divisionist legacies are significant for harbors as they can, for example, perpetuate short-sighted technological solutions to future-proof harbors against climate change by hardening seawalls, rather than interdisciplinary approaches of recent years that offer more innovative, long-term, systems-oriented responses. For example, included in this book is SCAPE Studio's Living Breakwater in chapter 21 and the floating breakwater in Marcus's chapter 19. Importantly, *the field* and a field guide user's relationship to the *field* is unsettled and enlivened through attention and interdisciplinary provocations. Through our discussion of field guides as a form of interdisciplinary pedagogy within the context of harbors, we hope to have contributed to the growing field of the Blue Humanities.[5] We have also argued that the interdisciplinary and coproductive process of producing a field guide fosters *careful noticing*.

The impending rapid transformation of the Bay into a growth center is an opportunity to consider ways in which critical field guides could be used to facilitate public engagement with issues (e.g., development) and public spaces (e.g., hospitals, parks, harbors). For example, field guides could be useful tools for addressing wicked problems within public policy, such as water infrastructure in many coastal and rural areas of Australia. They could also function within local council processes of community consultation to help facilitate an understanding of community—what's important to the public and what the public want council members to notice and care for, or, vice versa, for council to foster noticing and care within the public.

Processes of productively unsettling dominant narratives in relation to harbors and oceans (including narratives about gender, sustainability, nature, labor) are well established in humanities and social science literature (as many of the entries in this book attest to; see also Hamilton 2017; Neimanis 2017; Probyn 2016). The critical field guide offers a different kind of tool and process, one that is public facing, interdisciplinary, and provides opportunities for students and publics to engage with pressing issues relevant to harbors and the seas with which they merge.

ACKNOWLEDGMENTS

The walkshop and field guide were made possible with funding from the Sydney Environment Institution. The project also received in-kind support from the Faculty of Transdisciplinary Innovation at the University of Technology Sydney and the Sustainable Fish Lab. The Sustainable Fish Lab was funded by an Australian Research Council Discovery Project grant (Professor Elspeth Probyn) and housed in the Department of Gender and Cultural Studies at the University of Sydney. We would also like to thank all of the people who took part in the walkshop and contributed to the field guide.

NOTES

1. Jakelin Troy was also a participant in the walkshop, which we use as a case study further along in this chapter.

2. Puig de la Bellacasa builds on Bruno Latour's concept of a matter of concern (see Latour 2004).

3. A Superfund site is a name given to sites in the United States that are deemed toxic by the federal government and are funded to clean up the pollutants, under the Superfund program.

4. "In Aboriginal English, a person's land, sea, sky, rivers, sites, seasons, plants and animals; place of heritage, belonging and spirituality; is called 'Country'" (Australian Museum 2019, n.p.).

5. For a description of the Blue Humanities, see Buchanan 2018.

REFERENCES

Australian Museum. 2019. "Glossary of Indigenous Australia Terms." Accessed November 19, 2019. https://australianmuseum.net.au/learn/cultures/atsi-collection/cultural-objects/glossary-indigenous-australia-terms/.

Banks, J., Luke H. Hedge, Caroline Hoisington, Elizabeth M. Strain, Peter D. Steinberg, and Emma L. Johnston. 2016. "Sydney Harbour: Beautiful, Diverse, Valuable and Pressured." *Regional Studies in Marine Science* 8: 353–61. https://doi.org/10.1016/j.rsma.2016.04.007.

Bossell, Kassandra. 2017. "The Sea Sleeps and Wakes as Land, Remembering Its Past and Future as Sea." In *Speculative Field Guide to Blackwattle Bay*, edited by Kate Johnston and Susanne Pratt, 46–49. Sydney, NSW: Sydney Environment Institute.

Buchanan, Ian. "Blue Humanities." *A Dictionary of Critical Theory*. Oxford University Press. February 15, 2018. Accessed April 24, 2019. http://www.oxfordreference.com/view/10.1093/acref/9780198794790.001.0001/acref-9780198794790-e-764.

Bugnot, Ana, Ross Coleman, Will Figueira, and Ezequiel Marzinelli. 2014. "Patterns of the Non-Indigenous Isopod *Cirolana harfordi* in Sydney Harbour." *PloS One* 9, no. 1: 1–7. https://doi.org/10.1371/journal.pone.0086765.

Burdick, Jake, Jennifer A. Sandlin, and Michael P. O'Malley. 2014. *Problematizing Public Pedagogy*. New York: Routledge.

Dafforn, Katherine A., Mariana Mayer-Pinto, Ana B. Bugnot, Ross A. Coleman, Rebecca L. Morris, and Emma L. Johnston. 2016. "Guiding Principles for Marine Foreshore Developments." Report prepared for UrbanGrowth NSW. University of New South Wales, Sydney: 1–53.

Davies, Peter J., and Ian A. Wright. 2014. "A Review of Policy, Legal, Land Use and Social Change in the Management of Urban Water Resources in Sydney, Australia: A Brief Reflection of Challenges and Lessons from the Last 200 Years." *Land Use Policy* 36: 450–60. https://doi.org/10.1016/j.landusepol.2013.09.009.

Eager, Kate. 2017. "Mapping Blackwattle Bay." In *Speculative Field Guide to Blackwattle Bay*, edited by Kate Johnston and Susanne Pratt, 9–13. Sydney, NSW: Sydney Environment Institute.

Elliot-Ranken, Mark. 2017. "Who Lives on the Harbour?" In *Speculative Field Guide to Blackwattle Bay*, edited by Kate Johnston and Susanne Pratt, 34–37. Sydney, NSW: Sydney Environment Institute.

Fargher, Caitlin. 2017. "Follow Deep the Oyster Shell." In *Speculative Field Guide to Blackwattle Bay*, edited by Kate Johnston and Susanne Pratt, 18–21. Sydney, NSW: Sydney Environment Institute.

Farnsworth, Elizabeth J., Miyoko Chu, W. John Kress, Amanda K. Neill, Jason H. Best, John Pickering, Robert D. Stevenson, et al. 2013. "Next-Generation Field Guides." *BioScience* 63, no. 11: 891–99. https://doi.org/10.1525/bio.2013.63.11.8.

Floating Studio for Dark Ecologies (FSDE). 2017. "Dark Ecologies Field Guide to Newtown Creek." newtowncreekfieldguide.com. Accessed January 25, 2019.

Giroux, Henry A. 2004. "Public Pedagogy as Cultural Politics: Stuart Hall and the 'Crisis' of Culture." *Cultural Studies* 14, no. 2: 341–60. https://doi.org/10.1080/095023800334913.

Givens, Jean A. 2006. "Reading and Writing the Illustrated *Tractatus de herbis*, 1280–1526." In *Visualizing Medieval Medicine and Natural History*, edited by Jean A. Givens, Karen M. Reeds, and Alain Touwaide, 115–46. Burlington, VT: Ashgate.

Hamilton, Jennifer. 2017. "All the World's a Drain: Review of *Bodies of Water*." *Sydney Review of Books*, November 15, 2017.

Hawthorne, William, and Anna Lawrence. 2006. *Plant Identification: Creating User-Friendly Field Guides for Biodiversity Management*. London: Earthscan.

Hickey-Moody, Anna C. 2013. "Little Public Spheres." *Performance Paradigm* 9, no. 1: 4–15. https://www.performanceparadigm.net/index.php/journal/article/view/129/0.

Johnston, Kate, and Susanne Pratt. 2017. *Speculative Field Guide to Blackwattle Bay*. Sydney, NSW: Sydney Environment Institute.

Kanouse, Sarah. 2010. "A Post-Naturalist Field Kit for Saint-Henri." Accessed January 23, 2019. http://readysubjects.org/portfolio/a-post-naturalist-field-kit-for-saint-henri/.

Latour, Bruno. 2004. "Why Has Critique Run Out of Steam? From Matters of Fact to Matters of Concern." *Critical Inquiry* 30, no. 2: 225–48.

Masters, Jamaya. 2017. "Becoming Shucker." In *Speculative Field Guide to Blackwattle Bay*, edited by Kate Johnston and Susanne Pratt, 38–41. Sydney, NSW: Sydney Environment Institute.

Mattern, Shannon. 2016. "Cloud and Field." *Places Journal* August. Accessed January 25, 2019. https://doi.org/10.22269/160802.

McDonald, Joseph. 2017. "Listening to Country." In *Speculative Field Guide to Blackwattle Bay*, edited by Kate Johnston and Susanne Pratt, 22–25. Sydney, NSW: Sydney Environment Institute.

Metzger, Jonathan. 2014. "Spatial Planning and/as Caring for More-Than-Human Place." *Environment and Planning A* 46, no. 5: 1001–11. https://doi.org/10.1068/a140086c.

Neimanis, Astrida. 2013. "Feminist Subjectivity, Watered." *Feminist Review* 103: 23–41. https://doi.org/10.1057/fr.2012.25.

———. 2014. "Alongside the Right to Water, a Posthumanist Feminist Imaginary." *Journal of Human Rights and the Environment* 5, no. 1: 5–24. https://doi.org/10.4337/jhre.2014.01.01.

———. 2017. *Bodies of Water: Posthuman Feminist Phenomenology*. London: Bloomsbury Academic.

New South Wales Government. 2017. *Study Requirements for Bays Market District: Nominated State Significant Precinct, the Bays. April 2017*. NSW Government Planning and Environment.

Pater, Ruben. 2013. "Drone Survival Guide: Twenty-First-Century Birdwatching." Accessed January 23, 2019. http://www.dronesurvivalguide.org/.

Pearson, Natalie. 2017. "Preserving Our Maritime Pasts." In *Speculative Field Guide to Blackwattle Bay*, edited by Kate Johnston and Susanne Pratt, 14–17. Sydney, NSW: Sydney Environment Institute.

Plumwood, Val. 1993. "The Politics of Reason: Towards a Feminist Logic." *Australasian Journal of Philosophy* 71, no. 4: 436–62.

———. 2008. "Shadow Places and the Politics of Dwelling." *Australian Humanities Review* 44: 139–50.

Pratt, Susanne, and Kate Johnston. Forthcoming. "Speculative Harbouring: Wading into Critical Pedagogy and Practices of Care." *Journal of Public Pedagogies.*

Probyn, Elspeth. 2016. *Eating the Ocean*. Durham, NC: Duke University Press.

Puig de la Bellacasa, Maria. 2011. "Matters of Care in Technoscience: Assembling Neglected Things." *Social Studies of Science* 41, no. 1: 85–106. https://doi.org/10.1177/0306312710380301.

Salvio, Paula M. 2014. "Reconstructing Memory through the Archives: Public Pedagogy, Citizenship and Letizia Battaglia's Photographic Record of Mafia Violence." *Pedagogy, Culture and Society* 22, no. 1: 97–116. http://dx.doi.org/10.1080/1468 1366.2013.877620.

Sandlin, Jennifer A., Michael P. O'Malley, and Jake Burdick. 2011. "Mapping the Complexity of Public Pedagogy Scholarship: 1894–2010." *Review of Educational Research* 81, no. 3: 338–75. https://doi.org/10.3102/0034654311413395.

Scharf, Sara T. 2009. "Identification Keys, the 'Natural Method,' and the Development of Plant Identification Manuals." *Journal of the History of Biology* 42, no. 1: 73–117.

Schmidt, Diane. 1999. *A Guide to Field Guides: Identifying the Natural History of North America*. Englewood, CO: Libraries Unlimited.

———. 2006. "Field Guides in Academe: A Citation Study." *The Journal of Academic Librarianship* 32, no. 3: 274–85. https://doi.org/10.1016/j.acalib.2006.02.014.

Smith, Bernadette. 2017. "Beyond Drains." In *Speculative Field Guide to Blackwattle Bay*, edited by Kate Johnston and Susanne Pratt, 50–53. Sydney, NSW: Sydney Environment Institute.

Springgay, Stephanie, and Sarah E. Truman. 2018. *Walking Methodologies in a More-than-Human World: WalkingLab*. London, UK: Routledge.

Stengers, Isabelle. 2005. "The Cosmopolitical Proposal." In *Making Things Public: Atmospheres of Democracy*, edited by Bruno Latour and Peter Weibell, 994–1003. Cambridge, MA: Massachusetts Institute of Technology Press.

Strain, Elizabeth, Rebecca Morris, and Melanie Bishop. 2017. "Case Study: Sydney Harbour: Enhancing Seawall Sustainability." *NSW Department of Primary Industries and Sydney Institute of Marine Sciences.*

Tsing, Anna. 2011. "Arts of Inclusion, or, How to Love a Mushroom." *Australian Humanities Review* 50: 5–21. http://australianhumanitiesreview.org/2011/05/01/arts-of-inclusion-or-how-to-love-a-mushroom/.

UrbanGrowth NSW Development Corporation. 2017. "About the Bays Growth Centre." Accessed February 15, 2019. https://www.ugdc.nsw.gov.au/growth-centres/the-bays-precinct/.

Wickson, Fern, Roger Strand, and Kamilla L. Kjølberg. 2015. "The Walkshop Approach to Science and Technology Ethics." *Science and Engineering Ethics* 21, no. 1: 241–64. https://doi.org/10.1007/s11948-014-9526-z.

Winkiel, Laura. 2019. "Introduction: Hydro-criticism." *English Language Notes* 57, no. 1: 1–10. doi: https://doi.org/10.1215/00138282-7309633.

Winter, Christine, Louisa King, and Jakelin Troy. 2017. "The Body Is Harbour." In *Speculative Field Guide to Blackwattle Bay*, edited by Kate Johnston and Susanne Pratt, 26–33. Sydney, NSW: Sydney Environment Institute.

Zorrilla, Ana, and Elizabeth J. Tisdell. 2016. "Art as Critical Public Pedagogy: A Qualitative Study of Luis Camnitzer and His Conceptual Art." *Adult Education Quarterly* 66, no. 3: 273–91. https://doi.org/10.1177/0741713616645666.

Chapter Four

Caring for the Tuna of the Western Indian Ocean

Where Politics and Ecology Meet

Mialy Andriamahefazafy, Christian A. Kull,
Pamima Leste, Patsy Theresine, and Safina Echa

EXPLORING THE POLITICS OF TUNA FISHERIES IN THE WESTERN INDIAN OCEAN

Tuna resources are the "blue gold" of the Western Indian Ocean (WIO). Although the WIO represents only around 5 percent of our oceans, it contributes between 12 and 20 percent of the global tuna catch (Obura et al. 2017, 4; Poseidon et al. 2014, 4). From the fishing activity itself to the sale of tuna locally or beyond national borders, the fishery puts in contact, congregates, and opposes actors such as small-scale and industrial fishers, intermediaries, processing companies and their workers, and governmental departments. It also involves actors farther away from the tuna itself such as distant-water fishing nations (DWFNs), which fish in coastal waters of the WIO through fishing companies based in Spain, France, Japan, or Korea. Other management stakeholders are also involved, including big nongovernmental organizations or the regional fisheries management organization for tuna, the Indian Ocean Tuna Commission (IOTC). Managing this shared resource often depends on the socioeconomic contexts in coastal countries as well as intricate sociopolitical interactions.

We use political ecology as a theoretical anchoring. Through this framework, we can present the role of broader political economic drivers in local uses of resources (see Blaikie 1989, 23; Peet, Robbins, and Watts 2011, 9–10), analyze the winners and losers (e.g., Ribot and Peluso 2003, 154; Robbins 2012, 87) and explore the role of biophysical characteristics of nonhumans in social practices of resource management (see Bakker and Bridge 2006, 18; Bennett 2010, 1–19; Robbins 2012, 23). Using the theory of access of Ribot and Peluso (2003) to look at the politics of access to the resources, we particularly focus here on the role of social relations, investigating the relations between stakeholders as well as the geopolitics between the countries involved. For the former, we shed light on the social interactions in local resource access that only a few studies in tuna fishery have looked at (see, e.g., Campling 2012a; Probyn 2016, 77–100). In global fisheries, small-scale fisheries have long been underestimated in their importance (Pauly 2018, 372). For the latter, we explore the role of governments and extraregional institutions such as the European Union in shaping the fishery and its management. By exploring those, we highlight how different actors care about the resource and the people in contact with it. We associate the concept of care to both individual welfare and welfare of the community (D'Alisa, Deriu, and Demaria 2015, 63–66), a community that in our case includes the tuna resources. Establishing care for natural resources has been recognized as dependent on the level of engagement with the resources (Probyn 2014, 291–92). It was also studied as being able to influence livelihood outcomes (Ellis-Jones 1999, 182–86) and enhancing connections between the environment and its people (Suchet-Pearson et al. 2013, 196).

Our understanding of care goes beyond the interpretation of free or unpaid work and combines *taking care of* and *giving care*, two broader ideas that cover public engagement and private tasks (Lawson 2007, 6–7; Tronto 1993, 101–26). *Taking care* involves the assumption of responsibility to respond to a need and choosing to take action toward that need. *Giving care* implies a commitment and work to satisfy a need of care in a direct relationship (D'Alisa, Deriu, and Demaria 2015, 64). By *politics of care*, we then mean the various ways through which actors choose and support these expressions of care, and the consequences of such expressions or the lack of them.

As one of the tuna species—yellowfin tuna—is currently assessed as overfished in the entire Indian Ocean (Indian Ocean Tuna Commission [IOTC] 2018, 14), the WIO is facing an overfishing crisis. There is, therefore, a need for actions to sustain the tuna resources of the region. Caring for the tuna requires taking responsibility and actions toward the improvement of the state of tuna resources as well as sustaining the benefits that stakeholders gain from the resources. We show different politics of care through which actors in the various segments of the fishery take and give care to the currently depleted tuna resources and the people dependent on the fishery. Furthermore, as tuna fisheries are often considered as mainly driven by capitalist interests and accumulation, we show that in the WIO, there is a diversity of economies at different scales. This diversity brings an equally diverse politics of care among actors.

We focus on three case studies: Madagascar, Mauritius, and Seychelles. They hold a central place in the tuna exploitation of WIO—each having an active industrial fishing port, tuna canneries, and local fishers. Local fishing in the three island nations have two components: the small-scale segment using small wooden *pirogues* or glass fiber boats with hooks or nets, and the semiindustrial segment, composed of larger boats measuring nine to twenty meters, with stronger engines and using longline as a fishing technique. Considering their distinctive socioeconomical contexts—Madagascar being much poorer than its two neighbors (United Nations Development Programme 2016)—tuna fisheries also play different role in the national economies. They are at the center of the economy of Seychelles whereas they contribute less in Madagascar, which is more focused on a cash-crop agriculture, and in Mauritius, a tertiary-based economy (Sellström 2015, 46–47). However, in all three countries, tuna is an important source of foreign revenue, job opportunities, and food security (Indian Ocean Commission [IOC] 2018; Obura et al. 2017, 15). This chapter is based on six months of fieldwork undertaken in 2017 and 2018 in the three islands. We interviewed stakeholders at landing ports and fishing villages as well as governmental offices. Interviews consisted in gathering actors' experience and perspectives on their means of access to tuna resources, the state of the resources, and their interactions with foreign fleets. We also observed landing of tuna in local and industrial ports as well as the twenty-second session of the IOTC that took place in May 2018.

In this chapter, we tell the stories of actors within the local fishery and the foreign industrial fisheries. These two levels of fishing are often seen as in conflict, with the industrial sector considered by local fishers as causing the depletion of tuna and other marine resources in the WIO. We proceed in the chapter with a reflection on the challenges the WIO faces in sustaining its tuna resources. We then conclude the chapter by pointing out the role of different actors in contributing to a politics of care that considers both the tuna resources and its people.

FROM SITUATED FISHING TO EXTENDED LIVELIHOODS

In the three countries studied, tuna fishing by local fishers is often considered negligible. This is mainly due to the low quantity of catch attributed to the small-scale fishery that is recorded in official reports (Government of Madagascar 2017; Government of Mauritius 2017; Government of Seychelles 2016). When interviewing government officials, common phrases in the three islands were "artisanal fishers do not really catch tuna here" (Anonymous, Fisheries department's representative in Madagascar, pers. comm.) or "it is a very small catch" (Anonymous, Fisheries department's representative in Mauritius, pers. comm.). Behind this limited catch, however, there is a broader story of connections among people within the local tuna fishery.

In the three island countries, tuna fishing is a highly social activity that allows members of the community to access the resources. Fishing trips are primarily undertaken in groups that include individuals from the same family or village who do not have a formal right to the fish themselves. In this instance, the participant receives a part of the catch that is divided between the crew members. This social relation is embedded within a reciprocity principle (Ribot and Peluso 2003, 172). Reciprocity here manifests itself by offering to help a relative, friend, or neighbor in the same fishing village. We also observed that access to labor opportunities is not formalized and rather horizontal. Trust and reputation building (as developed in Berry 1989, 46–49) are the access mechanisms to labor here. Tuna fishers, like in Mauritius, are well known in their village and often considered successful in their activity. Each fisher can also ask for work by approaching the boat owner, sometimes referred to as an equal comrade rather than as a boss. In Ramena village (Madagascar), a tuna fishing boat often has ten to fifteen fishers on board, mostly coming from the same village. Each fisher has a family of three to five people to take care of. As one interviewee declared, "In Ramena, fishing is a matter of survival. I accept anyone who is in need of work. There is always something to do in the fishing, whatever your skill" (Kongo, boat owner in Madagascar, pers. comm.). In the three islands, the social effects of tuna fishing on local livelihoods is largely untold.

The case of the semiindustrial sector in the Seychelles is worth mentioning here in terms of the labor opportunities it brings. In the past five years, the Government of Seychelles (2016) has strongly invested in developing its semiindustrial tuna fishery, catching around two hundred tons of tuna a year. With a national fleet of around thirty boats in 2017, demand for crew was not satisfied. As a result, fishers from Sri Lanka have come to work on Seychellois vessels. Local views on this flow of human resources have, however, been mixed. Some consider them as taking the jobs of local fishers at a lower cost. This view is carried by other small-scale fishers and some government officials. Others, mainly boat owners, see them as contributing to the development of the fishery. Sri Lankans are here seen as filling a labor gap because young Seychellois are less and less willing to take part in fishing. Here, tuna fishing brings in a flow of foreign labor, which, despite creating conflicting local views, is key to the development of a national tuna fleet.

The story of people linked to tuna also goes beyond the fishers themselves. The role of tuna in food security and providing work for other locals is significant. Tuna caught by the small-scale sector plays a nonnegligible role in feeding the local markets with protein sources. In the north of Madagascar, tuna caught locally has become a highly appreciated source of protein, considering the high cost of other protein sources. In the southeast of Madagascar, tuna are carried on foot from the coast to isolated villages fifteen to twenty kilometers inland by local intermediaries. The importance of these intermediaries is largely undocumented in the value chain of tuna in Madagascar. In Mauritius and Seychelles, local catches of tuna not only go to the local markets but also to neighboring hotels. In the two islands, tuna contributes to food security of local residents and is also considered a high-value product in restaurants. The same applies to the catch of the semiindustrial fleet in the Seychelles where the tuna is shipped straight away to Europe and the United States through local processing companies, for fresh consumption.

As we have seen, despite its limited scale, the connections of people behind the small-scale and semiindustrial tuna fishery are essential to local livelihoods and produce foreign labor exchanges. This diversity brings a specific politics of care for the tuna resources at the level of each country. Caring about the tuna is not only strongly linked to taking care of one's livelihood but also to giving care to the community by providing livelihood to other members of the community. To maintain the benefits that the fishery generates for the local economies, a continuous access to tuna by both fishers and other indirect actors is necessary. This access is presently achieved through a low-intensity exploitation: use of smaller boats and limited fishing effort. Although it could be argued that the situation shows a lack of development in the fishery, the social interactions and lower pressure on the resources demonstrate that this way of fishing does benefit local actors and the tuna of the WIO.

THE CONTRASTED BENEFITS
FROM THE INDUSTRIAL EXPLOITATION

The better-known aspect of tuna fishing in the WIO region is its industrial segment, undertaken by purse seine and longline vessels that are foreign-owned, mainly by European and Asian companies. This industrial exploitation is established through fishing access agreements under which access to the tuna resources of the WIO is granted by governments of the region in exchange for an access fee. The industrial tuna fishery has always been subject to various critiques, notably regarding equity of fishing access agreements and the sustainability of the exploitation (Gagern and van den Bergh 2013, 380–84; Gegout 2016, 2196–97; Le Manach et al. 2013, 5–7). DWFNs have responded to these critiques by advocating that the industrial tuna exploitation, through the access fee paid by DWFNs, represents a reliable source of revenue for national economies (Barnes and Mfodwo 2012, 11–20). They also argue that for the three islands studied, it is an important source of local labor opportunities (European Union 2017; Hanazaki 2017). Here, we discuss the relevance of these two narratives, mainly through the example of the EU exploitation in the WIO, and attempt to establish who actually benefits from this segment of the fishery and how the tuna resources of the WIO are affected.

First, we start with the revenue argument. It is recognized that fishing access agreements are a significant source of revenue for coastal countries (Barnes and Mfodwo 2012, 11–20; Campling and Havice 2014, 715). In the case of Madagascar for example, revenues from those agreements are directly injected to the national budget, which funds different governmental departments (Le Manach et al. 2013, 261). In Mauritius and Seychelles, this revenue has been used for the construction of different port infrastructures or processing factories, as well as contributing to various capacity-building initiatives such as monitoring, controlling, and surveillance (IOC 2013, 71). However, in the broader picture of benefits, the price of access to the tuna resources of the WIO represents a minimal fraction of what fishing companies are getting as profit. As an illustration, the price of access to tuna resources in the WIO by the EU in 2014 was estimated at around 2 percent of the profits made by the EU fleet from the fishery (Poseidon et al. 2014, 86–87). DWFNs here have a strong discourse of taking care of the host countries where their fleet catch tuna. The economic model of access, however, shows that fishing by DWFNs' fleets is largely to accumulate profits that ultimately provide limited care to the host countries and their people.

If we look at labor opportunities from the industrial sector, we can highlight those of stevedores and cannery workers based in host countries. An interesting set of actors who access tuna further up the value chain are the local stevedores at port who land the tuna from EU purse seine vessels in the WIO. Although the practice is common in the three islands studied, in Madagascar the role of these stevedores in the local economy is substantial. In ports like Antsiranana, stevedores are offered or buy bycatch from industrial vessels (Government of Madagascar 2017). Bycatch species here are small or damaged tuna that are not taken by the canneries or nontuna species that are caught in the nets of purse seiners. Every evening of landings, stevedores sell the fish outside the port to intermediaries and local residents who welcome this fish as a sought-after local source of protein. A typical evening event consists of around one hundred men and women waiting for the fish to come out of the port. The tuna also goes from the port to local markets or is transported farther inland to remote villages through intermediaries. This indirect access to the resources, from which local residents are benefiting, constitutes a valid contribution toward food security for a vulnerable country like Madagascar.

DWFNs here can be considered as giving care to locals without necessarily intending to.

Industrial tuna fisheries are also a source of labor within the three islands through the canneries. In 2010, more than eight thousand people were working in tuna canneries in Madagascar, Mauritius, and Seychelles (Gillett 2011, 12). In Mauritius, the tuna cannery is the largest single employer in the country (Cervigni and Scandizzo 2017, 16). In addition to employing a large part of the local population, the canneries in Mauritius and Seychelles also hire migrant cannery workers from Madagascar and other countries of the WIO, such as Kenya. This flow of migrant labor, although key to the tuna economies, also comes with common issues of social integration (Craig 2015, 58–63). Considered as low-skill workers, migrant tuna workers often struggle to integrate, despite working on a shared resource of the WIO. From these two cases, industrial tuna fisheries bring benefit to local economies through employment but also bring in less apparent issues of labor migrations from which the tuna workers in the canneries suffer from.

Another aspect that weighs in the balance of benefits is the effect of industrial tuna fishing on the resources of the WIO. The destructive influence of industrial fishing is widely acknowledged (Kroodsman et al. 2018, 905; Sumaila, Bellmann, and Tipping 2016, 175). Despite the narrative of national benefits provided by the industrial tuna fishing, the tuna resources of the WIO are affected by this type of exploitation. As an illustration, the overfishing status of yellowfin tuna since 2015 has been attributed to the increase of fishing effort in the past eight years, including by the industrial sector, which catches more than 50 percent of the tuna in the Indian Ocean (IOTC 2016, 114–16). Local perceptions on the state of the resources also indicate a strong effect of this industrial exploitation. In our three case studies, actors and local fishers especially cite a substantial decrease in tuna resources and other fishes. Highlighting the role of industrial fishing in this situation, local fishers we interviewed at ports or in their villages made fierce remarks such as, "They do not choose what to catch, small and big fishes" (Elvis Won, local fisher in Seychelles, pers. comm.), "The big boats take all the tuna before the fish can get to our coastal waters" (Karon, local fisher in Madagascar, pers. comm.), and "They catch too much, they have very good equipment for that" (Judex Renfle, local fisher in Mauritius, pers. comm.). Visiting a couple of landing industrial vessels confirmed those stories. Walking inside a purse seiner, we could see mature tunas, juveniles, and other marine species caught in large amounts, suggesting a bleak picture for the WIO tunas and marine resources (personal observation). Little care is given to the tuna resources here. The industrial exploitation, in its current way of catching the resources, contributes to the depletion of the resources.

What is the politics of care of the industrial fisheries here? We have seen that a number of local actors and national economies are benefiting from the industrial exploitation. There is to some extent an engagement of DWFNs in *taking care* of the economies of the WIO countries. We have, however, also illustrated that those benefits do not outweigh the larger profits obtained by the foreign fishing fleets nor the detrimental effects of the fishery. The tuna resources of the WIO are not sustained by this segment of the fishery.

THE CHALLENGING WAY
TOWARD SUSTAINING THE WIO TUNAS

As is the case in all oceans, tuna has a high value in the different segments of the fishery as well as up and down the value chain. The question is, then, what are the drivers that prevent stakeholders from ensuring the future of tuna resources in the WIO? For the case of the WIO, the answer is not easy, given the role of strong foreign interests as well as powerful capitalist firms in the fishery (Campling 2012b, 263–69). Three obstacles prevent real socio-ecological change in the way tuna is managed and a politics of care that puts the tuna and its people first.

The first obstacle is geopolitics. Coastal countries are strongly entangled with DWFNs, which provide various national benefits (Andriamahefazafy, Kull, and Campling 2019, 9). The level of economic dependency of countries here determines how much they can defend the need for a better management of tuna. During our observation of the 2018 IOTC meeting, we noticed that it was difficult, for example, for a government official from Madagascar, compared with those of Mauritius or Seychelles, to openly challenge actors such as the European Union or Japan, which are both fishing and providing consequent development aid to the country. In this context, interests in strong management measures toward the tuna are superseded by other national interests. Although attendance of the IOTC shows an engagement of governments in taking care of the tuna resources, showing real activities of care by taking actions that benefit the resources, remains a challenging decision to take.

The second obstacle is the exploitation system of tuna itself, considering the large portion of tuna caught by the industrial sector in the WIO, especially by purse seiners. This industrial exploitation has allowed the worldwide distribution of tuna, especially in cans. While reducing this type of exploitation and replacing it with less impactful fishing practice is essential for the long-term sustainability of tuna resources, addressing the question of satisfying the demand for tuna, consumed in most households, cafeterias, and restaurants becomes a political one: do we provide an affordable source of protein that was caught with a destructive method? Or do we reduce the consumption of tuna and promote the consumption of tunas that are only caught with non-destructive fishing methods? Similarly, how can we shift to a less damaging fishing practice while the capitalist way of exploiting the resources has been part of the history of tuna fisheries? To address those questions and favor the management of tuna, a complete change of paradigm is needed in tuna fisheries, including in the WIO. As long as industrial exploitation is considered more beneficial than smaller-scale use of the resources, the tuna resources and the local people that daily depend on the resources will continue to be affected. Although consumers in developed countries have an array of choice for their protein intake, those of coastal countries do not have this privilege and rely heavily on marine resources. Without a shift toward a lower-impact exploitation, the future of the tuna resources will continue to be uncertain. Adopting the idea of giving care to the resources can contribute to a degrowth of the sector within which real socioecological change is possible along with more just benefits for the custodians of the resources.

The third obstacle is that tuna as a migratory species brings its own specificities for the use of the resources. Coastal countries only have jurisdiction over the resources when tuna is in their waters; local fishers with their current means of fishing are not able to follow the fish that goes beyond national borders, in the high seas (Campling and Havice 2014, 717–18). The combination of tuna movement, the marine environmental conditions, and the pressure put on the tuna biomass determines how much tuna is available to fish, to process, or to sell. In addition to the strong effects of fishing practices on the biomass of tuna, tuna remains a wild species for which cycles of productivity and movement are not fully understood (Kaplan et al. 2014, 1744). Other ecological factors such as climate change also represent out-of-sight and unknown variables for the future of tuna productivity and quality of feeding grounds of the resources. Different to other resources that are either stationary or moving within an accessible range, the vast movement of tuna in the ocean requires a broader care for the ocean, beyond sight and borders.

INSIGHTS ON MOVING FORWARD

How can we then sustain the tuna resources of the WIO and its people? We have seen that tuna fisheries in the WIO provide benefits to an array of actors, of whom some are less seen but equally dependent on the resources. From the local fisher who relies on the resources for livelihood to the intermediaries, cannery workers, or the capitalist fishing firms that feed the international consumer markets for tuna, all actors aspire to sustain their access to the resources. These actors all have different interests and politics of care for the resources. For the case of the WIO and particularly in the three island countries studied, the main challenge lies with managing the industrial segment of the fishery and its effects on the tuna resources and local livelihoods. An improved care, involving robust actions for the improvement of the state of tuna resources, will require difficult trade-offs that are necessary to sustain the future of the resources. Nationally, there is a strong need to refocus the fishery toward those who are the most dependent on the resources, notably by giving more voice to the stories of livelihoods behind the local tuna fisheries. The question of whether to pursue industrial fishing or not can only be addressed by a systemic change at the global level, where the socioecological effects of the exploitation are fully revealed and prompt the adoption of less growth-oriented exploitation. It also requires a change in the supply-and-demand market where the real value of tuna is reconsidered and exposed to the consumer. Its reputation as cheap protein needs to be rebranded and include the socioecological costs of its fishing. Ultimately, stakeholders in the fishery need to realize that there is no win-win solution to sustain the tuna of the WIO. Caring for the tuna of the WIO demands putting the resources and their custodians first, at a potential loss for some economic and political actors.

ACKNOWLEDGMENTS

We acknowledge the financial support from the Institute of Geography and Sustainability of the University of Lausanne to complete this fieldwork.

REFERENCES

Andriamahefazafy, Mialy, Christian A. Kull, and Liam Campling. 2019. "Connected by Sea, Disconnected by Tuna? Challenges to Regionalism in the Southwest Indian Ocean." *Journal of the Indian Ocean Region* 15 (1): 1–20. https://doi.org/10.1080/19480881.2018.1561240.

Bakker, Karen J., and Gavin Bridge. 2006. "Material Worlds? Resource Geographies and the 'Matter of Nature.'" *Progress in Human Geography* 30 (1): 5–27. https://doi.org/10.1191%2F0309132506ph588oa.

Barnes, Collin, and Kwame Mfodwo. 2012. *A Market Price Valuation of Tuna Resources in the Western Indian Ocean—An Indicative Regional and Country/EEZ Perspective.* A WWF Report. February 2012. Switzerland: World Wildlife Fund.

Bennett, Jane. 2010. *Vibrant Matter: A Political Ecology of Things.* Durham, NC: Duke University Press.

Berry, Sara. 1989. "Social Institutions and Access to Resources." *Journal of the International African Institute* 59, no. 1: 41–55.

Blaikie, Piers. 1989. "Environment and Access to Resources in Africa." *Africa: Journal of the International African Institute* 59, no. 1: 18–40.

Campling, Liam. 2012a. "The EU-Centered Commodity Chain in Canned Tuna and Upgrading in Seychelles." PhD diss. London: University of London.

———. 2012b. "The Tuna 'Commodity Frontier': Business Strategies and Environment in the Industrial Tuna Fisheries of the Western Indian Ocean." *Journal of Agrarian change* 12, no. 2: 252–78. https://doi.org/10.1111/j.1471-0366.2011.00354.x.

Campling, Liam, and Elizabeth Havice. 2014. "The Problem of Property in Industrial Fisheries." *The Journal of Peasant Studies* 41, no. 5: 707–27. https://doi.org/10.10 80/03066150.2014.894909.

Cervigni, Raffaello, and Pasquale L. Scandizzo. 2017. *The Ocean Economy in Mauritius Making It Happen, Making It Last.* Edited by World Bank Group, conference editions. Washington, DC: World Bank Group.

Craig, Gary. 2015. *Migration and Integration: A Local and Experiential Perspective.* Paper IRiS WP 7-2015. Institute for Research into Superdiversity (IRiS). University of Birmingham.

D'Alisa, Giacomo, Marco Deriu, and Federico Demaria. 2015. "Care." In *Degrowth, a Vocabulary for a New Era*, edited by Giacomo D'Alisa, Federico Demaria, and Giorgos Kallis, 63–66. New York: Routledge.

Ellis-Jones, Jim. 1999. "Poverty, Land Care, and Sustainable Livelihoods in Hillside and Mountain Regions." *Mountain Research and Development* 19, no. 3: 179–90.

European Union. 2017. *EU SFPAs: Sustainable Fisheries Partnership Agreements Leaflet.* Edited by the European Union. Brussels: DG Mare.

Gagern, Antonius, and Jeroen van den Bergh. 2013. "A Critical Review of Fishing Agreements with Tropical Developing Countries." *Marine Policy* 38: 375–86. https://doi.org/10.1016/j.marpol.2012.06.016.

Gegout, Catherine. 2016. "Unethical Power Europe? Something Fishy about EU Trade and Development Policies." *Third World Quarterly* 37, no. 12: 2192–210. https://doi.org/10.1080/01436597.2016.1176855.

Gillett, Robert. 2011. Tuna for Tomorrow. *Working papers.* Edited by the Indian Ocean Commission. Ebene, Mauritius: Indian Ocean Commission.

Government of Madagascar. 2017. *Bulletin Statistique Thonier 2017 de l'Unité Statistique Thoniere d'Antsiranana (USTA).* Consulted at the USTA in October 2018. 75 p.

Government of Mauritius. 2011. *Annual Report of the Ministry of Fisheries and Rodrigues.* Archive consulted at the Albion Fisheries Centre in March 2017. Port-Louis: Ministry of Ministry of Ocean Economy, Marine Resources, Fisheries and Shipping.

———. 2017. *Compilation of Tuna Data.* Provided by the Ministry of Ocean Economy during fieldwork in April 2017.

Government of Seychelles. 2016. *2016 Fisheries Statistical Report.* Edited by the Seychelles Fishing Authority (SFA), pp. 6–8. Victoria: Seychelles Fishing Authority.

Hanazaki, Mayu. 2017. "L'OFCF du Japon Rénove les Installations Des Pêcheurs pour l'Association TAZARA à Toamasina." *Communiqué de presse de l'Ambassade du Japon à Madagasca*. Edited by Ambassade du Japon. Madagascar.

Indian Ocean Commission (IOC). 2013. *Rapport Annuel 2013—Indianocéanie*. Edited by the Indian Ocean Commission. Ebene, Mauritius: Indian Ocean Commission.

———. 2018. *Premier Plan d'Action Régional sur la Pêche Thonière Côtière pour le Sud-Ouest de l'Océan Indien*. Press release. Edited by the Indian Ocean Commission. Ebene, Mauritius: Indian Ocean Commission.

Indian Ocean Tuna Commission (IOTC). 2016. *Report of the 19th Session of the IOTC Scientific Committee—Seychelles, 1–5 December 2016*. Document reference: IOTC-2016-SC19-RE. Victoria, Seychelles: IOTC.

———. 2018. *Report of the 21st Session of the IOTC Scientific Committee Seychelles, 3–7 December 2018*. Document reference: IOTC-2018-SC21-R[E]. Victoria, Seychelles: IOTC.

Kaplan, David M., Emmanuel Chassot, Justin M. Amandé, Sibylle Dueri, Hervé Demarcq, Laurent Dagorn, and Alain Fonteneau. 2014. "Spatial Management of Indian Ocean Tropical Tuna Fisheries: Potential and Perspectives." *ICES Journal of Marine Science* 71, no. 7: 1728–49. https://doi.org/10.1093/icesjms/fst233.

Kroodsman, David A., Juan Mayorda, Timothy Hochberg, Nathan Miller, Kristina Boerder, Francesco Ferretti, Alex Wilson, et al. 2018. "Tracking the Global Footprint of Fisheries." *Science* 359: 904–8.

Lawson, Victoria. 2007. "Geographies of Care and Responsibility." *Annals of the Association of American Geographers* 97, no. 1: 1–11. https://doi.org/10.1111/j.1467-8306.2007.00520.x.

Le Manach, Frédéric, Mialy Andriamahefazafy, Sarah Harper, Alasdair Harris, Gilles Hosch, Glenn-Marie Lange, Dirk Zeller, and Rashid U. Sumaila. 2013. "Who Gets What? Developing a More Equitable Framework for EU Fishing Agreements." *Marine Policy* 38: 257–66. https://doi.org/10.1016/j.marpol.2012.06.001.

Obura, David, Marty Smits, Taz Chaudhry, Jen McPhillips, Douglas Beal, and Camille Astier. 2017. *Reviving the Western Indian Ocean Economy: Actions for a Sustainable Future*. Edited by World Wildlife Fund International. Gland, Switzerland.

Pauly, Daniel. 2018. "A Vision for Marine Fisheries in a Global Blue Economy." *Marine Policy* 87: 371–74. https://doi.org/10.1016/j.marpol.2017.11.010.

Peet, Richard, Paul Robbins, and Michael J. Watts. 2011. "Global Nature." In *Global Political Ecology*, edited by Richard Peet, Paul Robbins, and Michael J. Watts, 1–48. London: Routledge.

Poseidon Aquatic Resource Management, Marine Resource Assessment Group, Nordenfjeldske Development Services, and COFREPECHE. 2014. *Review of Tuna Fisheries in the Western Indian Ocean*. Framework contract MARE/2011/01—Lot 3, specific contract 7. Brussels: DG MARE.

Probyn, Elspeth. 2014. "The Cultural Politics of Fish and Humans: A More-Than-Human Habitus of Consumption." *Cultural Politics an International Journal* 10, no. 3: 287–99.

———. 2016. *Eating the Ocean*. Durham, NC: Duke University Press.

Ribot, Jesse C., and Nancy Lee Peluso. 2003. "A Theory of Access." *Rural Sociology* 68, no. 2: 153–81.

Robbins, Paul. 2012. *Political Ecology. A Critical Introduction*, 2nd edition. Malden, MA: Wiley-Blackwell.

Sellström, Tor. 2015. *Africa in the Indian Ocean: Islands in Ebb and Flow*. Leiden: Koninklijke Brill NV.

Suchet-Pearson, Sandie, Sarah Wright, Kate Lloyd, and Laklak Burarrwanga. 2013. "Caringas Country: Towards an Ontology of Co-Becoming in Natural Resource Management." *Asia Pacific Viewpoint* 54, no. 2: 185–97. https://doi.org/10.1111/apv.12018.

Sumaila, U. Rashid, Christophe Bellmann, and Alice Tipping. 2016. "Fishing for the Future: An Overview of Challenges and Opportunities." *Marine Policy* 69: 173–80. https://doi.org/10.1016/j.marpol.2016.01.003.

Tronto, Joan. 1993. *Moral Boundaries: A Political Argument for an Ethic of Care*. New York: Routledge.

United Nations Development Programme (UNDP). 2016. *Human Development Report 2016: Human Development for Everyone*, 212–14. New York. Available at http://hdr.undp.org/en/2016-report/.

FISH AS FOOD

Consuming and Sustaining

Chapter Five

The Multiple Meanings of Fish

Policy Disconnections in Australian Seafood Governance

Sonia Garcia Garcia, Kate Barclay, and Rob Nicholls

In Australia, fisheries management jurisdictions at the state and federal level regulate fisheries according to sustainability objectives contained in legislation and related formal regulatory measures such as harvest strategies and management plans. The current system for governing fisheries was shaped in the 1990s against the backdrop of concerns about dramatically overfished stocks such as gemfish and orange roughy, which increased public awareness of environmental issues and employment-reducing restructuring of major commercial fisheries. Contemporary fisheries management objectives thus came to be focused on sustainability. In this vision of sustainability, fish are imagined as biological stocks and the amounts of fish of particular species is the key indicator regarding availability of the resource for current and future generations. Economic objectives are narrowly considered and social aspects of sustainability are not—fish are not treated as sources of food, culture, or livelihoods. Once fish enter the postharvest supply chain, the policy imaginary changes and they are regulated as a food commodity. In this space, governors for consumer protection manage products to ensure they are safe to eat and set the conditions for sale, including what information needs to be available to consumers. Seafood supply is guaranteed through trade flows, so sustainability is not regulated in the postharvest part of the supply chain. Sustainability is treated as a consumer value to be left to the initiative of the private sector, for example, through branding seafood product with ecolabels.

This chapter addresses the effects of this disconnection between the meanings of fish as a natural resource and fish as food, between the goals of the governors in the harvest and in the postharvest space. We have analyzed qualitative data—policy documents, interviews, and event observations—for insights into the consequences of this policy disconnect, as well as possible tools to address it. The chapter begins by exploring the main features of the present regulatory environment. First, the regulatory burdens of fisheries management—the costs of regulation imposed on those subjected to compliance with it—are different for domestically caught seafood and for the imported seafood that sits beside it on the supermarket shelf or menu. This creates a risk that the final point-of-sale business may choose to attribute an incorrect place of origin if the expected compliance risk is outweighed by the margin improvement flowing from misleading labeling. Second, the available voluntary tools for labeling seafood as having been sustainably harvested, such as Marine Stewardship Council (MSC) certification, are targeted to high-value, economically efficient fisheries. MSC certification is too expensive for the small-scale operators that constitute a large part of the Australian fishing sector. Finally, a history of conflict between Australian professional and recreational fishers and the conservation movement has damaged public perceptions of professional fisheries. This has been compounded by a lack of effective public communication about the relatively strong management of commercial fishing by Australian governments since the 1990s, leading to the situation in which the fishing industry fears the loss of public trust in their activity. These three situations arise from inaction by government on fisheries sustainability after the seafood is harvested to enable a level playing field for sustainable produce. The absence of this level playing field risks advantaging imports over domestic seafood, and product from unregulated fisheries over sustainably caught fish.

An attempt has been made to address this policy disconnect through country-of-origin labeling, based on the assumption that, because Australian fisheries are fairly well regulated, this would act as a proxy for sustainability requirements in the retail sphere. Another possible way to address the disconnect is through requiring importers to demonstrate that their seafood shipments were legally caught. Regulatory bodies in the European Union and the United States have addressed possibly unfair market competition between imported seafood, which may not have been produced under stringent fisheries management, and domestically produced seafood that has been subject to such management through requiring traceability documentation to demonstrate the fish comes from a regulated fishery.

GOVERNING FISH, GOVERNING SEAFOOD

In the first half of the twentieth century, the main goal of fisheries management in Australia was the development of fisheries as an economic sector, driven by state and federal governments to maximize the opportunity offered by apparently large stocks along the Australian coast (Clark 2017, 101).

Rapid development of commercial fisheries from the 1950s led to plummeting stocks of southern bluefin tuna and gemfish in the 1980s, and orange roughy in the early 1990s (Clark 2017, 106–8). These fishery collapses occurred against a background of increased scrutiny from environmental groups and conflict for resource access with the recreational fishing sector. In Australia, recreational fishing groups have successfully lobbied to have professional fishing excluded from certain waterways and continue to push for further restrictions on professional fishing (King and O'Meara 2019). These conflicts over resource access—episodes of collapsed fisheries and media coverage of overfishing as a global problem—have damaged the public image of professional fishing and brought about a profound revision of government regulatory objectives. Regulation has shifted from industry expansion to the pursuit of biological sustainability through the monitoring of stocks, restricting entrants and preventing overfishing:

> Fish are a renewable, but not inexhaustible, resource. They are subject to the well-recognized potential for a "tragedy of the commons," where the unregulated efforts of individual fishers deplete the resource. Governments must therefore limit catches to sustainably manage resources and, where there is competition between fishers, determine how access is to be shared. (Productivity Commission 2016, 3)

This focus on sustainability crystallized in the adoption of Ecologically Sustainable Development (ESD) principles; the definition and core objectives of ESD for Australia are contained in the *National Strategy for Ecologically Sustainable Development* endorsed by the Council of Australian Governments in 1992, integrated in the *Environmental Protection and Biodiversity Conservation Act 1999* (Commonwealth) and in the different Fisheries Acts of the states. The three components of sustainability—social, economic, and environmental—were taken into account in the high-level objectives of the legislation. However, the primary effect of the legislation addressed the biological component of sustainability, complemented in some cases by economic objectives (Barclay 2012). The operational objectives in management plans and harvest strategies rely strongly on the conception of fish as a stock whose existence in sufficient numbers is the fundamental goal of fisheries management:

> The Status of Australian Fish Stock Reports are a series of assessments of the biological sustainability of a broad range of wild-caught fish stocks against a nationally agreed framework. The reports examine whether the abundance of fish and the level of harvest from the stock are sustainable.
> The 2018 reports focus solely on the status of fish stocks. The status classifications do not consider broader ecosystem impacts of fishing or social and

economic considerations that some consumers may be interested in. (Fisheries Research and Development Corporation [FRDC] 2018a, paras 1, 9)

The economic component of ESD is present in some fisheries legislation, such as the aim to maximize the "net economic returns to the Australian community" (Commonwealth of Australia 1991, 2) of the harvest. Economic measurements of the value of Australian fisheries remain unsophisticated (Pascoe et al. 2016) and are undefined in many jurisdictions. Governments have not clearly articulated who should benefit from the exploitation of fishery resources: "Is the objective of the fishery to provide employment, food, reward entrepreneurship, generate income for the community, provide recreational utility or some other goal?" (Emery et al. 2017, 143). The vagueness of economic objectives, coupled with the disregard for the social component of sustainability (Barclay 2012, 38), means that fish stocks are managed to be available for future generations, but beyond that the goals are unclear.

The governance arrangements oriented to maintain fish stocks cease once the fish leaves the wharf and enters the supply chain as seafood. In the postharvest space, seafood is categorized as a food product, and governance arrangements aim at ensuring market competition and that food is safe to eat. This affects both imports and domestic product. Imports are subject to border controls for food safety and biosecurity but not to the demonstration of sustainability, in line with a trade regime in which the former constitute an acceptable trade barrier but environmental provisions have been traditionally problematic. Apart from the border control of biosecurity, the main regulatory responsibilities for seafood in the postharvest space are transferred to food authorities and consumer protection frameworks. In these frameworks, the government's ESD objectives are not pursued in the legislation or policies of the food safety and consumer protection authorities. In this framework, food safety is the object of mandatory information, followed by preventative health. Other consumer values issues, including environmental values, are left up to industry-initiated regulation.

> Regulatory action in relation to food safety, preventative health and new technologies should primarily be initiated by government and referenced in the Food Standards Code. Regulatory action in relation to consumer values issues should generally be initiated by industry and referenced to consumer protection legislation. (Blewett et al. 2011, 48)

The private sector thus emerges as a governor to fill the regulatory gap regarding sustainability, that is, to certify the sustainability of the seafood in the marketplace and to communicate this attribute to consumers. Internationally, some companies along supply chains use a variety of certification programs and associated marketing tools provided by independent organizations to demonstrate sustainability claims and differentiate their product to consumers (Auld 2014, 1). Such programs have contributed to the environmental regulation of fisheries and constitute one of the prime examples of global nonstate regulatory mechanisms (Hatanaka, Bain, and Busch 2005, 355). Some Australian fisheries have been active in seeking certification. In March 2000, the Western Rock Lobster Fishery was the first fishery in the world to become MSC certified and, in 2017, the first to have been recertified for a fourth time (MSC 2018). In 2012, the Northern Prawn Fishery was the first tropical prawn fishery in the world to be MSC certified, achieving one of the best MSC scores ever (Hadjimichael and Hegland 2016, 131). Western Australia is also the site of a collaboration between MSC and the state government to enable simultaneous preassessment of different fisheries by regions (Bellchambers et al. 2016). However, certifications cover a fraction of the seafood sold in Australia. This partial coverage of sustainability concerns in the postharvest space through private sector tools fails to solve the regulatory gap and has a number of effects.

THE EFFECTS OF POLICY DISCONNECTIONS

The absence of government regulation for sustainable fisheries in the post-harvest space has a number of implications. It produces a regulatory gap in that Australian fisheries are regulated for sustainability whereas imports are not subject to the demonstration of sustainability. The different sustainability requirements for domestic and imported seafood may result in a price advantage for less sustainably sourced seafood, thus reducing economic opportunities of fisheries subject to regulatory measures, whether domestic or foreign. The regulatory gap is especially relevant in a market characterized by a strong dependency on imports—in 2015–2016, imports made up 67 percent of the total consumption of seafood in Australia (Hogan 2017, 34).

The imbalance in the regulatory requirements is exacerbated by the partial success of private governance in the certification and communication of sustainability credentials. Certification of the sustainability of seafood remains restricted to the large-scale, industrialized, economically efficient fisheries and large supply chain actors. This presents a problem for the small, family-owned fishing businesses that make up the majority of Australian fisheries (FRDC and Ridge Partners 2015, 71). As an interviewee pointed out, "The problem for Australia for going into the market and demonstrating sustainability is that you've got the haves and have nots" (research funder, pers. comm., November 14, 2017). The have nots—those domestic fisheries subject to strong fisheries regulation but unable to demonstrate their sustainability credentials in the marketplace—thus compete in the retail space without the capacity to use voluntary tools to communicate their sustainability credentials. This means they compete directly with less regulated fisheries with lower regulatory costs. The result is a price disadvantage for seafood from more sustainable fisheries.

Another problem arising from lack of government involvement on sustainability outside the regulation of fisheries per se is that governments have not effectively communicated to the public that most Australian fisheries have since the 1990s been well managed in terms of preventing overfishing. The damage of the public image of professional fishing brought about by the collapse of fisheries in the 1980s and 1990s, plus ongoing media coverage of overfishing as a global problem has led to a low degree of trust in both government and the fishing industry well documented in studies (Mazur, Curtis, and Bodsworth 2014, 12). The failure to communicate to the public that Australian fisheries are by and large sustainably managed has contributed to the industry's lack of social license to operate.

Social license can be defined as "the level of acceptance or approval continually granted to an organization's operations or project by the local community and other stakeholders" (Mazur, Curtis, and Bodsworth 2014, 38). Lack of social license means a lack of goodwill and can give rise to a range of problems, key among which is maintaining access to the resource. Wild fish stocks are a common resource to which governments grant access. If the professional fishing industry does not have goodwill with the constituency, they are more likely to lose out when recreational fishing or conservation groups call for government to exclude professional fishing from certain estuaries and coastal areas. Social license is a key preoccupation for the Australian fishing industry: "For Seafood Industry Australia [a national peak body representing the fishing industry] I would reflect on the members' advisory forum that we had yesterday and the key issue the number one issue that came out of that is social licence" (Lovell 2017, n.p.).

The loss of public trust in the fishing industry is seen as a shared failure by government and industry to transmit to the community the sustainable management of the resource.

> [W]e worked twenty to thirty years to change and to improve our processes and to work on sustainability. We need to be acknowledged for that. We want to be acknowledged for that. And science and research has acknowledged us but that hasn't led into sufficient communication back to the community. (seafood industry organization representative, pers. comm., February 8, 2018)

Industry recognizes that it is their responsibility to communicate about their sustainability credentials to the public, and some initiatives have attempted to do that in recent years. There is, however, limited capacity, especially for smaller companies.

> I think the fishers know they have to do it [improve social license], but they don't know how and they're too busy fishing and too busy surviving because they, you know, what was coming out of yesterday's meeting was this fear of access, keeping their access to fishing is their main priority. (seafood industry consultant, pers. comm., March 8, 2018)

Government actors recognize that government is also responsible for communicating about the effectiveness of fisheries management to help rebuild the social acceptability lost with overfishing in the past. They have, however, been slow to address the problem, and still tend to remain focused on the technical aspects of fishery management. The Fisheries Research and Development Corporation (FRDC) that manages research funding for fishing and aquaculture industries was only enabled by the legislation to invest in marketing from 2018 onwards (FRDC 2018b).

> Some of our issues with social license come from our lack of focus on educating the public and the community about how well we manage our fisheries. (fisheries manager, pers. comm., March 23, 2018)

> The Western Australian state commitment to MSC is fantastic but the missing links are definitely the chain of custody involvement in the project and the communication aspects. That kind of money was ring-fenced towards fishery improvements and certification but not necessarily thinking about how to bring that message back to consumers or to communities. (nongovernmental organization representative, pers. comm., November 16, 2017)

The disconnection between efforts to manage the harvest of fish sustainably and the lack of tools to address sustainability concerns in downstream processes is thus the result of two main factors: the lack of involvement by public governors in the space and the partial success of private governors in filling the regulatory gap. The lack of a regulatory level playing field between domestic and imported seafood is aggravated by a lack of effective communication about the sustainability credentials of Australian fisheries to the general public. This leads to a lack of goodwill and uncertain access to fisheries resources.

CONSUMING LEGAL, REPORTED, AND REGULATED FISH

The problems detailed in this chapter have arisen from the lack of government involvement in sustainability beyond the act of fishing. This policy gap has had an effect further down the supply chain, with seafood industry representatives seeking government regulation in the retail sphere. In Australia, the main industry demand has been to lift the exemption contained in the Food Standards that allows food service outlets (fish and chip shops, restaurants, etc.) to sell seafood without specifying its country of origin. This strategy relies on perceived premiums for fresh local domestic product and aims to avoid product substitution and to explain potential price differentiation between Australian and imported products. Because of the relative stringency of Australian fisheries management, labeling seafood as Australian has also been seen as a proxy for sustainability. The low public awareness of these efforts, discussed previously, is a major flaw in this strategy.

The extension of Country of Origin labeling for seafood in the food service sector has so far only been implemented in the Northern Territory using the avenue enabled through the *Fisheries Act 1988* (Northern Territory) that regulates the requirements for fish retailer licenses. Although the demand was initially granted a positive recommendation of Senate Committee inquiry into labeling requirements citing the need of a level playing field for domestic and imported product (Commonwealth of Australia 2014, 27), the Productivity Commission has more recently recommended against the extension (Productivity Commission 2016, 41–42). The Commission provided a clear reminder that environmental provisions belong to the private domain: "Consumer health and safety interests would not be enhanced by such a policy change, and there are practical impediments to implementation. If such arrangements are desired to better meet consumer preferences, industry should apply them voluntarily" (Productivity Commission 2016, 267).

In recent times, a possible instrument to address sustainability concerns in a form compatible with legal frameworks for international trade, food safety, and consumer protection has been located in the control of the legality of seafood. The attention to the lawful sourcing of fish is linked to the emergence of illegal, unreported, and unregulated (IUU) fishing as a global issue over the past two decades. IUU fishing is considered to constitute a major threat to the sustainability of fisheries and its prevention is included as an indicator in the Sustainable Development Goal 14 (Food and Agriculture Organization [FAO] 2016, 80).

The fight against IUU has highlighted the economic dimension of fishing and the importance of preventing IUU products from entering the supply chain through port state and trade-related measures (Witbooi 2014, 293). Trade-related measures aim to prevent seafood originating from IUU fishing from entering markets by requiring traceability documentation covering the transit of fish from the ocean to the supply chain. Such measures may be multilateral, as in the case of catch documentation schemes (Agnew 2000), or unilateral, as implemented by the European Union (Elvestad and Kvalvik

2015) and the United States (Simões and Dolle 2016). These measures have been accepted under the World Trade Organization regime (Tsamenyi et al. 2010, 30–31). Trade scholars conclude that, despite the challenges they pose to the trade regime, trade-related measures "may have become one of the few practical options to urge uncooperative flag, production, and reexporting states to contribute to and actively engage in the global campaign against IUU fishing" (He 2017, 197).

In the European Union and the United States, anti-IUU trade-related measures have been justified in terms of fisheries management—pursuing environmental, economic, and social sustainability of fisheries—on the grounds that (1) IUU fishing may distort market conditions and undermine the economic opportunities of legitimate, well-managed fisheries; (2) there is a need to establish equal controls and a level playing field for imported and domestic product; and (3) there is a threat posed by IUU fishing to the sustainability of the oceans (Department of Commerce—National Oceanic and Atmospheric Administration 2014; European Union [EU] 2008, 1). Although the long-term effect of these measures remains to be assessed, they have been evaluated as "likely to become prevalent and embedded in parts of national, regional, sub-regional and international fisheries governance arrangements to ensure sustainable and responsible fishing practices" (Tsamenyi et al. 2010, 31). This signals the success of a new approach to justify environmental provisions in the trade regime by shifting such concerns from the realm of consumer values to the prevention of market distortions. Measures to prevent IUU thus constitute a new proxy for sustainability that aims at reconciling the conservation of the oceans with its sustainable exploitation—by ensuring that fishing is regulated and seafood harvested legitimately.

Australia has been one of the pioneers in the adoption of multilateral trade–related measures to prevent IUU fishing for its export fisheries (Agnew 2000). However, the adoption of EU-style measures for the domestic market have been objected to by most stakeholders, including fisheries managers:

> Traceability and labelling is attracting increasing attention in international fisheries management. Some countries are seeking more information on where and how seafood was caught and whether it is consistent with international, regional and domestic fisheries regulations. Unilateral market measures taken by an importing country can be trade restrictive in that they do not necessarily recognise equivalent or better arrangements put in place by other countries with differing approaches. Some, including the EU and the US, have already implemented market state certification requirements that have caused additional requirements for some Australian seafood exporters. (Commonwealth of Australia 2014, Submission 11, 4)

This objection rests on the key acknowledgment that traceability and labeling are matters of interest for fisheries management. As explored in this chapter, this is not the case in Australia, where traceability and labeling are firmly located in the food safety and consumer frameworks. Instead, fisheries management is disengaged from the postharvest regulatory space and the industry has focused on country of origin as a proxy for sustainability. So far, the potential of anti-IUU measures on imports through traceability schemes has failed to attract the attention of the majority of producers, as well as of the Australian fisheries management authorities.

The lack of sustainability requirements in the postharvest space continues to produce a disconnection in Australia between the efforts to preserve fishery resources and their sustainable exploitation as a food product. Measures to prevent seafood that has not been legally caught in a well-regulated fishery entering the market through traceability documentation would address environmental concerns related to seafood. The existing regulatory framework in Australia fails to accommodate this. The involvement of fisheries management to address concerns in the postharvest space seems a necessary precondition to bridge this gap. However, it remains to be seen whether possible competitive advantages for Australian fisheries will lead the industry to pursue improvements in the traceability of all seafood sold, and whether the Australian fisheries managers will be willing to venture out of fisheries regulation and into trade regulation.

ACKNOWLEDGMENTS

Sonia Garcia Garcia's PhD research is funded by the University of Technology Sydney (UTS) President's Scholarship and the UTS International Research Scholarship.

REFERENCES

Agnew, David J. 2000. "The Illegal and Unregulated Fishery for Toothfish in the Southern Ocean, and the CCAMLR Catch Documentation Scheme." *Marine Policy* 24, no. 5: 361–74. doi: 10.1016/S0308-597X(00)00012-9.

Auld, Graeme. 2014. *Constructing Private Governance*. New Haven, CT: Yale University Press.

Barclay, Kate. 2012. "The Social in Assessing for Sustainability: Fisheries in Australia." *Cosmopolitan Civil Societies: An Interdisciplinary Journal* 4, no. 3: 38–53. doi: 10.5130/ccs.v4i3.2655.

Bellchambers, Lynda M., D. J. Gaughan, B. S. Wise, G. Jackson, and W. J. Fletcher. 2016. "Adopting Marine Stewardship Council Certification of Western Australian Fisheries at a Jurisdictional Level: The Benefits and Challenges." *Fisheries Research* 183: 609–16. doi:10.1016/j.fishres.2016.07.014.

Blewett, Neal, Nick Goddard, Simone Pettigrew, Chris Reynolds, and Heather Yeatman. 2011. *Labelling Logic. Review of Food Labelling Law and Policy.* Canberra: Commonwealth of Australia.

Clark, Anna. 2017. *The Catch: The Story of Fishing in Australia.* Canberra: National Library of Australia.

Commonwealth of Australia. 1991. "Fisheries Management Act," 60 edn, vol. 162/1991. https://www.legislation.gov.au/Series/C2004A04237.

———. 2014. *Current Requirements for Labelling of Seafood and Seafood Products.* Canberra: Senate Printing Unit, Department of the Senate, Parliament House.

Department of Commerce—National Oceanic and Atmospheric Administration. 2014. "Recommendations of the Presidential Task Force on Combating Illegal, Unreported and Unregulated Fishing and Seafood Fraud," 79 FR 75536, 79 (243): Office of the Federal Register, National Archives and Records Administration, 75536-41, https://www.govinfo.gov/app/details/FR-2014-12-18/2014-29628/summary.

Elvestad, Christel, and Ingrid Kvalvik. 2015. "Implementing the EU-IUU Regulation: Enhancing Flag State Performance through Trade Measures." *Ocean Development and International Law* 46, no. 3: 241–55. doi: 10.1080/00908320.2015.1054745.

Emery, Timothy J., Caleb Gardner, Klaas Hartmann, and Ian Cartwright. 2017. "Incorporating Economics into Fisheries Management Frameworks in Australia." *Marine Policy* 77: 136–43. doi: 10.1016/j.marpol.2016.12.018.

European Union (EU). 2008. *Council Regulation (EC) No. 1005/2008 of 29 September 2008 Establishing a Community System to Prevent, Deter and Eliminate Illegal, Unreported and Unregulated Fishing.* Official Journal of the European Communities, O.J. 286.29.10.2008.

Fisheries Research and Development Corporation (FRDC). 2018a. *What are the Status of Key Australian Fish Stocks Reports 2016.* Accessed November 11, 2019. http://www.fish.gov.au/Overview/Introduction/What-are-the-Status-of-key-Australian-fish-stocks-reports-2016.

———. 2018b. *Seafood Marketing.* Accessed June 26, 2018. http://www.frdc.com.au/Services/Seafood-marketing.

Fisheries Research and Development Corporation (FRDC) and Ridge Partners. 2015. *Australian F&A Sector Overview 2014. A Report Supporting the Development of Working Together: the National Fishing and Aquaculture Rd&E Strategy.* Deakin West, ACT: Fisheries Research and Development Corporation.

Food and Agriculture Organization (FAO) of the United Nations. 2016. *The State of World Fisheries and Aquaculture 2016. Contributing to Food Security and Nutrition for All.* Rome: FAO.

Hadjimichael, Maria, and Troels J. Hegland. 2016. "Really Sustainable? Inherent Risks of Eco-labeling in Fisheries." *Fisheries Research* 174: 129–35. doi: 10.1016/j.fishres.2015.09.012.

Hatanaka, Maki, Carmen Bain, and Lawrence Busch. 2005. "Third-party Certification in the Global Agrifood System." *Food Policy* 30, no. 3: 354–69. doi: 10.1016/j. foodpol.2005.05.006.

He, Juan. 2017. "The EU Illegal, Unreported, and Unregulated Fishing Regulation Based on Trade and Market-Related Measures: Unilateralism or a Model Law?" *Journal of International Wildlife Law and Policy* 20, no. 2: 168–97. doi: 10.1080/13880292.2017.1346351.

Hogan, Lindsay. 2017. *Food Demand in Australia: Trends and Food Security Issues.* Canberra: ABARES, Department of Agriculture and Water Resources.

King, Tanya J., and Dayne O'Meara. 2019. "'The People Have Spoken': How Cultural Narratives Politically Trumped the Best Available Science (BAS) in Managing the Port Phillip Bay Fishery in Australia." *Maritime Studies* 18, no. 1. doi: 10.1007/s40152-018-0097-5.

Lovell, Jane. 2017. "Panel: Australia's Fisheries Management, Ensuring Seafood for Future Generations. Jane Lovell, Caleb Gardner, Colin Tannahill, Stan Lui." *Seafood Directions Conference 2017*, Sydney. Accessed June 10, 2018. http://www.seafooddirectionsconference.com/pages/conference-arch-seafood-directions-2017-66.html.

Macfadyen, Graeme, Gilles Hosch, Nina Kaysser, and Lyes Tagziria. 2019. *The IUU Fishing Index, 2019. Country profile, Australia.* Poseidon Aquatic Resource Management Limited and the Global Initiative Against Transnational Organized Crime. Accessed February 21, 2019, http://iuufishingindex.net/profile/australia.

Marine Stewardship Council (MSC). 2018. *Australian Western Rock Lobster.* Accessed June 2018. https://fisheries.msc.org/en/fisheries/australian-western-rock-lobster/about/.

Mazur, Nicole, Allan Curtis, and Andy Bodsworth. 2014. *Let's Talk Fish. Assisting Industry to Understand and Inform Conversations about the Sustainability of Wild-catch Fishing. FRDC Report 2012/301.* Canberra: Fisheries Research and Development Corporation.

Pascoe, Sean, James Innes, Renae Tobin, Nathalie Stoeckl, Samantha Paredes, and Kieron Dauth. 2016. *Beyond GVP: The Value of Inshore Commercial Fisheries to Fishers and Consumers in Regional Communities on Queensland's East Coast. FRDC Project 2013-301.* Canberra: Fisheries Research and Development Corporation.

Productivity Commission. 2016. *Marine Fisheries and Aquaculture, Final Report.* Canberra: Australian Government, Productivity Commission.

Simões, Bruno, and Tobias Dolle. 2016. "The Global Combat against IUU Fishing: The United States Proposes a New Seafood Traceability Program." *European Journal of Risk Regulation* 7, no. 2: 421–25. doi: 10.1017/S1867299X00005833.

Tsamenyi, Martin, Mary Ann Palma, Ben Milligan, and Kwame Mfodwo. 2010. "The European Council Regulation on Illegal, Unreported and Unregulated Fishing: An International Fisheries Law Perspective." *International Journal of Marine and Coastal Law* 25, no. 1: 5–31. doi: 10.1163/092735210X12589554057604.

Witbooi, Emma. 2014. "Illegal, Unreported and Unregulated Fishing on the High Seas: The Port State Measures Agreement in Context." *International Journal of Marine and Coastal Law* 29, no. 2: 290–320. doi: 10.1163/15718085-12341314.

Chapter Six

What Is a Fresh Fish?

Knowledge and Lived Experience in the United Kingdom and Portugal

Monica Truninger, João Afonso Baptista, David M. Evans, Peter Jackson, and Nádia Carvalho Nunes

The concept of freshness is a paradoxical one. There are widespread positive associations with purity and nature; healthy and wholesome produce; quality, taste, and distinction (Freidberg 2009; Jackson et al. 2019). Yet, there are also unintended environmental and social consequences in the production and circulation of food freshness. In this sense, there is a *dark side* to the concept of freshness, including the energy burden of the cold chain, increased water stress in already vulnerable areas, food waste, and the labor exploitation (e.g., women, migrants, low paid) in industrialized food production practices where freshness is often deemed the pinnacle of food quality (Evans 2016). To make fresh food (and, in this case, fish) available all year round relies on global supply chains and technological innovations.

This chapter is based on comparative research conducted in the United Kingdom and Portugal, which included interviews with major retailers, their suppliers, and smaller producers, as well as interviews with consumers supplemented with ethnographic methods.[1] Our work focused on the production and consumption of poultry, fruit, vegetables, and fish to explore the *enactment* of freshness (Jackson et al. 2019) across commodity chains. In this chapter, we refer mainly to the data on fish, examining eaters' knowledge of fish freshness when acquiring, appreciating, and appropriating fish. Although the analysis is clearly more focused on consumption, we will not disregard the interface with production, namely fish retailers' practices.

We firstly offer an overview of the scarce literature on consumers' knowledge of fish freshness and present our own conceptual perspective of fish knowledge inspired by a relational approach to knowledge. We then describe the research methods used and present the main results by looking at the different ways of knowing what a fresh fish is. We conclude with some comments on fish knowledge and freshness from a relational framework. In this vein, knowing what a fresh fish is involves understanding it as produced through practices, encounters, relations, and lived experiences of eaters and retailers with fish bodies.

WHAT IS A FRESH FISH? BODIES, TUNINGS, AND RELATIONAL KNOWLEDGE

Scientific studies focusing on objectively assessing and measuring fish freshness from a production lens abound, and findings are often taken up by the fish industry sector (Nielsen, Hyldig, and Larsen 2008). Less abundant is the literature on consumers' fish knowledge. However, over the last few years this body of work is growing and includes contributions from various fields such as marketing, social psychology, marine sciences, nutrition, and agricultural economics. It tends to center on accounting for levels of consumers' objective knowledge (e.g., based on factual statements that consumers have to identify as true or false), subjective knowledge (a self-assessment of consumers' knowledge about a topic), measuring consumers' information about wild fish and farmed production, consumers' knowledge about fish labeling schemes, fish safety and levels of chemicals in fish (e.g., mercury), and other matters related to pollution and environmental issues (Pieniak, Vanhonacker, and Verbeke 2013). There is a bourgeoning literature on consumers' perceptions, attitudes, and values around fish quality and freshness, mostly driven by those aforementioned disciplinary fields (Nielsen, Hyldig, and Larsen 2008). Consumers generally value fish freshness (more than frozen fish). Health and nutritional reasons, taste, a sense of well-being, and diet variation are the main reasons for buying fresh fish, with less positive aspects related to the time and skills needed for fish preparation (Frewer and Trijp 2007). In some countries, such as the United Kingdom, there is a preference for buying filleted, packaged, and prepared fish rather than buying whole fish. In Portugal, on the other hand, there is still a preference for buying certain fish species whole (European Market Observatories 2018).

This chapter looks at the links between eaters, knowledge, and fish freshness by taking a particular conceptual and theoretical route. Instead of engaging with matters regarding objective or subjective knowledge of fish freshness, we position knowledge as the effect of *lived* experiences (Carolan 2011) and freshness as a quality enacted differently throughout the fish supply chain. One of the reasons for this detour is that assessments of objective or subjective knowledge do not capture the richness of contexts, practices, and relations where knowledge is used. In our previous work, we explored "the ontologies of freshness as a quality of food that is enacted in multiple ways and with variable effects" (Jackson et al. 2019, 81). We identified four ways of enacting freshness, which are the outcome of material and discursive practices of producers and consumers when handling food: uniform and consistent, local and seasonal, natural and authentic, and sentient and lively. In this chapter, although we keep this typology at close analytical range, we want to shed light on the knowledge arrangements that emerge when fish freshness is enacted in different ways by the active bodies of eaters and retailers. These bodies continually measure, monitor, and assess *freshness* in the course of performing fish-related practices, from shopping and selling to meal preparation (Evans 2016, 110). They may do this by resorting to "the use of codified knowledge (such as date labeling) or more tacit and embodied forms of know-how (such as feeling for changes in texture or sniffing for changes in smell)" (Evans 2016, 110). Or, one may also add, freshness may be assessed through local knowledge that is generated in the marketplace and embedded in the close relationships between consumers and producers.

Thus we argue that knowledge of fish freshness is not fixed, objective, or absolute, and it is more than socially constructed by humans. It is understood as the outcome of the complex nonlinear entanglements of humans and fish bodies (Probyn 2016). As Carolan states, "what we call mind, thought, cognition, and knowledge are all *effects* of active bodies, of bodies-in-the-world" (2011, 1, italics in original). Knowledge of what is a fresh fish emerges through sociomaterial arrangements of varied heterogeneous elements: the living bodies of fish, fish corpses, frozen fish, and human bodies; eaters and retailers; local market fishmongers and supermarket shelves; thermometers and fridge-freezers; sea-salted ice and tap-water ice; frozen on board and other capture methods; and certifications and labels. Such arrangements are activated through the lived experience of retailers and eaters when assessing the quality of fish freshness. More than identifying types of knowledge that emerge through the different enactments of freshness, we are interested here in understanding how retailers and eaters are variously tuned in to fish freshness, that is, how their bodies are affected by the different enactments of fish freshness. Our research suggests these enacted realities can stand alone or coexist with each other in either harmonious or conflicting terms.

RESEARCH METHODS

This chapter is based on comparative research that the authors conducted in the United Kingdom and Portugal, including interviews with major retailers, their suppliers, and smaller producers, as well as interviews with consumers (twenty-five in Portugal and twenty in the United Kingdom), supplemented with ethnographic methods (shopping with consumers and conducting kitchen tours, cupboard and fridge rummages, cooking observations, combining audio and visual recording).[2] Participants in each country spanned in age from those in their twenties to those over seventy, and have diverse social, economic, and ethnic backgrounds. The interviews were transcribed, selectively translated (from Portuguese to English) and coded, following the ethical guidelines and protocols that were sanctioned by our respective universities. We also conducted a series of tasting events that were inspired by the work of Mann and colleagues (2011). These experiments were designed to capture eaters' embodied and sensory tunings with fresh fish, including things that are difficult to put into words (e.g., facial expressions, gestures, sounds), but which can be captured through visual methods (e.g., video cameras and photos) or observed directly by the researcher (Jackson et al. 2019, 81). We have explored elsewhere nonverbal material from the tasting events by paying particular attention to facial expressions, embodied gestures, and gustatory *mmms* when assessing and appreciating food freshness (Jackson and Meah 2019; Jackson et al. under review).

FISH BODIES, EATERS AND RETAILERS: WAYS OF KNOWING FRESHNESS

Embodied and Technical Knowledge

In both countries several eaters of fresh fish learned to tune in to freshness through their senses: eyes, fingers, hands, noses are all involved in assessing fish freshness. Crucially, fish also participate in this collective human-fish bodies tuning to freshness. The appearance, smell, and texture of fish skin and scales, together with colors of the eyes and gills, all leave traces of the qualities of freshness that are sensorially experienced by active bodies. According to Carolina (twenty-three years old, Portugal), "If the fillet isn't firm, if the scale and the eyes aren't shiny, it's not worth buying" (Carolina, interview, June 8, 2017). Maria (forty-seven years old, Portugal) explains that the smell of sea is crucial to know if a fish is fresh or not:

> I especially like to buy in Edgar's that has a fish that gets in your eyes and smells like sea, if it smells of sea it is fresh. . . . You get inside the market and immediately you sense the smell of sea, it smells very good. He has very good fish. (Maria, interview, March 14, 2017)

Among British eaters, the smell of sea (and not of fish) can be something that gives away how fresh shellfish is. During one of our tasting dinners in the United Kingdom, in between courses, Ruby (in her twenties) says that she could smell the squid when she was cutting it: "It smelled like the sea . . . which is something that I'd look for, you don't want it to smell like fish" (Ruby, tasting event, July 7, 2017).

One (human) body without the other (fish) body cannot be tuned in to freshness; here, knowledge is the effect of collective sensory tunings to fish freshness by active bodies. The freshness of fish is assessed by how it looks, how it feels, and how it smells, but also by highlighting the lively materiality embedded in the qualities of fish (enacting sentient and lively freshness). For example, Carolina (twenty-three years old, Portugal) mentions that the eyes of the fish can't have blood and they need to look "alive" (Carolina, interview, June 8, 2017). Maria (forty-seven years old, Portugal) also turns to the metaphor of vitality, explaining that "the eyes have to be fresh, the gills have to be red—it means the blood is still very alive" (Maria, interview, March 14, 2017). In a Portuguese tasting event, this feature of liveliness in a dead fish emerged again when participants at the dinner started a conversation on *what is a fresh fish*:

> Luísa: I don't know how exactly, but I immediately identify a fresh fish in the market stalls.
> Carmelinda: How do you do it?
> Luísa: The scales, the eyes; when everything is nice . . . I don't know how to explain it accurately. . . .
> Anabela: A person senses it instantly!
> Carmelinda: But how?
> Luísa: The fishes have to be *alive*, . . . (tasting event, Lisbon, April 18, 2018)

The vitality of matter (Bennett 2009; Evans 2018) is a sign that eaters not only look for on fish corpses, but importantly is also mentioned by both small and larger fish retailers. In late summer of 2016, we visited the head of *Cabaz do Peixe* (Fish Basket), an association that sells fresh fish directly to consumers, based at Sesimbra fishing harbor. António, a retired fisherman, told us that to be fresh a fish needs to seem *alive*. In his own words, "It is when a dead fish shows to be alive" (António, interview, September 22, 2016). António mentioned that skills and knowledge are mobilized to make dead fish alive, through *reviving* the fish from the time it is caught at sea until it is sold in the market. He spoke about how fish is handled throughout the fish supply chain; the characteristics of the ice for conserving the caught fish; hygiene practices on the boat; the different ways of storing boxes; temperature limits for preserving the fish; and labels that identify the method, the zone of capture, and the *fishing art* employed. How the fish dies crucially affects how the fish appears to be alive and fresh in the fishmonger's shop. In this sense, the knowledge that is mobilized in the work of death is fundamental to enact the vitality of fish flesh and freshness. This is also explained by José, who is responsible for fish and seafood training of technical operators who work in the supermarkets' fishmonger counters of a big food retail chain in Portugal. Although farmed fish has a different treatment at the source of production, it is prepared and gutted so that it increases its shelf life in the market place: "We can control better the life of the fish," says José (interview, October 19, 2016); wild fish has a shorter shelf life and needs to be sold quickly. José then explains the amount of work and knowledge that is mobilized and activated to make a dead fish look alive. Skills and knowledge revolve around maintaining a certain consistency and standard of shininess and liveliness of the fish throughout the day to entice consumers to buy it.

> *José:* It has some roughness, the scale also has that colour still *alive*. The work with the lights, the ice that we put on top that humidifies the fish and gives that brightness. Over time this glow begins to fade. . . .
>
> *Interviewer:* How long? One day?
>
> *José:* It depends on the work we put into it. For example, it will lose brightness if we don't put ice on top of it to lower the temperature and give it longevity, and it needs to be the right ice as well. . . . Before we used to have the micro sprinkles, but there was an issue with the salted water . . . as these are salted water fishes they need salted water, if we use normal tap water this will affect the composition of the skin and speed up fish deterioration. (José, interview, October 19, 2016)

In this quote, one can see an overlap of different ways of knowing freshness that are harmonious and not conflicting. Embodied and technical knowledge and skills are mobilized to secure consistency in the delivery of fresh fish in large supermarkets, which meets desired quality standards and makes fish available in its optimal state (e.g., showing brightness and aliveness of the skin, avoiding skin's deterioration). In the enactment of technical and sensory freshness multiple knowledges emerge: from assessing colors and textures of the fish through the senses (embodied knowledge) to applying technologies to maintain consistency (e.g., micro water sprinkles, ice with the right salt content) throughout the day.

KNOWING AT A DISTANCE AND LOCAL KNOWLEDGE

In Britain, fish is often made available in supermarkets preprepared and filleted in a package. When prepared this way, some eaters mentioned they could hardly use their sensory skills to assess fish freshness. As stated by former deep-sea fisherman Steve (sixty-two years old, United Kingdom), "Supermarkets don't allow you to pick fish up" (interview, February 9, 2017). Instead, UK participants had to rely on codified knowledge inscribed on the package (e.g., fish sources, how it was caught, use-by dates, and other certifications associated with environmental impacts). Quality attributes were made known through labels that mediated knowing fresh fish at a distance. In this context, eaters had to rely on the information provided by supermarkets, to decode that information, and to trust it.

> If I get tuna, I try and get line-caught tuna, sustainably, if it's tinned, sustainably, er, dolphin-friendly or whatever on the tin, umm. . . . I suppose, really, you have to take supermarkets at their word because they say they have fresh fish in every day. (Jo, forty-eight years old, interview, June 21, 2017)

However, Stella (sixty-three years old, United Kingdom) mentioned how difficult it is to know whether a fish is fresh at the supermarket, either packed or uncovered at the counter: "I wouldn't know whether it's fresh or not. I don't think you can . . . in a supermarket you just make an assumption that they are selling [fresh fish]" (Stella, interview, April 18, 2017).

The difficulty of knowing if supermarket fish is fresh is also shared by Steve (sixty-two years old, United Kingdom), but in his case this knowledge is the effect of his experience as a former deep-sea fisherman. In a walkalong interview with Steve, he shared his lived experience as a trawlerman, where he stayed out at sea close to the Arctic for twenty-one days, starting fishing from the fourth day onward, and coming back ashore to sell the fish:

> So, the oldest fish on that boat was seventeen days old, kept on ice, and this was the delicious freshest fish because it was Hull. Whereas when you were aboard the trawler, every single tea-time you had fish that was there. . . . Now they get most of their fish from Iceland, in containers . . . and it's kept until they've got a load, then it's sent to Grimsby, and then it's filleted and sent off, so even that's not entirely fresh. . . . You are not certain [pause] how fresh your fish is, no matter what they say, the supermarkets. (Steve, interview, February 9, 2017)

Thus, some British eaters were quite aware of the intricate global routes fish take, and how distantly they are caught, before they arrive at the supermarket fishmonger counter. For some, the fish purchased in small retailers gave them more guarantees that it was fresher, given some believed it was brought ashore a day or two ago.

> Freshness is very important with fish, you never get it but you want it as fresh as possible [subject explains that if you go to a Tesco Local and get some fish there which has been sitting around for a few days, having already been sent from a

larger store or distribution center] it's alright, you can eat it, it won't poison you, but it has no special taste. If you buy a fresh piece of fish, even from the fishmonger down Sharrow Vale Road for example, they were probably in the sea a day or two ago, they were fresh yesterday. (Ted, seventy-three years old, interview, July 13, 2017)

In the last part of this quote Ted brings about a different way of knowing *what a fresh fish is.* In this enactment of freshness, time is important to take into consideration, wherein skills and knowledge about the local shops that sell fish recently brought ashore are mobilized. This local knowledge is an instantiation of learning to be affected by recently caught fish and equating that short time span to freshness and also to a particular taste and texture of the fish, which is not the same if the fish is old or has been frozen. Jo (forty-eight years old, United Kingdom) reports from her experiences of going on holidays in Lanzarote:

You can literally see them bringing it from the sea and putting it onto the lorry . . . that's how fresh it is. . . . And you can tell the difference in the taste . . . when it's been frozen, the flavor changes I think. . . . It's softer and sweeter when it's fresh. (Jo, interview, June 21, 2017)

In Portugal, a large proportion of participants preferred to shop in local fish markets, particularly the participants recruited in the Algarve, a tourist desti-nation that has kept its local fish markets and auctions, with a close relation-ship with the sea (e.g., former small fishing villages, the triangle of beach-sun-surf, fish restaurants with the catch of the day). For many Portuguese eaters who participated in our study the timespan from catch to purchasing is crucial to assess fish freshness. Inês (thirty-three years old, Lisbon) reports that "fish has to be from the day of the catch, otherwise it starts to turn soft" (Inês, interview, June 6, 2017). For her, freshness is opposed to having spent a large amount of time in cold storage: "It means it wasn't frozen, hasn't been in cold chambers, which is one of the things that happens in big hypermar-kets, they are in cold chambers and are frozen and then thawed . . . until they get to the fish stall where you buy it as a fresh fish" (Inês, interview, June 6, 2017). Similarly, for Roberto (fifty-four years old, Algarve), "a fresh sardine is the one that was caught early in the morning and eaten at noon" (Roberto, interview, June 16, 2017).

Several interviewees can identify signs of freshness in fish bodies (gills, skin, eyes) but still feel safer by having a close relationship with someone more experienced to measure freshness on their behalf. They delegate this knowledge and skill to someone else by developing a relationship of trust with the fishmonger, usually at the local fish market. Almerinda (forty-seven years old, Lisbon) places her trust on the expertise of the fishmonger to choose the freshest fish: "I get there, and I sometimes ask the girl 'So, what do I want today?'" (Almerinda, interview, May 16, 2017). She explains that, by asking this, she hopes the seller will give her the freshest fish. She adds that she developed a relationship of trust and believes the seller will not deceive her. Vanda (forty-four years old, Algarve), who purchases mostly frozen fish on account of being comparatively cheaper and fitting better into her everyday food practices, also resorts to trusted fishmongers when occasionally buying fresh fish: "I know it's trustworthy because I already know some people . . . there is some key-people, there is this sort of relationship" (Vanda, interview, June 13, 2017). Others, such as Bruna (thirty-one years old, Algarve), claimed that they have gained skills and knowledge about fish because they are from a coastal region like the Algarve. "With fish, I, being from the Algarve, know a few techniques . . . the redder the blood, the redder the gill, the fresher the fish is" (Bruna, interview, June 14, 2017). Joaquim (sixty-seven years old), also from the Algarve, considers himself "an artist" in fish selection (Joaquim, interview, June 12, 2017). In these cases, knowing fish freshness is an effect of the engagement with lived experiences of a particular place, like the Algarve, which is famous for the quality of its fish. That is, it is an effect of the practices of observing fish coming to the shore, of going to the fish markets and seeing whole fish at the fishmonger counters, and of chatting with the fishmongers to learn more about fresh fish. The infused sea-related stories and practices rooted in a particular locality affect the ways eaters tune in to freshness. How they learn about fish freshness is affected by sensing fresh fish bodies caught early in the morning, interactions with fishmongers in local fish markets, the tastes and textures of fresh fish in eaters' mouths when dining at the table of restaurants or at home.

This chapter has explored the different ways British and Portuguese eaters know what a fresh fish is. We kept retailers of fresh fish, either small or large, at close analytical range given our alignment with a relational approach to knowledge. In this vein, knowledge of fresh fish goes beyond cognitive representations that perceive objective and universal signals of fish freshness (technically and scientifically measured) or beyond subjective knowledge of what one thinks a fresh fish is. Instead, we engage with the literature that understands knowledge as an effect of relations between heterogeneous elements. Such relations are embedded in the practices and lived experiences of both eaters and retailers. Thus, the relations that eaters establish with the retail contexts for fish selection are important to derive what they know or do not know about fish freshness.

We have examined different ways of knowing what a fresh fish is. Sensory and embodied knowledge used by both eaters and retailers (large or small) to tune into freshness in fish, meant that the color of the eyes and the gills, the shininess of the skin, and its texture were observed, smelled, touched, and tasted. Sometimes, different ways of knowing a fresh fish were juxtaposed in harmonious ways, when combinations between elements extended beyond the sensory and vitality signs inscribed on fish bodies, to technologies, technical skills used in handling fish onboard, types of ice and storage boxes, water content, and micro water sprinkles. All this network of relations between heterogeneous elements resulted in a combination of embodied and technical knowledges when tuning into a fresh fish that had to comply with a particular enactment of freshness: one that delivers consistently the desired quality standards of freshness to eaters (technical freshness).

Our research also demonstrates that eaters rely on particular information clues to know fish freshness at a distance, especially when buying fish at large retail shops. Certificates about zones of capture, use-by dates, and environmental labels are inspected, especially among our British participants who were more tuned in to this sort of codified knowledge in which ethics and environment are highlighted. This constituted important ways to know fish quality and freshness at a distance, especially in a supermarket context where some were quite aware of the difficulties of knowing what fresh fish is in such a place. Such doubts were not necessarily an instantiation of ignorance on the part of eaters (an interpretation that may come from the measurement of objective knowledge in a questionnaire). On the contrary, they expressed an awareness of the complexities of the global fishing industry, to the point of having to rely on supermarkets' information without being totally convinced that they were getting the freshest fish in the market. In this sense, a fresh fish was associated with the local and with a short time span from catch to sale, something that both British and Portuguese eaters valued. There was some investment in learning about the local fish suppliers that have the freshest fish and that sell the catch of the day. This was more visible in the Portuguese context, where fish is such an important part of eaters' diet, and fresh fish bought in local fish markets is especially appreciated. However, not all were certain about what a fresh fish was, and often relied on the skills and knowledge of others, namely the fishmongers with whom they have established a long and trusting relationship. Knowledge was an effect of the encounters with the fishmongers. It was interpersonal and localized, marked by territorial tunings to freshness, as we demonstrated with the Algarve participants, where fresh fish knowledge was deeply ingrained in spatial practices and localized lived experiences. All this produced a different way of knowing what a fresh fish is that is more local and interpersonal than the instituted codified knowledge used through labels, certifications, and other informational cues that are embedded in the ways of knowing at a distance. Thus our research demonstrates that knowledge emerged in various ways and was a product of the practices, encounters, relations, and lived experiences of eaters and retailers with fish bodies, instead of being reduced to human mental and cognitive capacities, apprehended and deployed in a stable and fixed manner.

ACKNOWLEDGMENTS

This chapter was elaborated under the research project titled *Enacting Freshness in the UK and Portuguese Agri-food Sectors* funded by the Economic and Social Research Council (ESRC) award number ES/N009649/1.

NOTES

1. The project *Enacting Freshness in the UK and Portuguese Agri-Food Sectors* was funded by the Economic and Social Research Council of the UK (ES/N009649/1).
2. Dr. Angela Meah was a research associate who undertook most of the UK fieldwork and also participated in the analysis and writing up of the data. We are grateful for her excellent work in the collection of rich and high-quality data.

REFERENCES

Bennett, Jane. 2009. *Vibrant Matter: A Political Ecology of Things*. Durham, NC: Duke University Press.

Carolan, Michael. 2011. *Embodied Food Politics*. Farnham: Ashgate.

European Market Observatories for Fisheries and Aquaculture Products. 2018. *The EU Fish Market*. Brussels, Belgium: European Market Observatory for Fisheries and Aquaculture Products.

Evans, David. 2016. "Biological Economies and Processes of Consumption: Practices, Qualities and the Vital Materialism of Food." In *Biological Economies Experimentation and the Politics of Agri-Food Frontiers*, edited by Richard Le Heron, Hugh Campbell, Nick Lewis, and Michael Carolan, 69–102. London: Routledge.

———. 2018. "Rethinking Material Cultures of Sustainability: Commodity Consumption, Cultural Biographies and Following the Thing." *Transactions of the Institute of British Geographers*, 43, no 1: 110–21. https://doi.org/10.1111/tran.12206.

Freidberg, Susanne. 2009. *Fresh: A Perishable History*. Cambridge, MA: Harvard University Press.

Frewer, Lynn, and Hans van Trijp, eds. 2007. *Understanding Consumers of Food Products*. Cambridge: Woodhead.

Jackson, Peter, David Evans, Mónica Truninger, Angela Meah, and João A. Baptista. 2019. "The Multiple Ontologies of Freshness in the UK and Portuguese Agri-Food Sectors." *Transactions of the Institute of British Geographers* 44, no. 1: 79–93. https://doi.org/10.1111/tran.12260.

Jackson, Peter, David Evans, Monica Truninger, João Baptista, and Nádia Nunes. Under Review. "Tasting Events: A Methodological Experiment for Doing Food Research." *Social and Cultural Geography*.

Jackson, Peter, and Angela Meah. 2019. "Taking Humour Seriously in Contemporary Food Research." *Food, Culture and Society* 22, no. 3: 262–79. https://doi.org/10.1 080/15528014.2019.1580535.

Mann, Anna M., Annemarie M. Mol, Priya Satalkar, Amalinda Savirani, Nasima Selim, Malini Sur, and Emily Yates-Doerr. 2011. "Mixing Methods, Tasting Fingers: Notes on an Ethnographic Experiment." *HAU: Journal of Ethnographic Theory* 1, no. 1: 221–43. https://doi.org/10.14318/hau1.1.009.

Nielsen, Jette, Grette Hyldig, and Erling Larsen. 2008. "'Eating Quality' of Fish—A Review." *Journal of Aquatic Food Product and Technology* 11, no. 3–4: 125–41. http://dx.doi.org/10.1300/J030v11n03_10.

Pieniak, Zuzanna, Filiep Vanhonacker, and Wim Verbeke. 2013. "Consumer Knowledge and Use of Information about Fish and Aquaculture." *Food Policy* 40, June: 25–30. https://doi.org/10.1016/j.foodpol.2013.01.005.

Probyn, Elspeth. 2016. *Eating the Ocean*. Durham, NC: Duke University Press.

Chapter Seven

Late Nights and Live Tanks

Entanglements of Caring at Golden Century

Nancy Lee

Golden Century is a Hong Kong–style live seafood restaurant in Sydney's Chinatown and has been an enduring fixture in my life. Brightly lit and constantly surrounded by a queue of diners, it looms over bustling Sussex Street with its tanks full of slow-moving fish and hills of bivalves. Golden Century is many things to many people; it has also been many things to the many versions of me over the last three decades. The restaurant is something of an institution in Chinatown, operated by the same owners since 1989. Birthdays, Lunar New Year celebrations, babies' 100 days[1] celebrations, and dinners with family visiting from overseas have taken place here. As I've grown and changed over the years, so too have the ways I come to relate to those bubbling tanks that line its walls.

This chapter responds to Elspeth Probyn's suggestion that exploring the politics of caring in complex, "more than human" entanglements "is through detail, through relating stories and experiences, through paying attention to the intersections of human and fish lives" (2016, 34). I consider my own relationship to Golden Century and its place in Sydney's Chinatown as witness to an ever-changing Chinese community; its ubiquitous tanks full of coral trout and barramundi keeping a watchful eye over the evolving streetscape. Probyn suggests in *Eating the Ocean* that how we care or don't care is a question of habitus (2016, 36). Her centering of the emotional in understanding habitus is the hinge I use to articulate tensions between notions of sustainability, care, and culture from the perspective of Chinese-ness (my own) and from another group that has adopted Golden Century as its own, chefs in Sydney. I draw on Ien Ang's work in Australian diasporic Chinese politics and ideas of belonging, in particular from her work on Sydney's Chinatown and "nation as contested home" (Ang 2016, 259) to analyze Golden Century as a space of ritual. My analysis is also built on Beverley Skeggs's work on the production of value and her argument that "perspectives are made from classed positions" (2004, 173)—that is, who we are produces what we know.

GROWING UP CHINESE

I have always known seafood to be a status-signaling food in Chinese culture. Steamed fish and prawns featured on the dining table at every celebratory family meal, and whenever we had special guests. When I was older, mum and dad discovered one of our neighbors went diving; as a result, abalone became a regular fixture at special meals (the diving neighbor didn't know how to cook them). An idiosyncrasy of Chinese manners I find amusing is the host insisting they have prepared an extremely humble meal, when really the spread is quite extravagant. My mother would often brush aside impressed exclamations of the seafood with variations of, "It's nothing, we live close to water so why not make the most of it?" For the Chinese, seafood symbolizes prosperity in several ways—it is not particularly cheap, so serving seafood reflects the ability to purchase it. As well, the Cantonese word for *fish*, for example, is "yu" which is a homonym for abundance. During Lunar New Year, the types of food consumed are homonyms for desirable traits, including vitality (*xing*, represented by lettuce) and wealth (*fat choi*, represented by black moss). Food is a significant part of Chinese culture, and this was certainly true of my upbringing.

I have been eating at Golden Century since I was about five years old. As a child, I was mesmerized by the tanks full of fish; they were used as a distraction for restlessness at the dinner table. For as long as I can remember, Golden Century has been a place of prestige and so-called authentic Chinese food—before it became my go-to late night post-clubbing venue in Sydney (it's open until 4:00 a.m.), it was the place my mum took visiting guests, the place to go for a proper fancy dinner. Part of its prestige for my mum back then was the fact that you can pick and choose your fresh seafood on the spot. For those wanting to show off to their friends, choosing from the tank is a form of conspicuous consumption—it's literally flapping fresh, and it is expensive. You make your order, and the waitstaff bring it over for your approval, both of quality and size (they are sold by weight), after which it is taken into the kitchen to be cooked. Excess at the Chinese dinner table is a huge compliment to the guest—here, platters of seafood are themselves a display of wealth. When my uncles visit from overseas, we take them to Golden Century, even with so many options in Chinatown, not to mention in Sydney. Mum will order the pippies in XO sauce and the mud crab with ginger and shallots as a gesture of extravagance to our guests. When you want to show face to special guests, you take them to Golden Century. This tradition endures in my family even as Chinatown has changed rapidly, with the increase in mainland Chinese migrants reflected in the diverse food options populating what used to be a predominantly Cantonese area.

Probyn says, "It's hard, though not impossible, to cuddle a fish" (2016, 26). She is right—fish are not cute and cuddly—but one thing I always notice when I go to Golden Century is that the fish seem sad. They don't really move—there is no space to do so—they kind of just hover and stare out into the dining room, their bright tropical colors dulled in captivity. Until recently I was happy not to think too hard about sad fish, tangled up as it is with the cultural ambivalence the Chinese have when it comes to animal welfare.

Growing up, I learned that food is food. Food isn't animals, animals are food. Culturally and historically, the Chinese (and perhaps I generalize) are not particularly concerned about the sentience of what they eat, and certainly not when it comes to fish. But younger generations in China are gradually becoming more and more concerned about animal welfare, evidenced by, for instance, the recent rise in pets and the amount of money invested in looking after them (in the 1980s, Mao denounced pet-owning as bourgeois) (Zhang, Pei, and Woo 2018). China is also one of the few countries catching up on the abolition of animal testing (Smith 2018; Bloomberg 2018). The idea of animal welfare is presumably not top-of-mind in Hong Kong–style seafood restaurants that rely on live tanks as a key drawcard.

A few years ago, I was at Golden Century with my then partner (non-Chinese). We were watching a table near us order lobster and what looked like ocean trout. The waiter took a net and scooped out a lobster, along with a long fish. He put them into two plastic bags and carried them over to the table. A man, who was the host of the table, said something to the waiter about the lobster. Nodding, the waiter set the fish on the floor, and marched back to the tank to swap the lobster. My boyfriend and I watched, agape, as the fish flapped around desperately on the floor. The diners at the table, and others around them, glanced dispassionately at the fish flailing next to them until it suddenly thunked out of the bag and onto the carpet. Still, everyone watched calmly. Shortly after, the waiter strolled back and seamlessly flicked the fish back into the bag as the host gave the new lobster his tick of approval. The waiter carried both to the kitchen and conversation at the table carried on like nothing had happened.

That night stuck with me. As a Chinese person, I feel like I should have somehow understood their ambivalent, passive reaction enough to dismiss the whole episode. But I didn't. I found it distressing, and I was at a loss for what to do in that particular situation. Skeggs argues that "class is in continual production" and that "value becomes established through exchange, but in this exchange the relationship to the commodity itself generates different forms of personhood" (2004, 9). Consumption of live seafood produces a kind of value that is quite specific to Chinese culture; although I have grown up instinctively understanding it, I find it difficult to translate.

CULTURE AND HABITUS

At this point, it is perhaps pertinent to admit that I do not consider myself "Asian in any substantial, cultural sense" (Ang 2004, 148). Ien Ang's prolific work on Australian Chinese diaspora is useful to provide some context of what it's like being Chinese without really *being Chinese*. Ang argues, "white racism becomes a source of self-identification" (2004, 149); that is, it is an internalization of what others project. In many cases this can be in the form of simplistic stereotypes. Although I grew up with a traditional Chinese mother who drilled into me the importance of education and reputation and status, I lack knowledge about the nuances of Chinese culture. (I think I learned more about Chinese culture from reading the *Crazy Rich Asians* trilogy [Kwan 2013, 2015, 2017] than I ever have just existing as a Chinese person.) The stereotypes I've internalized over the years fill in the gaps. They reinforce the dread I feel when I see the slow, sad looking fish at Golden Century. I am Chinese and often, I also believe the negative stereotypes of Chinese people as uncaring and self-serving, especially when it comes to animals.

At this point I will turn back to Probyn's argument that "the body eats into the social as our bodies are simultaneously eaten by practices that are instantiation of class, gender, ethnicity" (2016, 36). Skeggs suggests, "Different bodies carry unequal values depending on their position in space, on their cultural baggage" (2004, 17). My Chinese-ness has given me some insight into the cultural value of Golden Century—namely, the symbolic value and status transferred by its live seafood. I also know the food is very good—I order their popular XO pippies (pippies stir-fried in a spicy sauce of chili, soy sauce, and dried seafood) for my birthday each year; American chef David Chang is quoted on their menu saying the XO pippies at Golden Century might be the "greatest dish in the world." For this chapter, I decided to put those assumptions to the test. I requested an interview with the owner of Golden Century, Eric Wong. Speaking to Wong highlighted to me the different ways the mythology of Golden Century had been constructed, from its relative lack of media hype, to its exaltation by Sydney chefs.

LIVE TANKS

We meet at the restaurant over some jasmine tea just before the thrum of dinner service kicks in. Wong has an efficient, approachable-but-not-quite-warm demeanor. He and his wife have built the behemoth of a restaurant that is the Golden Century I grew up with. Before Wong and his family migrated to Australia in 1989, they had decided to buy the business from its owners, who were retiring. "We took over Golden Century in May 1989. Two weeks [after we arrived]. We arrived in April and took over Golden Century on the first of May. From that day we had to work hard" (interview, March 12, 2019). Back then, they were in a much smaller shopfront a few doors down the road. A year in, the age of the building began to show, with several building leaks pushing the restaurant to find another home. In the beginning, they started with one floor and a small fish tank, mostly keeping lobsters, sometimes abalone. Their customers were fine with eating seafood that had been frozen, but:

> because we came from Hong Kong, we had the experience of the seafood restaurants in Hong Kong: in Hong Kong the style is all the fish tank show the customer. The customer like to [choose], to [choose] their seafood themselves. Seafood, lobster, they want to [choose] themselves. And this is one point. Second is because the traditional Hong Kong seafood restaurant . . . it's fresh from the tank. Must be live. (interview, March 12, 2019)

The switch to live seafood was a gradual process, one that stemmed from the inherent value that was perceived in its freshness.

> Because when you're used to that . . . when you compare . . . no, that is not my standard. I can't . . . the Chinese traditional Hong Kong-style seafood [must be] that way. I must upgrade. So we try to discuss with the supplier . . . but most of them is not interested. I don't need to do that way because I can sell my product, I don't need to worry about how to . . . I don't need to do a lot of work to get the same business. (interview, March 12, 2019)

Golden Century worked on developing close partnerships with their seafood suppliers, and through them spoke to fishermen and convinced a few to transport the seafood live for additional payment. Wong recalls an anecdote from the early days of the restaurant turning to live produce.

> I said, "I am happy to pay you more if you can keep the prawns alive, I can pay more to you." And he said, "Ok, I'm happy." Because the same job, for them it's the same job. He said, "I don't know how to keep the live prawn." And we said, "Easy, you just use the plastic bag, and put some . . . pump air through the plastic bag, and that will keep them alive, that was our experience in Hong

Kong." And then when they arrived, bad luck. Say, ten kilos, there was only one kilo alive. Why? Because when they drive back, all the prawns cut through the plastic, the plastic bag was too thin. So all the . . . the plastic bag was broken, so there was no more air. (interview, March 12, 2019)

They ended up taking the prawns despite the hiccup and devised a better storage solution to prevent future leaks. The Wongs imported their knowledge and rituals from Hong Kong with the belief that their Cantonese customer base would support it. One species that Golden Century is well-known for is the king crab. Back in the 1990s, the five-kilogram king crabs were sold at markets already cooked, for about $5 per kilogram. Wong tells me the king crabs of Australia are unique—they are king-sized in body as well as pincers, whereas the Alaskan king crab, for example, have much smaller bodies. A king crab from Golden Century will make three courses: salt and pepper crab, stir-fried with ginger and shallot, and with noodles or sweet and sour sauce. The cost of king crabs has risen to $130 per kilogram from significant increases in demand. Wong sees Golden Century as playing a role in this:

When a visitor comes from overseas, after they order they must take a photo and they send to all their friends—this is the king crab I had at Golden Century in Australia. So they [share this] very fast to overseas people. They understand that only Australia have big king crab like this. (interview, March 12, 2019)

Wong takes pride in his business and the produce they serve, although he acknowledges the very best Cantonese food is still being cooked in Hong Kong and Guangzhou. In his thirty years on Sussex Street, Wong has also seen the significant changes in migration from China, from his own migration as a relatively affluent Hong Konger, to the growing population of migrants from the mainland China community. Walking through Chinatown in 2019, it is no longer the mini Hong Kong I knew in my formative years—the sheer number of Ma La Tang outlets (a Sichuan-style, do-it-yourself hot pot popular with students in Beijing) and queues for bubble tea with cheese foam are testament to a rapidly changing enclave. Wong concurs: "When we arrived, I can say, 90 percent of our clients [were] Cantonese. They speak Cantonese. Now, in Chinatown, say . . . 60 percent is speaking Mandarin. Only 40 percent speak Cantonese" (interview, March 12, 2019). The changing migrant community has changed the fortunes of Cantonese restaurants in Chinatown. "The big functions, or the very important events still happen in Cantonese restaurants. But everyday eating or friends gathering, some of them, the younger ones, they go to the Sichuan restaurant. It's more casual and more suitable for their style" (interview, March 12, 2019).

Today, Golden Century takes up three floors (a third floor was added and was being decorated at the time of writing; the space will be used for private functions). Each service period sees seventy floor staff and forty kitchen staff spread across the two floors currently in operation. In 2011, they opened a high-end Cantonese restaurant in Sydney's Star Casino called The Century, with plans to open a third venue in the tourist hub of Darling Park. Wong estimates that the restaurant sells three hundred to four hundred kilograms of pippies a week; they keep up to three hundred kilograms of king crab in stock at a time. The restaurant serves up to 450 guests a day, using live seafood that lives in twenty-four tanks. The Wongs could be seen as an immigrant success story, albeit already affluent immigrants who left Hong Kong before it was handed back to China by the British.

With the distressed flapping fish from a previous visit in mind, I asked Wong how the seafood is killed. Listening to the recording, I sound hesitant—I'm not really sure I wanted to know the answer. I ask, "The seafood is a big reason people come here. How . . . it must be a lot of work to look after the fish and seafood. Is . . . is there a special way that the seafood is killed when it gets cooked? What do they do to the fish?" Wong responds, "Because we understand by the law of Australia, we must follow the law, otherwise we should be fined. . . . We have the lobster sashimi. We used the old way to do it . . . is not allowed by Australian law." I ask what he means by "old way," and he answers:

> The old way, it means we just take the meat out, and then cut for the customer. But because we understand and . . . we've been told by the government already, and they sent us some information. We have to put in the ice. We have a bucket of ice; when we kill the lobster we put in the bucket of ice. And then after that we do the sashimi for the customer. (interview, March 12, 2019)

Treating the lobster humanely is not driven by motivation to be humane, but rather to adhere to regulation that allows selling their lobsters. But really, I want to know about the fish. He says, "The fish is the old way, just hit on the head." I probe a little more, probably wanting to know if he feels remorse: "And that is . . . that is . . . does the government send information about the fish?"

Wong is calm, matter-of-fact. "I understand the lobster and then the mud crab. These two are the most serious. Because the lobster sashimi, at the start, some of the lobster still moved at the table. After the government law passed, everyone can't do it like that" (interview, March 12, 2019). At this point, I am projecting my trepidation on Wong, convinced he is hedging and avoiding the question. I move back to the fish, asking what is used to kill them, and he tells me they use their cleaver handles. I ask if customers have ever inquired about how the fish are killed. Without so much as a pause, Wong replies, "No." Still pushing on the topic of fish, I ask what strategy is used to choose what fish to have on display. "How do you choose what seafood to keep, do you think about what the customers want, sustainability, how to cook them? How do you choose what goes in the tanks?" He responds,

Actually we have the deal with the supplier and then with the fisherman. I said to them, every time when they catch the new fish, I don't mind what product [it is]. I still want them to provide to us, I don't mind to pay them, and after that we can taste, we can cook and taste, and then we can get advice, get the opinion from our customer. (interview, March 12, 2019)

Wong's response is not quite what I expected. I expected him to talk about the fish that would confer the most status and be most impressive to Chinese diners. Instead, what I hear is quite sensible, indeed adhering to one of the mantras of seafood sustainability—to trust the advice of his fishmonger and to be open to trying different fish. He tells me about the process of tasting the fish and trying different methods of cooking. If the flesh is quite soft, it is more likely to be deep-fried, for example. It is at this juncture that I begin to feel the weight of my "cultural baggage." Wong speaks English extremely well but does not have the vocabulary to get into the nuances of the ideas around sustainability. Similarly, I would describe my Cantonese as fluent, but I do not have the language skills to explain what I want to know about the restaurant's philosophy on sustainability, particularly when it comes to the species of fish and how they are treated. I don't have the language skills or share a mutual conception of sustainability to ask him if he feels sorry for the fish, or to express my own uncertainty about the welfare of fish.

The technique Wong describes for preparing the fish is indeed humane, at least according to the Royal Society for the Prevention of Cruelty to Animals (RSPCA Knowledge Base 2019) in Australia. Their website states: "Humane killing requires that the fish is stunned (rendered instantaneously insensible) before being bled out. Fish should remain in water until immediately prior to stunning." The fish at Golden Century do not remain in water immediately prior to stunning. They are carried over to the customer for inspection in a plastic bag, before being taken into the kitchen. It is not clear if the fish is killed immediately after being brought into the kitchen, either. This particular step of giving the customer a choice in the ritual prevents the fish from being stunned immediately after being lifted from the water. That is, the Chinese-ness of having face, of being seen to control the quality of what you serve your guests (as well as being able to approve the weight and size, and therefore the cost), is the undoing of what might otherwise be a humane process. Wong claims he has never been asked by a customer about the way the fish is prepared. I am inclined to believe him. In some ways, speaking to Wong has reinforced my assumptions of Chinese-ness. In others, it has contradicted them. My concern, as it turns out, is not merely for the well-being for the fish, but also for my own ability to continue to enjoy eating at Golden Century without consternation; to continue to hold onto my affection and pleasant memories of the restaurant with a clear conscience.

LATE NIGHTS

Whereas Chinese families eat at Golden Century because it continues rituals of prestige and tradition, chefs in Sydney have adopted it as their late-night venue, giving it a different kind of identity that deemphasizes the conspicuousness of the tanks. The restaurant is open until 4:00 a.m. and has long been a haven for Sydney chefs as somewhere they can go for supper after cleaning down from dinner service. It's something of a cult venue, removed from the polished, fine dining rooms they are leaving for the night.

Instagram has become the key method for chefs to share their work and build their audience, where photos of food have all but replaced the role of the food critic. Additionally, social media has changed the ways we understand celebrity, and the methods of celebrity production (Marwick 2013). Social media is a significant platform on which a specific type of pseudo intimacy is facilitated, and on which celebrities can actively construct their own identities (Lee 2014).

This brings me to the influence of Instagram and how it, in its own way, produces another identity and form of value for Golden Century. Sydney chefs have coopted Golden Century as their space through posting on Instagram. Through letting people into their world postservice via close-ups of king crab and noodles, chefs signify their stake in it. Whereas Chinese migrants in part value Golden Century for its prestigious dining room (as far as Chinese restaurant dining rooms go) and its pecuniary symbolism, chefs value it as a kind of insiders' supper club. In 2016, it was celebrated as such through a special "midnight feast" promoted by well-known chefs.

Restaurant reviewer Myffy Rigby (2016) wrote:

> Golden Century is the Studio 54 of supper clubs. A crazy mix-up of royalty, politicians, celebrities, hospitality workers and beautiful strangers in various states of disarray, all mashed together in pursuit of the best possible time at a restaurant that barely sleeps. . . . Over the years it has become a multi-generational rite of passage for young chefs to eat here with their mentors.

The oversized menus still feature the 2016 "midnight feast," with chef recommendations including steamed parrot fish with ginger and shallots, endorsed by several well-known Sydney chefs, including renowned chef Tetsuya Wakuda: "This is a classic Cantonese dish which I order every time I dine at GC's . . . no one does it better." For chefs, eating at Golden Century means being in the know, being part of the in-crowd within the industry. Although the restaurant is still open to the public late at night, dining there after doing a dinner service is to signify that you are part of a brethren of chefs and part of that ritual. Chef Darren Templeman tells me, "I've been cooking in Sydney for about seventeen years now. Golden Century is one of the best places to go for a feed after midnight. Nothing changes, the same wait staff have been around since I first started eating there" (interview, November 10, 2017). Hong Kong chef Jowett Yu calls it "a shrine for chefs, a beacon of light" (pers. comm. August 18, 2017). Mitch Orr is a young chef known for his southeast Asian spin on traditional Italian staples who sings the praises of Golden Century's roast pigeon in the pages of its menu. He tells me,

> For chefs GC's is almost a rite of passage. We all started going there early in our careers with our head chefs. Getting that invite and going with your work mates and head chef is an early and important bonding moment for a young chef. It's the place you go to celebrate a good week, a hard as fuck week, the winning of awards, staff members leaving. . . . From that, for my generation, it has grown into the place we choose to reconnect with each other. We always end up there after awards nights or when friends are in town. It takes us back to a place when we were younger, we share old stories of how it used to be as well as vent on how things are now. (pers. comm., July 13, 2019)

After midnight, the restaurant gears up for its supper rush. Although peak hour is still dinner time—around 6:00 to 7:30 p.m.—the crowds that arrive after midnight are far more eclectic than the Chinese families that gather for dinner. "Industry crew, to drunks, to taxi drivers, to a mix of characters from the suburbs and maybe a couple of unsavory types" is how Orr describes it (pers. comm., July 13, 2019). Another chef I speak to, Dan Johnston, finds this late-night mix refreshing: "It's not the same people you always see at the new and trendy restaurants. At Golden Century it's always a mix" (interview, July 12, 2019).

Golden Century is exalted by chefs to such an extent that I wonder what they make of the live tanks, particularly when their cultural influence is increasingly used to implore us to think more deeply about food systems (e.g., Jamie Oliver, Hugh Fearnley-Whittingstall, etc.). I ask Johnston if the tanks are a concern. "Look, it is a bit bizarre, even worrying, although maybe not for all the creatures. A tank of pippies doesn't really bother me, but snow crabs or barramundi would bother me, for example." However, he also notes,

As someone who has always been super conscious of where food comes from, and being in the industry, it's hard to judge without knowing a lot more about everything. Take Murray cod, for example. I don't know what a Murray cod farm looks like. I don't know what a *good* Murray cod farm looks like, you know? Golden Century is a place where I let go and get involved in the culture and the food. . . . Chinatown is one of those parts of Sydney that has cleaned up a bit, but I like that there are still pockets around town that aren't shiny like the rest of Sydney . . . that people have come here, and do what they do the same way they would have done before. . . . They haven't changed and molded to our expectations. I think that's amazing. (interview, July 12, 2019)

This perspective might be seen as non-othering; Johnston acknowledges that he doesn't have all the information to make a judgment on how the tanks are managed, while consumers in general lack information in even more aspects of the food chain. Orr takes a slightly different view:

I love the tanks. . . . Seeing the tanks, seeing the quality of product in the tank, is a direct link. . . . There's no hiding that what you're about to eat was alive five minutes ago, so you better savor it and respect it. If you can't handle that process you probably shouldn't be eating that animal. If you can't eat fish with bones in it, or meat on the bone because it's too real for you, you shouldn't be eating it. Suck every last remnant of crab meat from that crab, eat the cheeks and all that gelatinous, delicious meat in the head of the fish. (pers. comm., July 13, 2019)

From Orr's perspective, the tanks provide a kind of teaching moment; this is amplified given that he sees Golden Century as a place to "connect and bond" with other chefs. He suggests, "Only wanting to eat the 'best' cuts and not acknowledge the whole animal, no matter how big and small, is what people should be uncomfortable with." For Orr, consuming an animal after seeing it alive can contribute to better consumption practices. "We shouldn't be eating things that come [as] pre-packaged portions in plastic. It's not good for our understanding and respect for the animal, and it's not good for the environment" (pers. comm., July 13, 2019). Another chef, O Tama Carey, dined at Golden Century often with workmates when she was younger, but these days visits far less frequently. "It's not a nice way to store the fish," she says of the tanks. The pippies and the clams are fine, but not the fish . . . it's quite hypocritical [for some chefs to say otherwise]" (interview, July 19, 2019). She adds,

But it's like that funny thing where it's cool for fine dining chefs to go eat fried chicken or burgers or something else not good for you after service. But you have to draw the line somewhere [i.e., decide not to eat the fish from the tanks]. It's a very personal thing. That said, I've had some excellent times there, and it is an institution—that can't be denied.

Golden Century is enjoyed by chefs as a place to gather and share their experiences in the restaurant business; a byproduct of this is that it offers a space in which they are reminded to acknowledge different cultural rituals and value different practices, involving both consumption and production.

My belonging to Golden Century is fragmented by my feelings of trepidation toward the tanks. Maybe, by nurturing such a fondness for Golden Century and holding it on a pedestal for my Chinese-ness, I am complicit too. Do I care? Is it enough that I personally refuse to eat the sad fish, but happily order XO pippies each birthday? Chefs I've spoken to, on the other hand, consider the approach of Golden Century as a specific cultural practice from which to learn. Probyn notes:

> [H]ow and why we care about things, people, and places is a continual process whereby "caring" can hurt or reassure or be joyous, as new knowledge, new ideas, different practices intersect with the primary habitus.

In this chapter I have attempted to reflect on a complex entanglement of culture, ritual, and optimism to find some kind of clear-cut answer to how and why I care about what I perceive to be a lack of care when it comes to human–fish entanglement at Golden Century, or how I can care better—as though such a thing should be possible. Through exploring my relationship to Golden Century through the narrow lens of fan and concerned citizen at the same time, I find that I have to tell my own stories to myself to make things more palatable. Perhaps on some level, the chefs I've spoken to do the same. As Probyn reminds us, "There is no innocent place in which to escape the food politics of human–fish entanglement" (2016, 5). As a place of Chinese-ness, a place of familiarity, status, and nostalgia for so many; the details, stories and experiences of Golden Century are why I chose to care in the first place.

NOTE

1. Traditionally, Chinese families host a feast with close family members when babies pass their first one hundred days.

REFERENCES

Ang, Ien. 2004. "Between Asia and the West: The Cultural Politics of Food." *Life Writing* 1, no. 1: 147–54.

———. 2016. "At Home in Asia? Sydney's Chinatown and Australia's 'Asian Century.'" *International Journal of Cultural Studies*. 19, no. 1: 257–69.

Bloomberg. 2018. "Ending China Animal Tests Is Salve for Big Beauty." *Bloomberg*. January 17. https://www.bloomberg.com/news/articles/2018-01-16/ending-china-animal-tests-is-salve-for-big-beauty-quicktake-q-a.

Kwan, Kevin. 2013. *Crazy Rich Asians*. New York: Doubleday.

———. 2015. *China Rich Girlfriend*. New York: Doubleday.

———. 2017. *Rich People Problems*. New York: Doubleday.

Lee, Nancy. 2014. "Celebrity Chefs: Class Mobility, Media, Masculinity." PhD Diss., University of Sydney.

Marwick, Alice E. 2013. *Status Update: Celebrity, Publicity and Branding in the Social Media Age*. New Haven, CT: Yale University Press.

Probyn, Elspeth. 2016. *Eating the Ocean*. Durham, NC: Duke University Press.

Rigby, Myffy. 2016. "Legend of Golden Century Burnished by Generations of Gun Sydney chefs." *Good Food*, October 29. https://www.goodfood.com.au/eat-out/legend-of-golden-century-burnished-by-generations-of-gun-sydney-chefs-20161026-gsb64o.

Royal Society for the Prevention of Cruelty to Animals. 2019. "Knowledge Base: What Is the Most Humane Way to Kill a Fish Intended for Eating?" RSPCA, May 1. https://kb.rspca.org.au/knowledge-base/what-is-the-most-humane-way-to-kill-a-fish-intended-for-eating/.

Skeggs, Beverley. 2004. *Class, Self, Culture*. London: Routledge.

Smith, Matt. 2018. "Australia Eyes Cosmetics Boom as China Eases Mandatory Animal Testing." *Australian Financial Review*. 22 June. https://www.afr.com/news/world/asia/australia-eyes-cosmetics-boom-as-china-eases-mandatory-animal-testing-20180621-h11pbt.

Zhang, Lusha, Li Pei, and Ryan Woo. 2018. "China's Expected Surge in Pet Spending Draws Disbelief, Criticism." *Reuters*, September 10. https://www.reuters.com/article/us-china-pets/chinas-expected-surge-in-pet-spending-draws-disbelief-criticism-idUSKCN1LQ1AY.

Chapter Eight

Catfish: Halal, Green, or Disgusting?

Investigating Practices of Traditional Farming and Care in Indonesia

Arum Budiastuti

This chapter discusses how halal fish is defined within backyard catfish rearing as a *green* practice. *Halal tayyib* is an Islamic concept for Muslims to lead a wholesome life, including in food consumption. *Halal* is generally defined as the lawfulness of food as stipulated in the Quran and *hadith* (the Prophet's tradition), such as adherence to ritual slaughter and prohibition on pork, alcohol, and animals regarded as *khabaith* (dirty, disgusting). *Tayyib* (good), on the other hand, is more open to interpretations because it relates to various dimensions. Despite its multiple meanings, *tayyib* is traditionally associated with health. Creepy crawlies, for example, may be halal by law but they might not be considered *tayyib* as some of them pose a health risk if consumed.

In Indonesia, catfish has been incorporated in Muslim diet for centuries, but now it elicits disgust and is considered nonhalal. One reason is that in some catfish farming practices, especially in rural areas, catfish are also fed with human and animal excrement. Feces is a common elicitor of disgust (Angyal 1941), so it is not surprising to find some people disgusted by the feces-eating fish. The practice of feeding fish with livestock or poultry manure, however, is common in integrated farming systems. In rural areas such farming practice exists in the form of a backyard catfish farm (i.e., catfish reared in domestic effluents or septic ponds dug in the backyard soil). The relations between human, catfish, and dung has existed for a long time and was a mainstream practice. Backyard catfish farming was never problematized as nonhalal and even praised as a sustainable practice as it leaves zero waste. So why now is it regarded as nonhalal? My view is that in the past few decades halal food has been largely discussed and understood merely as a material practice, leaving less space to think about halal differently. Halal meat studies and news coverage, for example, typically revolve around practices of slaughter and pork contamination. Little is said about halal food as an ethical practice, such as whether or not the meat is sourced or produced sustainably. As in my case here, objections toward backyard pond catfish farming are raised solely on the basis of *what* it eats, not how it lives and relates with other entities in its environment, or on the effects of such relations.

Although conversations on halal and sustainability, or what is often called *green* halal, have started to emerge (see Ali and Suleiman 2016; Razalli, Abdullah, and Yusoff 2012), more needs to be done to further explore practices beyond the food industry, like agriculture. In contemporary discourse of halal as a green practice, *tayyib* is translated as *wholesome* and encompasses all aspects of the physical, environmental, social, economic, and ethical footprints of a particular food (Iqbal 2015; Koenig 2008). Talking about *green* halal is thus referring to both *halal* and *tayyib* as an inseparable entity. Here I seek to explore this further by suggesting that halal as a *green* practice is affective and relational. In my discussion of backyard catfish farming, I focus on how relations are forged in the more-than-human world and demonstrate how traditional methods of fish cleansing and feeding render the fish halal materially (how to make catfish clean from haram substances) but also as an ethical entity (how such practice benefits the environment). My data were sourced from the fieldwork I conducted between July and September 2017 in Kemloko village, Blitar, East Java.

I begin my discussion by introducing some species of catfish cultivated in Indonesian aquaculture and providing a brief description on the development of catfish farming. I then tell a story about the long-standing relationship between humans, catfish, and other life forms in the traditional farming practice. To grasp these connections, I will employ Law and Mol's concept of "metabolic intimacy" (2008, 137), which discusses the flux of energy within a community of species: what is thrown away goes back into the human metabolic system. In their discussion about the practice of pigswill in a Burnside farm in the north of England, they showed how the pigs were fed with human leftovers from nearby restaurants and establishments, and the pigs in turn end their lives as objects of human consumption. Similar to the traditional human-pig relation described in Law and Mol's case, villagers in Kemloko village have had a long-standing metabolic tradition with plants and animals in their surroundings, including catfish. Whereas Law and Mol (2008) did not pay much attention to immaterial aspects (such as affects), I would like to bring them into the picture. To understand how physical affects are played out in these relations, I draw on Elspeth Probyn's article, "Queer Fish: Eating Ethnic Affect" (2017) to give another example of what Probyn has shown in her work: how affects serve to block or enable further connections in a more-than-human arrangement. In human-catfish connection, the love of catfish and respect for the environment have allowed this relationship to endure, although it mainly exists in villages. This affection also manifests in the way traditional catfish farmers think about and do halalness through feeding and sanitizing the catfish, techniques passed down through generations as local wisdom. By mobilizing the affective dimension, I also hope to offer another way of talking about human-catfish relations, which has been largely defined in the language of economy.

CATFISH: FROM THE RIVERS TO PONDS

Catfish (*Clarias sp.*) is found in freshwaters of tropical Africa and Asia and can live out of water as long as its skin stays moist. The fish is characterized by an elongated body, long dorsal and anal fins, a very thin skin, four pairs of circumorbital barbels, and pectoral fins (Sudarto, Teugels, and Pouyaud 2004). *Clarias sp.* can grow as big as an adult human, but the common size of catfish consumed is about twenty to thirty centimeters long when aged less than a year. In this section, I briefly describe the local catfish, *C. batrachus*, and the global aquaculture development, especially in Africa and much of Asia.

Indonesians have long been familiar with the native catfish, *C. batrachus*, which is called *lele* by the locals. It is a feral species definitively known only from river drainages in Java, Indonesia. Other species from mainland Asia and India are often confused under the same name, but they are separate undescribed species (Ng and Kottelat 2008). The native fish, however, is now very hard to find. The decrease of water areas and the change of paddy fields into housing and industrial areas have slowly eroded the fish's habitat. In addition, there has been an increasing activity of fish breeding and hybridization for aquaculture in Asia started in the 1980s, and Indonesia is no exception (Iswanto, Suprapto, and Marnis 2015).

The catfish species most commonly farmed today are strains of African catfish, *C. gariepinus*. Hybridization is the most common form of genetic improvement practiced in Southeast Asia, and it is performed to produce more profitable species for farming. For this purpose, African catfish is preferred because it is bigger and grows faster than the local fish: it takes only two or three months for the African species to grow to market sizes (150–300 grams), whereas the local catfish requires much longer to reach similar size.

The selection of African catfish as the main species for catfish culture throughout Asia and Africa is made possible through decades of aquaculture studies through research collaborations and support from world organizations such as the Food and Agriculture Organization (FAO) of the United Nations

since the 1950s (Lawonyawut 1996). Soon the research on the catfish proliferated, most notably in the 1980s, and was not only conducted by the
state but also by private companies and research institutions. In addition to
the rapid growth of *C. gariepinus*, the fish is preferred because it can thrive
in diverse environments, adapt its feeding ability depending on food availability, and most importantly, be successfully hybridized with native species and produce superior hybrids in countries where the exotic species was
introduced (Wachirachaikarn et al. 2009). The hybrid first introduced to Indonesia in 1985 under the government's food security program was known
locally as *Dumbo*, followed by the other strains, such as *Paiton*, *Masamo*,
Egypt, and *Sangkuriang* (Iswanto, Suprapto, and Marnis 2015). *Dumbo* is
a hybrid between African species and a Taiwanese native, *C. fuscus*, and
Sangkuriang was bred in Indonesia from F2 (second-generation) female
Dumbo and F6 (sixth-generation) male *Dumbo*. Although more hybrids
of African catfish are farmed and fill the market today, people call these
hybrids as *lele Dumbo* (regardless of the different strains) to differentiate
them from the native catfish.

This short history of African catfish making distant links exemplifies
one of the underpinnings of contemporary globalization, that is "the spread
of industrialisation throughout the decolonised world, under the ideology
of development" (During 2005, 84). Industrialization in catfish farming is
not only about a large-scale production of catfish for domestic and export
markets, but also the exchanges of technology and knowledge to maximize
catfish aquaculture industry wherever it takes place, all of which are under
the logic of development. Food security was the language used by the FAO
in the 1980s and 1990s to develop catfish aquaculture in Africa and Asia to
provide fish supply to the world and boost economic growth in the regions.
About 90 percent of the world's aquaculture output are from developing
countries (Hishamunda et al. 2009, 102), and global fish demand is expected
to increase by forty-seven million tons in the early 2020s compared with the
mid-2010s (Cai and Leung 2017).

Whereas standardized modern fish farms are becoming the norm in Indonesian aquaculture today, traditional fish farms exist mainly in rural areas.
Backyard catfish ponds are an example of traditional aquaculture, which is
"mainly integrated with other human activity systems such as agriculture and
animal husbandry, cottage agro-industry and sanitation" (Edwards 2012). I
use this definition throughout this chapter to differentiate the practice from
modern catfish farming, which uses modern technology and relies heavily on
pelleted feed and chemical fertilizers.

BACKYARD CATFISH FARM PRACTICE:
PRESERVING METABOLIC INTIMACY

My fieldwork visit was arranged by one of my research interviewees (Dani, twenty-two years old), who invited me to see his family's backyard catfish pond in Kemloko village, located in Nglegok district, Blitar regency, East Java (Figure 8.1). Dani introduced us to his sixty-eight-year-old grandfather, Ahmad, who took care of the pond. The catfish pond was about $8 \times 3 \times 2m^3$, occupying the central part of the backyard area. Next to it were two drop toilets with pipes connected to the pond, and a chicken coop made of bamboo. The old man has taken care of the pond since his father passed away, and the pond has been there for generations.

In this type of farming, feeding is about making use of whatever is available and plentiful from the environment. Ahmad fed his catfish kitchen refuse and food leftovers, and sometimes papaya leaves and sweet potato bushes growing near the pond. He also gave his fish small critters, like mice caught around paddy fields, making rice farmers happy. Catfish is omnivorous, never rejecting any food. Ahmad said keeping catfish this way is an advantage:

Figure 8.1.   Traditional backyard catfish pond in Blitar, East Java.
Photo by Arum Budiastuti.

his household never produced any organic waste as everything goes to the fish and the chickens. Keeping poultry and catfish together also has its own merit. In the case of dead chicken or geese, he just needs to boil it to clean the feathers and give it to the catfish. An Islamic teaching that Ahmad strongly believes in and practices is that people should be grateful and not to be wasteful. Nature has provided everything people need, so it is their obligation to return the favor and be mindful to the natural environment. Preserving the balance will eventually lead to prosperity. In Ahmad's example, caring for his plants and animals gives him food security in return.

As I noted previously, the more-than-human relationship in Ahmad's pond exemplifies what Law and Mol (2008) call *metabolic intimacy*. This metabolic tradition is a highly praised cultural value in Kemloko village. When animals eat leftovers, they are not in direct metabolic competition with people because the practice is based on the economy of plenty (Law and Mol 2008, 141). It is "a technique that fed some of the excess back into the metabolic circuits. The use of waste food does not require extra fertilizers, pesticides or herbicides either. Just a bit of extra energy" (Law and Mol 2008, 141). This is true for Ahmad. He even said that he did not *feed* the catfish, as feeding means *buying* fish meal for the catfish, which he did not do.

Fish feed is an issue in catfish farming, especially in modern catfish systems because most of the industrial feed ingredients, like corn and soy, are imported mostly from South America (Rimmer et al. 2013, 272). Instead of being turned into fish feed, soy "would have been perfectly suitable to feed the rural population of Argentina or elsewhere," as Law and Mol (2008, 140) put it. In Indonesia, soy is the main ingredient for making *tempeh* (soy cake), one of Indonesia's staple foods. Using soy as fish feed puts catfish in direct competition with humans. This is one reason why local scientists and farmers have experimented with locally sourced ingredients for cheaper alternative feed. But this alternative feed is usually not yet mass-produced, so the middle- and large-scale fish industries still depend on imported ingredients.

While urban catfish farmers often complain about the rising price of industrial fish feed, traditional catfish farmers flourish as they work based on the economy of plenty. They make use of whatever is available in their surroundings to feed the fish and avoid waste at the same time. In Kemloko village, for example, chicken offal is abundant because chicken farm industries are thriving in the area. As Ahmad told me, many farmers even integrate their chicken farms with catfish farming, erecting chicken coops above catfish ponds. "That is a beneficial practice," he said. "Nothing is wasted. Even the chicken manure can be used to feed the fish. They just need to make sure to change the water every now and then. If not, the water will be *hot*." Ahmad used the word *hot* to talk about the bad water condition for fish culture caused by the build-up of nitrogen from the chicken excrement. Nitrogen from fecal matter can drain the oxygen out of the water, and it is a serious problem in the fish industry (Probyn 2016). Whereas oysters have been used elsewhere to cleanse the water from nitrogen excreted through the more-than-human practices of agriculture (Probyn 2016, 22), here catfish gobble up the poop and minimize the problem. The integration of poultry and fish farming in Asia can increase the production intensity and minimize land, labor, and water use for both fish and poultry. One hectare of static water fish ponds, for example, can "process" the waste of up to 1,500 poultry, producing fish in quantities of up to ten metric tons per hectare without other feeds or fertilizers (Little and Satapornvanit, 1996, 426).

Traditional backyard ponds preserve not only metabolic intimacy, but also local knowledge and wisdom that have been practiced for generations. As shown previously, Ahmad highly values the old views and practices regarding feeding inherited from his parents. That is why he could not understand urban people's objection toward catfish fed with refuse and the view that such fish is not halal. He remarked,

> So, some people think that catfish is not halal? What do you mean? All fish is halal . . . what people say . . . don't you think that fish has its own way of digesting its food? Like . . . we give manure to plants, and people eat the leaves or the fruits. Isn't catfish the same thing [as the plants]? We also raise the fish in a good way. . . . There is also a certain treatment to eliminate the smell. Before taking them to the kitchen, we just put the fish in clean water for two or three days and the smell is gone.

Ahmad defines halalness in a broader sense. First of all, he believes that fish is inherently halal. The feed given to fish does not make it nonhalal. Second, the catfish is halal because it is reared in a "good way," which refers to Ahmad's method of care by keeping the balance with the environment. Furthermore, Ahmad mentioned a particular method to clean the fish. By applying fish quarantine, he could eliminate the bad odor commonly found in fish reared in domestic effluents.

Fish quarantine is basically a rule in Islamic teaching regarding *jallalah* (literally *dirty* animals that regularly feed on *haram* substances like feces or pig offal). Referring to *hadith* (the narration of the Prophet or his companions), the majority of Muslim jurists declare that eating such animals is not permissible, unless the animals are quarantined and given good food for a certain period (Muflih et al. 2017, 438). The quarantine period depends on the size of animals: forty days for camels, thirty days for cows, seven days for goats, and three days for chickens or smaller animals (Saidin, Rahman, and Abdullah 2017, 71). Because there is no specific rule for catfish, some researchers performed a polymerase chain reaction test on catfish fed with pig offal to find out whether traces of the pig's DNA are present after quarantine (Wan Norhana et al. 2012). Their findings showed that "the minimum quarantine period for catfish fed with pig offal is 1.5 days" (Wan Norhana et al. 2012, 1268). The catfish indeed digests the *bad* food completely after quarantine.

Traditional catfish farmers knew this, but not through formal education, scientific research, or religious training. It is a traditional knowledge passed on for generations. Ahmad, for instance, learned about the fish quarantine from his father, and it has been a common practice in his village. Islam has been a dominant religion in Java since the fifteenth century, so it is not surprising to find Islamic teachings embedded in people's way of life, including in aquaculture, and passed down as local wisdom. Farmers like Ahmad have enacted *halal* and *tayyib* beyond a religious stipulation; it is their way of doing fish and sustaining life. Ahmad confessed that he did not know precisely when rearing catfish that way began, but he was proud of the practice as he preserves the legacy of his parents and the community. Ahmad's affection and pride have enabled his backyard fish pond to survive amid the increasing popularity of modern catfish farming in his village.

Not all traditional farmers, however, are proud of the old practice. I learned about this when I visited another house having a backyard pond. The owner, Darman, was somehow reluctant to show it to me and took us instead to more modern catfish ponds his son built recently. "My backyard pond was nothing. I will show you the special one," he said. The *special* ponds he was talking about were actually smaller concrete tanks. Each tank has a set of pipes built above the tank for water input and another pipe erected from the bottom of tank, which can be laid down horizontally to throw away waste (fish feces and uneaten feed). As we have seen, catfish can thrive in any water condition and this new method of catfish rearing includes water quality control as one of its protocols. Darman's son, who just graduated from a prestigious state university in Surabaya, learned about the new practice of catfish farming from a workshop held in the city. He soon asked Darman's permission to build some concrete tanks in an empty space close to the family's old backyard catfish pond. This new practice is now gaining popularity, especially among younger people, who participated in similar workshops.

Darman seemed very proud of what his son was doing. He happily showed us the tanks, although he did not know much about it. The son was not present when we were visiting the tanks, so we only listened to Darman's enthusiasm in talking about his son. For a moment I thought my visit was fruitless. I did not get to see the backyard pond he owns, let alone talk about it, and saw a different farming practice instead without much information. Only a moment later did I realize that I had picked up something else: Darman's display of affect. Darman blushed when I first asked for his backyard pond, and he politely brushed off my attempt to see the pond. He was ashamed of his backyard practice. That was why he showed me the *special* (modern) catfish tanks and talked about his successful son instead.

In my attempt to comprehend the role of affect in a more-than-human relationship, Darman's shame is important to note. He has cared for the backyard pond for a long time, but the connection might come to an end because of embarrassment about his old practice compared to the modern farming his son introduced. "Shame marks the break in connection," as Probyn writes (2005, 13). Shame and interest are inseparable, as "shame illuminates our intense attachment to the world, our desire to be connected with others" (Probyn 2005, 14). This explains the blush on Darman's face when I enquired about his backyard pond. His passion for the traditional practice does not seem to be appreciated by his son who said that the modern farming is better than the backyard practice; hence the word *special* is used. If the shame continues and deepens, the metabolic tradition found in traditional backyard farming practice might cease to exist. The break could also affect the long-held practice of avoiding waste because the feed used in the modern catfish farming is mainly industrial pellets. The turn to modern ponds might also eliminate traditional knowledge and wisdom, like the metabolic intimacy and fish quarantine, which are entirely absent in the new practice.

I have talked about affective and relational practice in backyard catfish farming. Caring for catfish in traditional backyard ponds has allowed the farmers "to create, hold together and sustain life's essential heterogeneity" (de la Bellacasa 2012, 198). Although backyard fish farming has its own challenges, it deserves further attention and discussion. This old farming practice blended local wisdom and traditional knowledge in favor of a more-than-human relationship, which is inherently more sustainable. In Africa and elsewhere in Asia, traditional practices of rearing catfish (incorporating animal or human excreta as additional feed) are complementing the modern practice of farming (e.g., Van Heerden and Frey, 1994; Bouelet Ntsama et al., 2018). Yet the public rejection of such practices could undermine the existence of traditional catfish farming.

Urban dwellers' opinions that backyard pond catfish is repulsive and not halal is an example of such aversion. The sentiment, however, is ungrounded because halalness is practiced through fish quarantine; the *disgusting* catfish can be made clean and halal. City people may not know about the halalling method because their link to metabolic tradition is broken. As shown previously, fish quarantine does not only render catfish halal materially (by eliminating the *haram* substances from the fish body), but also holds ethical value of halal in broader sense: it relates to a feeding practice that accentuates and sustains a more-than-human relation and life.

This is to say that halal food consumption as a wholesome practice contains irreducible relations and it should be a matter of concern in discussing halal as a *green* practice. A form of life-sustaining, multispecies relationality I set forth in this chapter is just a minuscule example of how this can be done. Given that conversations on halal have transformed the global food industry and relations to some extent, *green* halal is a potential avenue to put the more-than-human to the forefront and to address the increasing issue of environmental crisis as responsible Muslims and human beings.

REFERENCES

Ali, Mohd Helmi, and Norhidayah Suleiman. 2016. "Sustainable Food Production: Insights of Malaysian Halal Small and Medium Sized Enterprises." *International Journal of Production Economics* 181: 303–14. https://doi.org/10.1016/j.ijpe.2016.06.003.

Angyal, Andras. 1941. "Disgust and Related Aversions." *Journal of Abnormal and Social Psychology* 36 (3): 393–412.

Bouelet Ntsama, Isabelle Sandrine, Betrand Ayuk Tambe, Julie Judith Tsafack Takadong, Gabriel Medoua Nama, and Germain Kansci. 2018. "Characteristics of Fish Farming Practices and Agrochemicals Usage Therein in Four Regions of Cameroon." *Egyptian Journal of Aquatic Research* 44 (2): 145–53. https://doi.org/10.1016/j.ejar.2018.06.006.

Cai, Junning, and PingSun Leung. 2017. "Short-Term Projection of Global Fish Demand and Supply Gaps." 607. Rome: Food and Agriculture Organization–United Nations.

De la Bellacasa, María Puig. 2012. "'Nothing Comes without Its World': Thinking with Care." *Sociological Review* 60 (2): 197–216. https://doi.org/10.1111/j.1467-954X.2012.02070.x.

During, Simon. 2005. *Cultural Studies: A Critical Introduction*. New York: Routledge.

Edwards, Peter. 2012. "Rural Aquaculture: Reflections on Small-Scale Aquaculture." *Aquaculture Asia* 17 (1): 3–12.

Hishamunda, Nathanael, Neil B. Ridler, Pedro Bueno, and Wilfredo G. Yap. 2009. "Commercial Aquaculture in Southeast Asia: Some Policy Lessons." *Food Policy* 34 (1): 102–7. https://doi.org/10.1016/j.foodpol.2008.06.006.

Iqbal, Noor Fatima Karima. 2015. "From Permissible to Wholesome: Situating Ḥalal Organic Farms within the Sustainability Discourse." *Islamic Sciences* 13 (1): 49–57.

Iswanto, Bambang, Rommy Suprapto, and Huria Marnis. 2015. "Morphological Characterization of the African Catfish (*Clarias gariepinus* BURCHELL, 1822) Strains Introduced to Indonesia." *Indonesian Aquaculture Journal* 10 (2): 91–99.

Koenig, Leah. 2008. "Reaping the Faith." *Gastronomica* 8 (1): 80–84. https://doi.org/10.1525/gfc.2008.8.1.80.This.

Law, John, and Annemarie Mol. 2008. "Globalisation in Practice: On the Politics of Boiling Pigswill." *Geoforum* 39 (1): 133–43.

Lawonyawut, Khamchai. 1996. "Hybridization and Genetic Manipulation in Clarias Catfish." Stirling, Scotland: University of Stirling.

Little, David, and Kriengkrai Satapornvanit. 1996. "Poultry and Fish Production—A Framework for Their Integration in Asia." In *Livestock Feed Resources within Integrated Farming Systems*, edited by A. W. Speedy, C. Dalibard, and R. Sansoucy, 425–53. Rome: Food and Agriculture Council–United Nations. http://www.fao.org/livestock/agap/FRG/conf96.pdf/CONTENTS.HTM.

Muflih, B. K., N. S. Ahmad, M. A. Jamaludin, and N. F. H. Nordin. 2017. "The Concept and Component of Contaminated Animals (Al-Jallalah Animals)." *International Food Research Journal* 24 (December): 436–40.

Ng, Heok Hee, and Maurice Kottelat. 2008. "The Identity of *Clarias batrachus* (Linnaeus, 1758), with the Designation of a Neotype (Teleostei: Clariidae)." *Zoological Journal of the Linnean Society* 153: 725–32. https://doi.org/10.1111/j.1096-3642.2008.00391.x.

Probyn, Elspeth. 2005. *Blush: Faces of Shame*. Minneapolis: University of Minnesota Press.

———. 2016. *Eating the Ocean*. https://doi.org/10.1215/9780822373797.

———. 2017. "Queer Fish: Eating Ethnic Affect." In *Visuality, Emotions and Minority Culture*, edited by John Erni, 27–34. Berlin: Springer. https://doi.org/10.1007/978-3-662-53861-6_3.

Razalli, Mohd Rizal, Suzzaini Abdullah, and Rushami Zien Yusoff. 2012. "Is Halal Certification Process 'Green'?" *The Asian Journal of Technology Management* 5 (1): 33–41.

Rimmer, Michael A., Ketut Sugama, Diana Rakhmawati, Rokhmad Rofiq, and Richard H. Habgood. 2013. "A Review and SWOT Analysis of Aquaculture Development in Indonesia." *Reviews in Aquaculture* 5 (4): 255–79. https://doi.org/10.1111/raq.12017.

Saidin, Nurulaina, Fadilah Abd Rahman, and Noriham Abdullah. 2017. "Animal Feed: Halal Perspective Animal Feed: Halal Perspective." In *International Conference on Humanities, Social Sciences and Education (HSSE'17)*, edited by Angela Roman, Wijdan Eltijani Elsiddig Abbas, Anna Stoynova Andonova, and Avarvarei Bogdan-Vlad, 68–74. London: International Association of Humanities, Social Sciences and Management Researchers. https://doi.org/https://doi.org/10.17758/URUAE.UH0317018.

Sudarto, Guy G. Teugels, and Laurent Pouyaud. 2004. "Description of a New Clariid Catfish, *Clarias pseudonieuhofii* from West Borneo (Siluriformes: Clariidae)." *Zoological Studies* 43 (1): 8–19.

Van Heerden, D. J., and B. J. Frey. 1994. "Human Health Aspects of the Metals Zinc and Copper in Tissue of the African Sharptooth Catfish, *Clarias gariepinus*, Kept in Treated Sewage Effluent and in the Krugersdrift Dam." *Water SA* 20: 205–12.

Wachirachaikarn, Anyalak, Wikrom Rungsin, Prapansak Srisapoome, and Uthairat Na-Nakorn. 2009. "Crossing of African Catfish, Clarias Gariepinus (Burchell, 1822), Strains Based on Strain Selection Using Genetic Diversity Data." *Aquaculture* 290 (1–2): 53–60. https://doi.org/10.1016/j.aquaculture.2009.01.036.

Wan Norhana, M. N., G. A. Dykes, B. Padilah, A. A. Ahmad Hazizi, and A. R. Masazurah. 2012. "Determination of Quarantine Period in African Catfish (*Clarias gariepinus*) Fed with Pig (*Sus* Sp.) Offal to Assure Compliance with Halal Standards." *Food Chemistry* 135 (3): 1268–72. https://doi.org/10.1016/j.foodchem.2012.05.083.

Chapter Nine

Free Fish Heads

A Case Study of Knowing and Practicing Seafood Differently

Emma L. Sharp

It was a Saturday. After a phone call to a name on the Free Fish Heads list, and a guarantee they would be home, I transported the fish frames thirty kilometers in a chilly bin in the boot of my car from a boat ramp in East Auckland to a state house in South Auckland. More than frames—it was fish heads, fish wings, flesh leftover from a light filleting. When I arrived and approached the fence with a bag of fish parts, I was welcomed heartily, with a smile:

Eater 1: "The guys in there are all hungover. This will make a good tonic for them."

Me: "What will you do with it?"

Eater 1: "I'll put in lots of greens—silverbeet, watercress, carrots, potatoes, that kind of thing. Boil it up." She smacked her lips. She walked across her backyard, with big strides, her colorfully printed dress swinging. She snapped off big leaves of silverbeet, and gathered them in her arms and brought them to the fence. "Here, you want them? You came a long way, eh?"

In this chapter, I explore how Free Fish Heads materializes new practices of care in food provisioning associated with diverse and community economies (Hill 2015). This is in the context of an era that struggles to remake the social relations of food economies, in line with rising ecological, cultural, economic, and social concerns with the politics and ethics of food. In the seafood industry, practices of care are netted up together with politics of marine systems' sustainability, such as those related to feeding human and animal populations (Tibbets 2004), biodiversity concerns (Dick et al. 2012), cultural rights (Bargh 2016), fisherperson welfare (Stewart, Walshe, and Moodie 2006), and the humanity of fishing practices (Stafford 2013).

It is possible to identify multiple acts of caring embedded in *alternative* practices of procuring seafood: alternative to practices that are highly exploitative. Individually and collectively, different practices and knowledges around caring for our seas and their relations are present, and undertaken routinely as "small p" political practice. Otherwise neglected interdependencies of aspects of diverse culture, methods of conservation, types of nourishment, approaches to livelihoods, and recreation, are taken up through care-full practices that re-value fish and their relations. In this chapter, I examine the possibilities for a more caring future economy presented by fish heads and other parts, by building on two sets of ideas to do with remaking economy. The first are theories of care that emphasize thinking and acting in ways that treat human and more-than-human relations and activities as assemblages of interconnected bodies rather than bodies acting in isolation (Tronto 1993). The second set of ideas regards Tsing's notions of ruin and salvage read through diverse economic theory (Gibson-Graham 2008), which guides us to the hopeful economic futures that might be made possible through altered practice now, and creative response to collapse.

CARE-FULL ECONOMIES OF RUIN AND SALVAGE

Puig de la Bellacasa (2011) extends Tronto's (1993) reading of care practice by directing attention to what is *neglected* in theorizing the relationships between animate and inanimate actors. She argues that to truly *care* we must pay special attention to the diverse, marginalized practices that are typically understood or theorized as uncommon or *weak* (Gibson-Graham 2014). Knowing care care-fully and taking a caring approach to knowing, highlights the care-full practices that expose cracks in what could be perceived as a hegemonic and unsustainable food system of biological commodification, and this has been similarly noted in less watery food production (see Campbell et al. 2009). Instead, we might read this system as a suite of diverse economic (Gibson-Graham 2008) practices of care. The cracks, or experiments, appear as hopeful practices of difference now, and hopeful potential for what might be done differently in the future. We see in food scholarship a proliferation of caring concerns in smallholder pastoralism for cattle farmers, cows, paddock soils, boundary rivers, and consumer safety and experience (Scholten et al. 2013) given attention to environments, economies, and communities. Diverse practices in seafood provisioning that similarly recognize multiple and contested values and spaces might be redirected more firmly toward a deeper and more interconnected care for our seas.

Anna Tsing's recent book, *The Mushroom at the End of the World*, provides a different platform for a care-full economy. She explores sites where natural processes and bodies that are embedded in diverse human and nonhuman social relations are amassed and converted into wealth through interactions between diverse and capitalist economies (Tsing 2015). Her ideas on salvage accumulation are premised on mining natural capital from sites of ruin that are at the same time marked by different types of care. She vividly describes how *matsutake* mushroom bodies are valued, proliferated, cared for, and harvested in denuded plantation forests, in relation to noncapitalist forms of ownership, labor, exploitation, and accumulation and the capitalist production of other-valued mushrooms elsewhere. Building from Tsing's notion of a ruin economy I theorize a *salvage economy*, in which caring economic practices must be present and might be cultivated. In what follows, I use the case of fish-waste salvage through the work of Free Fish Heads to work notions of care into the idea of a care-full salvage economy.

SEAFOOD IN THE ANTHROPOCENE: A FISHY PROBLEM

Consumer and environmental advocacy groups are attempting to foster new economic relationships between humans and oceans that displace industrial overexploitation with more sustainable use of the seas. This work is backed by evocative films like *The End of the Line* (Murray 2009) and formal reports that detail the decimation of the world's fisheries and fishing-based livelihoods such as those by the Food and Agriculture Organization (FAO) of the United Nations (e.g., FAO 2014). Overfishing and methods such as bottom trawling in the oceans and industrial agriculture on land have used new technologies to ruin multiple oceanic and coastal sites and forms. Overfishing and questionable regard for biophysical and ecological damages have generated economies and ecologies of ruin (MacDiarmid et al. 2012). Economic and ecological ruin are registered in declining welfare of fish and other sentient sea life (e.g., Sneddon et al. 2018) as well as the loss of livelihoods of small local fishers and their communities (Stewart, Walshe, and Moodie 2006). In Aotearoa (the Māori name for New Zealand) New Zealand, some measure of the ongoing loss and altered quality of life is evidenced in declines of endangered and charismatic marine megafauna (Vance 2019).[1]

Reflecting on these industrialized and ruinous practices, one Aotearoa New Zealand commercial fisher interviewed recently pointed to how they view care: "[T]he whole ethics of what you do on the water changes when you are not stressed out just trying to feed a family" (Peart 2018, 101). At the other end of provisioning relations, consumers must confront the presence of industrialized fish in their supermarkets, and (via fishmeal and fish skins) in their cheese and face masks. Proliferating certification standards and lavish promotional campaigns that celebrate environmental credentials reveal that even multinational fishers recognize the disconnect between production and appetite, knowledge, cultural practice, and the political will embedded in them and embodied in consumers. There is a space for care-full practices in an oceanic salvage economy.

FISH AND FISHING IN AOTEAROA NEW ZEALAND

Aotearoa New Zealand manages the fourth largest fishing zone in the world—a four-million-square kilometer exclusive economic zone—and industrial fishing here is a more than $1 billion per annum industry. It provides more than 4,300 full-time equivalent jobs and harvests 450,000 tons of fish each year (Williams et al. 2017). There are efforts to control ruin out of this intensive industrial activity. Aotearoa New Zealand has one of the most extensive quota-based fishery management systems in the world, has been (albeit contentiously) ranked highly internationally in terms of monitoring and surveillance (Pitcher, Kalikoski, and Pramod 2006). One measure of its success is the survival of the national fishing industry; another is the reduction in the extensive return of waste catch to the ocean (Simmons et al. 2015). Annual takes are determined by ongoing monitoring of catch, scientific models, and calculations of maximum sustainable yields. These generate total allowable catches and the rights to take quantities of fish from particular sites in particular years, dependent on the tradable quota that any particular fisher holds (annual catch can also be leased to fishers by quota owners). As a whole, the quota management system depends on being able to model fisheries, and measure and police all catch—customary, commercial, charter, and recreational. Despite this, official records are unreliable. A recent study points to actual catch in Aotearoa New Zealand in the period between 1950 and 2010 as being approximately 2.7 times the fourteen million tons reported (Simmons et al. 2015) meaning that science and governance frameworks can only go so far in sustaining our seas.

RECREATIONAL FISHING:
CONTESTED SPACES AND PRACTICES

So, what of recreational fishing in Aotearoa New Zealand? Recreational fishers are enrolled in an ethically and politically complex set of practices. There is complexity around monetary value, and values around the cultural ethics of different practices. Here, I highlight indigenous culture and sporting culture.

In terms of capitalist economic value, 2012 figures suggested that recreational fishers in Aotearoa New Zealand fished somewhere between NZ$10.1 and NZ$80.8 million[2] worth of seafood versus a commercial value of $40.4 million (Barbera 2012). The uncertainty in recreational value is a consequence of informal practice. Growing populations and interest in recreational fishing mean that inshore fisheries, particularly of the Hauraki Gulf and Marlborough Sounds, are subject to increasing harvest pressure and also growing conflict between recreational and commercial fishing.

Motivations to participate in recreational fishing, however, are multiple. A 1999 survey captured reasons of (in order of frequency of response) pleasure and fun, relaxation and leisure, recreation, food supply and fresh fish, outdoors and environment, solitude and getting away, and sport and exercise (Walshe and Ackroyd 2000). In terms of food supply and fresh fish provisioning, cultural practices in recreational fishing are not explicitly disaggregated in this survey, but in Aotearoa New Zealand, the ocean is vital to Māoritanga.[3] Māori fisheries settlements recognize the right to collect kaimoana[4] for subsistence, or to sell. The practices of these rightsholders are folded together with those of other actors, such as game fishers who take part in this popular recreational activity involving the pursuit of predator fish for sport, using larger fish for bait.

The Ministry of Primary Industries has proposed recreational fishing parks in Aotearoa New Zealand to enhance the enjoyment and value of recreational fishing, and therefore highlighting a particular brand of care. Their proposal is not necessarily specifically care-full toward protecting marine biodiversity. This has been picked up by the recreational fishing advocacy group Legasea (an arm of the New Zealand Sport Fishing Council), which is skeptical of the value of recreational fishing parks, indicating that they would exist within larger and open-to-full-exploitation fish stock areas. This suggests that by swimming over a boundary line, fish are altered in the way they are valued, from a dollar-per-kilogram value (in the dominant mode of thinking of fish as resource), to a culturally specific—even intrinsic—value. Critique of state governance models suggest there is room for other solutions that do not frame fish in strictly economic terms.

The manifold accounts of exploitation in extraction from the sea narrated through multiple relations and locations of ruin therefore make a strong case for a critical stance on seafood production by any means. Undeniably, particular sites of ruin occur through recreational, as well as commercial, contexts. In response, and in parallel, a niche of alternative practices have emerged with a salvage-type response to the problems of fishing and its associated waste. Actual practices fill gaps left by science and governance. In the face of intensive approaches to seafood production and consumption to meet our metabolic needs and cultural wants, Free Fish Heads practiced in Aotearoa New Zealand offers something different.

FREE FISH HEADS: AN EXPERIMENT

The alternative food experiment of Free Fish Heads does seafood *production* differently. It is a network that exists online, of fishers and consumers of fish all achieving one thing in common through their participation. According to the initiative's website, the "utilization of our fish is conservation of our fish" (Free Fish Heads 2019). Conservation is achieved when recreational fishers, upon filleting their catch on board or on the jetty, then retain the fish frames and heads to be given to another member of the Free Fish Heads network who has indicated that they have a use for them. Members register to the Free Fish Heads website or app (with no exclusionary factors but for the requirement of a working telephone and an ability to drop off or pick up fish heads, usually at a boat ramp). Free Fish Heads operates in sixteen regions around Aotearoa New Zealand with, at the time of writing, 14,108 participants registered, most in the Auckland region. Advertising of the initiative takes place variously through conduits such as Māori TV (Māori TV 2019), through the organizations Love Food Hate Waste (Love Food Hate Waste 2019) and Legasea (Legasea 2019), as well as via the founder's (now online only) sport-fishing television show and current website (*The ITM Fishing Show* 2019).

The Free Fish Heads initiative concentrates the contradictions and contestations of recreational fishing, bringing into tension the value of the catch, the value of sport, and the value of recreation. I ask: What practices of care are represented by this initiative? How do these practices and properties reconfigure the various actors in this food assemblage? I will look at acts of salvage and acts of care to explore the most visible actors in the assemblage—the fishers, the fish parts, and the eaters—to answer these questions.

A SALVAGE ECONOMY OF FISH HEADS

To think through Free Fish Heads as a salvage economy I start by reading it as a diverse economies framework, to decenter conventional valuing of fish parts. Recreational fishing, as nonmarket hunting, fishing, and gathering, is recognized as a diverse economic transaction (Gibson-Graham 2008). In this framing, the Free Fish Heads initiative's practices of salvage might also occupy this space of a non-capitalist, perhaps food rescue-type enterprise (Cameron and Wright 2014, 3). Its network circulates surplus fish parts: Those that are *produced* by harvesting them for free, and then gifted on. In most cases, there is no expected return for the discarded fish parts. Labor, in this circulation, might rather be considered to be in-kind work, for which there is some voluntary compensation for the fish parts by the receiver, but again, it is typically unpaid or volunteer work. Free Fish Heads as a salvage experiment highlights a different economic activity, backgrounded by an industrial, capitalist fishing model and a conventional recreational fishing model, which both perpetuate a ruin of a kind. It has similarities to other alternative food economies like dumpster diving (Sharp, Schindler, Lewis, et al. 2016) or food waste redistribution organizations (e.g., Turner 2018) in which surplus or unwanted produce is offered to others outside of that immediate chain, for consumption.

LANGUAGE OF A SALVAGE ECONOMY

In keeping with the care-full politics of diverse economies thinking (Gibson-Graham 2006), we need a different language for the exchanges that take place. In capitalist economic terms, Free Fish Heads might be seen as a *broker*. As a salvage experiment, the initiative might be better thought of through their practice as *facilitators* of access to parts of the sea assemblage that were previously less well-connected (Sharp, Friesen, and Lewis 2015). The *producers* of this food are also reconfigured in a salvage economy and might be named for their practice instead—as *fishers*. Fishers' salvage of fish parts destined for the waste stream designates these human actors as new start points in a food production assemblage. This (re)making of food in a salvage economy might then be considered in ontological and material terms rather than economic ones. The exchange ends with more-than-*consumers* in a salvage economy, named instead for their practices as *eaters*. Fish parts are not valued as per a capitalist economy. The objects are valued as more-than-*commodities*, and I argue, care-fully.

Fish heads and frames (or fish waste, to some) in these terms are reconfigured as salvaged objects of new value. The salvage solution is the pairing of those who see little value in fish heads with others who do. When transferred from fisher to eater, the *waste* parts are remade and revalued as edible. A different materiality of fish heads challenges normative practice and thinking: When is a fish a part of a marine ecosystem? A carcass? Waste? Food? The ontological production of food is ascribed when particular value is found in a food assemblage. This value is not always an accumulation of wealth in the strict terms of a capitalist economy. Rather, the salvage illustrated through Free Fish Heads sits more comfortably in diverse economic terms (Gibson-Graham 2008)—of salvage extraction, where care is the currency.

CARE-FULLY REMAKING FISH AS MORE-THAN-*COMMODITY*

Most fish eaters who discard fish heads and frames will never be convinced of the "joys of picking at a fish frame or sucking on a fish eye" as Free Fish Heads claims (Free Fish Heads 2019), layering a cultural value on fish parts. In fact, recently, commercial interests in Alaska noted this indifference toward fish offcuts and have advocated for their rebranding as valuable commodities to whet appetites (Welch 2018). Further, the FAO (2014) noted concern for the large volumes of filleting byproducts. They are certainly monetarily valuable: In 2014, Norwegian data on wild fish parts discarded at sea estimated 220,000 tons (Olsen and Karunasager 2014) which, incredibly, equates to about half of Aotearoa New Zealand's industrial annual harvest (Williams et al. 2017). As per recent returns to terrestrial nose-to-tail eating, and despite the Western cultural convention to eat fish fillets preferentially, this shift in recognition of fish parts suggests that there could be a monetary (or other) market for what is commonly seen as waste. Free Fish Heads, however, advocates for the use of these neglected fish parts in creating a salvage economy of care in their distribution.

CARE-FULLY REMAKING
THE EATER AS MORE-THAN-CONSUMER

Although economic markets might be made by narrating fish heads as *good*, Free Fish Heads also claims that already "there are thousands of Kiwis that would gladly take fresh fish heads and frames and feast on the sweet succulent meat" (Free Fish Heads 2019). They make visible the people who care for fish heads. Hopeful recipients of fish heads listed on the Free Fish Heads website include a local marae (a Māori community meeting site) and a school. Although inferences cannot be made about the socioeconomics of Free Fish Heads' practices of a care-full salvage economy, information on attributes of fishers and eaters as gathered from participant data on the Free Fish Heads website does enable us to make interesting points about cultural dimensions of these practices. From literature including recipe books, it is clear that there are different cultural traditions of eating fish heads (e.g., Coles and Hallett 2012), including those of Māori (e.g., Allo 1970; State of Queensland 2015),

Indian (Cheah 2008), and Pacific (Chamberlain and Parkinson 2001) populations. Coincident with these traditions is the significant number of Māori, Indian, or Pacific names on the hopeful-recipient list of Free Fish Heads membership: more than half of the Auckland region's individual records displayed Māori, Indian, or Pacific surnames.[5] By care-fully redistributing the material capital of fish heads, Free Fish Heads also demonstrates care in redistributing social and cultural capital between givers and receivers, in turn transforming the ontological value of the fish as it moves between actors, in a food supply chain of cultural diversity. Ideals of food sovereignty are worked upon by extracting the value of fish parts out of a commodified system.

CARE-FULLY REMAKING
THE FISHER AS MORE-THAN-*PRODUCER*

There are practices that happen in tandem with fish-parts salvage. Juxtaposed against the initiative's philosophy, "utilization of our fish is conservation of our fish" (Free Fish Heads 2019) are practices that illustrate the contested space in which it operates. This contestation is embodied in the founder of the Free Fish Heads scheme, Matt Watson—a celebrated game fisherman and television personality, with a passion for the ocean and for hunting trophy fish. There are juxtaposed practices of his profession and the initiative for which he is a protagonist. In more than one episode of the founder's internationally renowned Sports Fishing TV show (*ITB Ultimate Fishing*) he feeds whole, freshly-caught skipjack and albacore tuna—"a couple of New Zealand's most underutilized fish species"—into a wood-chipper fixed onto the back of a boat, to generate a burley trail that attracts kingfish and snapper (Ultimate Fishing TV 2015, from 18.20 m). Fish parts here take on multiple representations beyond ecological care. There is a form of care practiced by the game fisher. A perhaps similar care is expressed by an independent harvester who claimed, "I enjoy hunting the fish out . . . with fishing, there are a thousand different things—currents, moons, time of year, depth, baits . . . when you line it all up it's a good feeling. Very satisfying" (Peart 2018, 3). Care in terms of the skill and fulfillment in navigating these oft unrecognized and interdependent elements of fishing might be translated to the previous example of sports fishing.

This vignette might resonate somewhat differently with a non–game-fishing ocean actor. Indeed, "care convokes trouble and worry for those who can be harmed by an assemblage or might be unable to voice their concern and need for care" (Puig de la Bellacasa 2017, 52). In this case, *those* might amount to the nonhuman—marine ecosystems, fish themselves—or human onlookers. As de la Bellacasa discusses, a caring obligation is not reducible to *feeling good* (in this case, about fish), nor is it only dependent on utility (whereas here, fish is a resource) (2017, 147). This does not neglect that if we are inattentive to nonhuman signals such as dwindling fisheries, we suffer consequences. Rather, a situated ethics of *caring with* the fish assemblage is more useful for sustaining our seas. To be sure, there is no singular ethics around the exchange or use of fish bodies, and conventional ethics may butt up against other politics. These are messy practices and politics, and if nothing else this account identifies the different carers and acts of care that are part of the assemblage and that have a stake in a care-full salvage economy of fish.

DOING SEAFOOD DIFFERENTLY

Although the sea is immense and gives an illusion of abundance, the reality is that multiple juxtapositioned human activities have contributed to devastations in the ocean. The case of Free Fish Heads demonstrates an awareness of the biological limits of our food production, in relation to fish and fisheries. It does more work than the "utilization" and "conservation" of our fish. The care-full salvage economy that it performs is just one part of contiguous cultural, economic, and social practices, where *waste* (or ruin) is revalued through the act of economic salvage from particular contexts that are not limited to capitalism. It reveals opportunities to foster new economic relationships that practice difference. It highlights a politics and ethics that decenters humans, while connecting important, neglected relations.

I consolidate my thinking about the future of fish by looking to the hopeful potentials of diverse economic practices of seafood to capture this complexity—the tensions, but also the opportunities afforded by practices of salvage from particular sites of ruin. The *doing differently* in this experiment of fish salvage traces the problematics and possibilities of working from the point of a ruin economy, to the care-full places it might take us.

ACKNOWLEDGMENTS

This project was supported by the New Zealand Sustainable Seas National Science Challenge (CO1X1515).

NOTES

1. A recent Greenpeace commissioned report published recorded casualties from commercial fishing in only one week of February of four Hector's dolphins (in trawl nets), five albatross (on long lines), and four New Zealand sea lions (in trawl nets) (Vance 2019).

2. Dependent on calculations used. Calculations are complicated by the fact that recreational fishers are not obligated to keep official catch records, and that figures dependent on which target fish are being modeled, among other factors.

3. Māori culture, traditions, and way of life.

4. Seafood.

5. Exact numbers are not offered here as there is obviously uncertainty around individual cultural backgrounds based on names. Further, there were noted to be some double-ups in names, and in particular a *marae* (Māori *iwi*/community) was listed which may also have individual registrations associated with it. What is clear, however, is that there is a significance in the trend of cultural valuing of fish heads, given a more than 50 percent registration of Māori, Indian, and Pacific membership while corresponding proportions of ethnicity in the Auckland region are 10.7 percent, 7.5 percent, and 14.6 percent respectively, according to 2013 census figures, which are the most recent figures available (Statistics New Zealand 2014).

REFERENCES

Allo, Jan. 1970. "Analysis of Midden from Two Sites on Motutapu Island, New Zealand." *Records of the Auckland Institute and Museum*, 83–91. http://www.jstor.org/stable/42906144.

Barbera, Mattia. 2012. "Towards an Economic Valuation of the Hauraki Gulf: A Stock-Take of Activities and Opportunities." *Auckland Council Technical Report TR2012/035*. Auckland: Auckland Council.

Bargh, Brian. 2016. The *Struggle for Māori Fishing Rights*. Wellington: Huia.

Cameron, J., and S. Wright. 2014. "Researching Diverse Food Initiatives: From Backyard and Community Gardens to International Markets." *Local Environment* 19, no. 1: 1–9.

Campbell, Hugh, Rob Burton, Mark Cooper, Matthew Henry, Erena Le Heron, Richards Le Heron, Nick Lewis, et al. 2009. "From Agricultural Science to 'Biological Economies'?" *New Zealand Journal of Agricultural Research* 52 (1): 91–97.

Chamberlain, Tony, and Susan Parkinson. 2001. "Seafood in our Meals." *Community Fisheries Training Pacific Series 2*. University of the South Pacific: Secretariat of the Pacific Community.

Cheah, Jeanie. 2008. "Enjoy Fish Head Curry on Banana Leaf." *Today's Manager* (December–January): 41. Academic OneFile. Accessed March 1, 2019. http://link.galegroup.com/apps/doc/A173422645/AONE?u=learn&sid=AONE&xid=27b9e016.

Coles, Benjamin, and Lucius Hallett IV. 2012. "Eating from the Bin." *Sociological Review* 60: 156–73.

Dick, Jonathan, Janet Stephenson, Rauru Kirikiri, Henrik Moller, and Rachel Turner. 2012. "Consequences of the Loss of Abundance and Biodiversity of Coastal Ecosystems in Aotearoa New Zealand." *Mai Journal* 1 (2): 117–30.

Food and Agricultural Organisation (FAO). 2014. "World Fish Trade to Set New Records." Accessed February 28, 2019. http://www.fao.org/news/story/en/item/214442/icode/.

Free Fish Heads. 2019. "Philosophy." Accessed October 10, 2018. https://www.freefishheads.co.nz/philosophy/.

Gibson-Graham, J. K. 2006. *A Postcapitalist Politics*. London: University of Minnesota Press.

———. 2008. "Diverse Economics: Performative Practices for 'Other Worlds.'" *Progress in Human Geography* 35 (5): 613–32.

———. 2014. "Rethinking the Economy with Thick Description and Weak Theory." *Current Anthropology* 5(9).

Hill, Ann. 2015. "Moving from 'Matters of Fact' to 'Matters of Concern' in order to Grow Economic Food Futures in the Anthropocene." *Agriculture and Human Values* 32 (3): 551–64.

The ITM Fishing Show. 2019. "Free Fish Heads." Retrieved February 28, 2019. https://www.thefishingshow.co.nz/free-fish-heads.

Legasea. 2019. "Free Fish Heads." Retrieved February 28, 2019. https://legasea.co.nz/support_partners/freefishheads/.

Love Food Hate Waste. 2019. "Free Fish Heads." Retrieved February 28, 2019. https://lovefoodhatewaste.co.nz/free-fish-heads/.

MacDiarmid, Alison, Andy McKenzie, James Sturman, Jennifer Beaumont, Sara Mikaloff-Fletcher, and John Dunne. 2012. "Assessment of Anthropogenic Threats to New Zealand Marine Habitats." *New Zealand Aquatic Environment and Biodiversity Report 93*.

Māori TV. 2019. "Free Fish Heads Website Gains Popularity." Retrieved February 28, 2019. https://www.maoritelevision.com/news/national/free-fish-heads-website-gains-popularity.

Murray, Rupert (director). 2009. *The End of the Line* (film). United Kingdom: Arcane Pictures.

Olsen, Ragnar, and Iddya Karunasager. 2014. "Challenges and Realistic Opportunities in the Use of By-Products from Processing of Fish and Shellfish." *Trends in Food Science and Technology* 36 (2): 144–51.

Peart, Raewyn. 2018. *Voices from the Sea: Managing New Zealand's Fisheries*. Auckland: Environmental Defense Society.

Pitcher, Tony, Daniela Kalikoski, and Ganapathiraju Pramod. 2006. "Evaluation of Compliance with the FAO (UN) Code of Conduct for Responsible Fisheries." *Fisheries Centre Research Report* 14 (2). Vancouver: University of British Columbia Press.

Puig de la Bellacasa, Maria. 2011. "Matters of Care in Technoscience: Assembling Neglected Things." *Social Studies of Science* 41 (1): 85–106.

———. 2017. *Matters of Care: Speculative Ethics in More than Human Worlds*. London: University of Minnesota Press.

Scholten, Martin, Imke de Boer, Bart Gremmen, and Kees Lokhorst. 2013. "Livestock Farming with Care: Towards Sustainable Production of Animal-Source Food." *NJAS Wageningen Journal of Life Sciences* 66: 3–5.

Sharp, Emma, Ward Friesen, and Nicolas Lewis. 2015. Alternative Framings of Alternative Food: A Typology of Practice. *New Zealand Geographer* 71, no. 1: 6–17.

Sharp, Emma, Ellen Schindler, Nicolas Lewis, and Ward Friesen. 2016. Food Fights: Irritating for Social Change among Auckland's Alternative Food Initiatives. *Kōtuitui: New Zealand Journal of Social Sciences Online* 11, no. 2: 133–45.

Simmons, Glenn, Graeme Bremner, Hugh Whittaker, Philip Clarke, Lydia Teh, Krystn Zylich, Dirk Zeller, et al. 2015. "Reconstruction of Marine fisheries Catches for New Zealand (1950–2010)." *University of British Columbia Working Paper Series, WP #2015-87*. Institute for Oceans and Fisheries: University of British Columbia.

Sneddon, Lynne, David Wolfenden, Matthew Leach, Ana Valentim, Peter Steenbergen, Nabila Bardine, Donald Broom, and Culum Brown. 2018. "Ample Evidence for Fish Sentience and Pain." *Animal Sentience* 21(17).

Stafford, Kevin. 2013. *Animal Welfare in New Zealand*. Cambridge, New Zealand: New Zealand Society of Animal Production.

State of Queensland. 2015. *Food and Cultural Practices of the Māori Community in Australia—A Community Resource*. Queensland: Metro South Health.

Statistics New Zealand. 2014. "2013 Census QuickStats about Culture and Identity—Data Tables." Statistics New Zealand. Retrieved February 28, 2019. https://web.archive.org/web/20140524102811/http:/www.stats.govt.nz/~/media/Statistics/Census/2013%20Census/profile-and-summary-reports/quickstats-culture-identity/tables.xls.

Stewart, James, Kim Walshe, and Beverley Moodie. 2006. "The Demise of the Small Fisher? A Profile of Exiters from the New Zealand Fishery." *Marine Policy* 30: 328–40.

Tibbetts, John. 2004. "Eating Away at a Global Food Source: The State of the Oceans." *Environmental Health Perspectives* 112 (5): A282(10).

Tronto, Joan. 1993. *Moral Boundaries: A Political Argument for an Ethic of Care*. London: Routledge.

Tsing, Anna. 2015. *The Mushroom at the End of the World: On the Possibility of Life in Capitalist Ruins*. Oxford: Princeton University Press.

Turner, Bethaney. 2018. "Playing with Food Waste: Experimenting with Ethical Entanglements in the Anthropocene." *Policy Futures in Education*. https://doi.org/10.1177/1478210318776851.

Ultimate Fishing TV. 2015. "Fishing . . . with a Woodchipper" (video). October 12, 2015. Accessed February 20, 2019. https://www.youtube.com/watch?v=Ui1-qlZjmwE&list=PLeQCp3MCY6VPfScbLEbadxElm9s2Mjsr0&index=17&t=0s.

Vance, Andrea. 2019. "Fishing Company Pulls Boat after Four Endangered Sea Lion Deaths." Accessed February 13, 2019. www.stuff.co.nz/environment/110574823/fishing-company-pulls-boat-after-four-endangered-sea-lion-deaths?fbclid=IwAR1bpYB9IXNkSm-flbdJmAQVxjTAxeVjYf8gVckQ9Muug-2goLONs8bI5B1M.

Walshe, Kim, and Jo Ackroyd. 2000. "Motivations and Perceptions of Seawater Recreational Fishers in New Zealand." *Ministry of Fisheries Project REC9802.* Auckland: Ackroyd Walshe Limited.

Welch, Laine. 2018. "Alaska Rebrands the Fish Head: ASMI Report Promotes Full Utilization." *National Fisherman.* Accessed November 8, 2019. https://www.nationalfisherman.com/alaska/alaska-rebrands-fish-head-asmi-report-promotes-full-utilization/.

Williams, J., F. Stokes, H. Dixon, and K. Hurren. 2017. "The Economic Contribution of Commercial Fishing to the New Zealand Economy." Wellington: Business and Economic Research Limited.

Section 3

GOVERNING AND REGULATING THE OCEANS

Chapter Ten

Out of Sight, Out of Mind

The Challenge of
Regulating High Seas Fisheries

Rosemary Rayfuse

When John Cabot arrived in the waters off Newfoundland in 1497, he found the sea so full of fish that they could simply be picked up out of the water in baskets. Right up to the 1600s English fishing captains reported cod shoals so thick that they could barely row a boat through them. Today, the cod fishery off Newfoundland has been fished to commercial (if not biological) extinction (Kurlansky 1998; Pringle 1997, 34). Worryingly, the story of the Newfoundland cod is not unique. Indeed, the whole history of fisheries has been one of rich harvests, followed by overfishing, followed by stock decline and collapse.

Of course, it is axiomatic that as a biologically renewable resource the supply of fish is potentially unlimited. To achieve this potential, however, appropriate management of the resource is necessary. Even Hugo Grotius, often referred to as the grandfather of the law of the sea, and the exponent, in 1608, of the principle of the freedom of the seas, acknowledged that restriction of fishing effort might be necessary if the resource were found to be exhaustible (Grotius 1608). He simply failed to conceive of it ever becoming so (Rayfuse 2004, 3). By the nineteenth century, however, the specter of exhaustibility of fisheries resources had reared its head. Instances of stock depletion caused by overfishing had begun to occur and by the end of that century the problem of stock depletion had become significant. States were forced to recognize that the continued unregulated exploitation of selected species would lead to their commercial extinction. The Grotian conception of the inexhaustibility of fisheries resources was dead (Rayfuse 2010, 201–2).

Throughout the twentieth century the story of fishing continued to be one of overexploitation. In its first estimates of world fishery production in 1945, the Food and Agriculture Organization (FAO) of the United Nations recognized that some stocks were already overfished. Fifty years later, in its

Report on the State of World Fisheries and Aquaculture (1996), the FAO concluded that over 60 percent of the world's fisheries were fully exploited, over exploited, or depleted (FAO 1996, 2). In its 2000 Report the percentage of world fisheries that were fully exploited, over exploited or depleted had risen to at least 75 percent, with these fisheries requiring "urgent action to stabilize or improve their situation." At least 28 percent of these fisheries required "forceful" action to reverse their decline (FAO 2000, 3).

Today the situation has not changed. In its latest State of the World's Fisheries and Aquaculture Report, published in 2018, the FAO reported that 33.1 percent of fish stocks are estimated as fished at a biologically unsustainable level and therefore overfished, 59.9 percent of stocks are fully fished, and only 7 percent of stocks are underfished (FAO 2018, 2 and 38–41). In other words, fully 93 percent of the world's fish stocks are either overexploited, fully exploited, or depleted. Nor have the adverse effects of overexploitation been limited to targeted stocks and species. Millions of tons (indeed 30 percent of all catches) of unwanted fish bycatch is caught, killed, and dumped back into the sea each year (World Wildlife Fund [WWF] 2017). Moreover, this bycatch is not limited to fish. Hundreds of thousands of seabirds, including endangered species of albatross, have been caught and drowned as bycatch in global fisheries (Birdlife International 2017). In addition, tens of thousands of sea turtles are reported caught in fishing operations each year and populations of whales, dolphins, and sharks are all in danger of overexploitation as incidental bycatch (Wallace et al. 2013). Worse still, destruction of benthic communities by inappropriate fishing techniques like bottom trawling has compounded the problem of destruction of marine habitats about which we still know very little but whose importance we are increasingly beginning to appreciate (WWF 2019). Increasingly, climate change and marine pollution are seen as major causes of concern (FAO 2018, 130–33).

As we know, fish provide an extremely valuable contribution to human food security, providing up to 20 percent of global animal protein intake, with some localized communities completely dependent on fish for their protein intake (FAO 2018, 2). According to the FAO, world per capita apparent fish consumption has increased from an average of 9.9 kilograms in the 1960s to more than 20 kilograms today (FAO 2018, 2). There is no doubt that this significant growth in fish consumption has enhanced people's diets around the world through diversified and nutritious food. However, as the increasing percentage of overfished stocks indicates, this has come at a cost.

There is one fundamental cause of the decline in fish stocks—anthropogenic overexploitation. The challenges of regulating fishing activities to reduce and prevent overexploitation are multiple and complex. In purely geographical terms, the oceans are vast, covering 70 percent of Earth's surface.

This enormous area is largely "out of sight" and, for the most part, "out of mind." In legal terms, the challenges are no less vast, involving as they do the need for comprehensive and effective international cooperation to ensure the conservation and management of global fish stocks. This chapter examines the international legal framework for the conservation of fish stocks focusing on the regulation of high seas fisheries and the role of the organizations through which international cooperation in their conservation is to occur. In doing so, it is hoped that this chapter will make some small contribution to the identification and understanding of what works, or does not work, and why.

THE INTERNATIONAL LEGAL
FRAMEWORK GOVERNING HIGH SEAS FISHERIES

The current international legal regime governing fisheries is grounded in the 1982 United Nations Law of the Sea Convention (LOSC),[1] which essentially divides competence for the regulation of fisheries between coastal states on the one hand and the international community as represented by regional organizations on the other. Within the two-hundred-nautical-mile exclusive economic zone (EEZ), the LOSC gives coastal states the right to conserve and manage fish stocks. On the high seas, which comprise the nearly two-thirds of ocean area lying beyond the jurisdiction of states, the nationals of all states enjoy the freedom to fish, although this is subject to the obligation on all states to cooperate in the conservation and management of high seas fish stocks (LOSC Articles 116–18; for discussion of the general LOSC framework, see Matz-Lück and Fuchs 2016, 491–515).

When the negotiators of the LOSC created this division between areas under and areas beyond national jurisdiction, it was assumed that conflicts between states over international fisheries exploitation would be resolved. Allowing coastal states to exercise their domestic jurisdiction over fisheries within their newly established two-hundred-mile fishing zones and EEZs was intended to ensure the proper conservation and management of these resources. It was also assumed that states would willingly cooperate in the development, adoption, and enforcement of fisheries regulations dealing with transboundary, straddling, and highly migratory fish stocks. Finally, it was assumed that the then existing economic paradigm of optimum utilization of fisheries resources would be ensured by the application of the management reference point of maximum sustainable yield (Rayfuse 1999, 110–13). With the benefit of hindsight, we know that each of these assumptions was ill-conceived. It is now a matter of historical fact that, within their EEZs, coastal states have by and large failed to properly manage their fishery resources, resulting in overexploitation. On the high seas, states have made limited progress in regulating fishing activities by vessels flying their flags and overexploitation has again been rife. This has led to the development of an ever-increasing body of international legal instruments, including the 1993 Agreement to Promote Compliance with International Conservation and Measures by Fishing Vessels on the High Seas, the 1995 Agreement for the Implementation of the Provisions of the UN Convention on the Law of the Sea relating to the Conservation and Management of Straddling Fish Stocks and Highly Migratory Fish Stocks Agreement (1995 Fish Stocks Agreement), and the 2009 Agreement on Port State Measures to Prevent, Deter and Eliminate Illegal, Unreported and Unregulated Fishing. There have also been a plethora of non–legally binding instruments including the International Plans of Action adopted by the FAO for Reducing Incidental Catch of Seabirds in Longline Fisheries (FAO 1999c), for the Conservation and Management of Sharks (FAO 1999a), for the Management of Fishing Capacity (FAO 1999b), and to Prevent, Deter and Eliminate Illegal, Unreported and Unregulated Fishing (FAO 2001), as well as the 1995 Code of Conduct for Responsible Fishing. In addition, there have been a number of United Nations General Assembly resolutions on a global moratorium on large scale pelagic driftnet fishing (United Nations [UN] 1989; UN 1990; UN 1991) and on destructive fishing practices and the protection of vulnerable marine ecosystems (UN 2007 and UN 2009). All of these instruments serve to establish and clarify the precise parameters of the conservatory obligations on states. They have also led to the establishment of a number of international organizations charged with conserving and managing high seas fisheries resources.

THE MECHANISM OF CHOICE: REGIONAL FISHERIES MANAGEMENT ORGANIZATIONS

Interstate cooperative efforts to control exploitation of high seas fisheries have their origins in the 1882 International Convention for the Regulation of the Police of the North Sea Fisheries outside Territorial Waters, the first treaty to establish a system of measures designed to stop overfishing. In the intervening years, international cooperation in high seas fisheries has been pursued through a number of bilateral and multilateral agreements. In recent decades, regional fisheries management organizations (RFMOs) have become the international community's mechanism of choice through which the duty to cooperate in the conservation and management of high seas fisheries resources is to be implemented. RFMOs are distinguishable from the broader corpus of regional fisheries bodies by virtue of their specific management mandate (as opposed to a purely scientific or advisory mandate).

RFMOs seek to regulate exploitation either of particular species throughout their area of distribution or of various species distributed throughout a particular geographic area. Management is conducted through the adoption of regulatory measures relating, among other things, to data acquisition, fishing practices and technologies, fishing effort and capacity, and control and enforcement. Currently, five RFMOs manage highly migratory species (mostly tuna): the International Commission for the Conservation of Atlantic Tunas, the Indian Ocean Tuna Commission, the Western and Central Pacific Fisheries Commission, the Inter-American Tropical Tuna Commission and its related Agreement on the International Dolphin Conservation Programme, and the Commission for the Conservation of Southern Bluefin Tuna. A further ten RFMOs have been established to manage non-tuna stocks within particular geographic areas: the North-East Atlantic Fisheries Commission, the Northwest Atlantic Fisheries Organization, the North Atlantic Salmon Conservation Organization, the South-East Atlantic Fisheries Organisation, the South Pacific Regional Fisheries Management Organisation, the Convention on Conservation of Antarctic Marine Living Resources, the General Fisheries Commission for the Mediterranean, the North Pacific Fisheries Commission, the South Indian Ocean Fisheries Agreement (SIOFA), and the Convention on the Conservation and Management of Pollock Resources in the Central Bering Sea (CCBSP). Not all of these bodies are self-standing organizations. The SIOFA and the CCBSP, for example, are "arrangements" in which decisions are taken by meetings of the parties. In addition, the CCBSP currently has no management role as the moratorium it establishes on fishing for pollock in the Central Bering Sea remains in place. A potentially new addition to the list, the Agreement to Prevent Unregulated High Seas Fisheries in the Central Arctic Ocean (CAOF Agreement) was adopted in 2018. The CAOF Agreement establishes a framework for the acquisition of science on which precautionary, ecosystem-based management measures can be based, if, and when, they become necessary in the future, although it remains to be seen whether this will ultimately lead to the establishment of a new RFMO.

Although there is some level of coordination and cooperation between the five tuna RFMOs, no formal process or mechanism exists for the institutionalization of cooperation between all RFMOs. Nevertheless, consistent approaches to common conservation, management, compliance, and enforcement issues are sought through cooperation at the inerorganizational level via observer representation at each others' meetings. This does not, however, stop member states from adopting inconsistent positions in different organizations depending on their particular perceived self-interest.

RFMOS AND THE LIMITS OF COOPERATION

Although the international community's mechanism of choice, RFMOs face a number of challenges in fulfilling their obligations (for a more fulsome discussion, see Rayfuse 2017, 817–30). Quite apart from the scientific and practical challenges of regulating fisheries, RFMOs are international, inter-governmental organizations established by treaties. Thus, at the most basic level, they face a number of challenges resulting from the inherent structure of international law with its emphasis on the sovereign equality of states and the particular emphasis in the law of the sea on high seas freedoms and flag state jurisdiction. The nature of this legal framework is such that it places considerable limitations on the potential efficacy of RFMOs.

The first challenge arises as a result of the application of the basic rule that treaties are only binding on their parties (1969 Vienna Convention on the Law of treaties, Article 34). As international law does not compel states to adhere to any particular treaty, they are free to choose whether to become party to a treaty. In other words, states whose vessels fish in areas of the high seas may not be members of RFMOs and their vessels are therefore not obliged to comply with the various conservation and management measures adopted by those RFMOs. These "free riders" can undermine conservation and management measures as well as any incentive for member state nationals to comply. In addition, free riders are under no obligation to supply data to RFMOs, so stock assessments may be unreliable, rendering the adoption of meaningful conservation and management measures impossible. Free riders also reap the economic benefit associated with not having to fund the operations of RFMOs and the implementation of their measures, giving them an unfair advantage over the vessels of member states.

Another challenge for RFMOs relates to their decision-making processes, many of which provide for "opt-out" or objection procedures whereby member states can escape the application of otherwise binding measures that have been adopted by an RFMO. Yet another way of avoiding the application of measures is through the use of "flags of convenience" or the practice of flagging or reflagging fishing vessels from member states to nonmember states to avoid the application of conservation and management measures adopted by a RFMO. Although RFMOs may invite nonmember states to join, the latter may have no desire to take on the financial and other obligations associated with membership, particularly where there may be no guarantee of a quota allocation.

Even in situations in which conservation measures apply, however, under the traditional law of the sea it is the flag state, and only the flag state, that has jurisdiction over its vessels on the high seas and can take any action in respect of them. The ineffective exercise of flag state jurisdiction through inadequate enforcement can thus undermine RFMO measures. Although progress has been made on developing the content of flag state responsibilities, there are no sanctions in international law for a flag state's failure to meet those responsibilities. There is thus no guarantee of member states' willingness or ability to ensure compliance with conservation and management measures by their vessels and nationals. For example, member states may fail to adopt legislation to implement measures that have been agreed on within the RFMO. Alternately, the sheer tyranny of geography, with vessels fishing on the high seas far out of sight, may make it difficult for flag states to control the activities of their vessels. They may thus fail to prosecute violations either because of lack of ability or resources to acquire legally acceptable evidence, or because of lack of interest.

Since the 1990s, the collective effect of these challenges has been summed up in the terminology of *illegal, unreported, and unregulated* (IUU) fishing. Briefly put, IUU fishing is fishing that "occurs in violation of—or at least with disregard for—applicable fisheries rules, [whether those rules have been] adopted at the national or international level" (FAO 2002, 6). Thus, although IUU fishing also relates to fishing activities carried out within the EEZ of coastal states, the focus here is on high seas IUU fishing where the term is used to encompass a wide range of fishing activities that undermine regulatory activities and conservation and management measures adopted by RFMOs. The whole point of the objection to IUU fishing is that it directly threatens effective conservation and management of fish stocks, thereby adversely affecting both fisheries and the people who depend on them. Nevertheless, although fighting the scourge of IUU fishing is an important part of the work of RFMOs, it is not their only task.

RFMOS AND THE REQUIREMENTS OF CONSERVATION

RFMOs are established for the express purpose of conserving and managing high seas fish stocks (for a detailed discussion, see Rayfuse 2016, 439–62). To that end, they are required to take measures, designed on the basis of the best scientific evidence available, to maintain or restore populations of harvested species at levels that can produce maximum sustainable yield, taking into consideration the effects on associated and dependent species (LOSC, Article 119). In modern parlance, RFMOs are to ensure the long-term sustainability and promote the optimum utilization of the resources they manage (1995 Fish Stocks Agreement, Article 5). Measures adopted by RFMOs (generally referred to as *conservation and management measures*) can roughly be divided into five broad categories: measures relating to stock assessment, management of fishing effort, allocation of fishing opportunities, compliance and enforcement, and protection of the wider marine environment.

Measures relating to stock assessment are aimed at providing RFMOs with the knowledge necessary to set total allowable catches (TACs) and determine participatory rights. These measures include data reporting requirements as well as the decision rules used in calculating stock status. Unfortunately, both under reporting and misreporting remain common occurrences, with RFMO annual reports evidencing a litany of calls for better compliance by members with their data reporting obligations. These measures have therefore become increasingly complex as RFMOs struggle to deal with the scientific uncertainty resulting from the absence or inadequacy of data resulting from nonreporting by members, nonmembers, or both. For this reason, RFMOs have moved to implement a precautionary approach to management that allows for estimates of unreported catch to be taken into account in their stock assessments.

Measures relating to the management of fishing effort include measures establishing closed seasons or closed areas (including in some cases total moratoria); restrictions on types of gear; or on the manner, time, and season of setting and use of gear; and other catch and effort restrictions designed to control where, when, and in what manner fishing activity takes place. Also included are measures aimed at managing capacity and effort. These measures can, however, only be effective if capacity limitations are in place. Despite ongoing efforts to reduce capacity, it continues to far outstrip supply, with the inevitable consequence of continued overfishing (Bell, Watson, and Ye 2012, 495–505; FAO 2008; FAO 2018, 6).

Management of capacity and effort is also related to the issue of allocation of fishing opportunities. Many RFMOs divide the TAC for a stock into national quotas, whereas others provide for an "Olympic style" fishery where the race for the fish is open to all but ends with the closure of the fishing season as soon as the TAC is reached. As between members, quota allocations are generally based on historic catch and capacity, taking into account scientific advice regarding stock status. However, negotiation of these quotas is often a fraught exercise in which political considerations trump scientific advice. One of the most difficult challenges for RFMOs relates to compliance and enforcement. RFMOs have adopted a plethora of measures aimed at persuading flag states to ensure their vessels comply with international rules and to address the enforcement challenges posed by flag states that are unwilling or unable to do so. These include measures calling for the adoption of "positive" or "negative" lists of vessels that are either authorized or not authorized to fish in the regulatory area, transshipment bans, the use of observers, port state controls, catch documentation schemes, and at-sea measures such as boarding and inspection. The efficacy of these measures is, however, undermined both by their lack of application to all vessels fishing for a particular stock or in a particular area and by the lack of political will on some flag states to take enforcement, investigation, and prosecution action against their vessels.

With respect to the broader environment, although predominantly concerned with the management of target species, RFMO measures also deal with management issues related to associated or dependent species and broader ecosystem concerns. Traditionally, the most basic measures adopted have related to bycatch of other fish species. However, the adverse effects of fishing on other (nonfish) species have increasingly occupied RFMOs, which have adopted measures aimed at preventing or reducing bycatch of species such as sharks, dolphins, turtles, and seabirds. Even more recently, RFMOs have broadened their concerns to include the adverse effects of fishing on vulnerable marine ecosystems and measures have been adopted aimed at controlling destructive fishing practices like bottom trawling and at the establishment of marine protected areas in which fishing activities are either limited or prohibited.

Regardless, however, of how effective RFMO measures are in dealing with the classical threats to global fish stocks of overexploitation and habitat destruction, the managerial efficacy of RFMOs is now seriously challenged by the new threat of climate change and associated ocean acidification. In its Fifth Assessment Report, released in 2014, the Intergovernmental Panel on Climate Change unequivocally confirmed that the oceans are warming and becoming more acidic. These changes are leading to increasingly rapid biological responses and ecological shifts, including species depletion, migration, and range shifts, all of which pose significant threats to global fish stocks (Pörtner et al. 2014, 411–84). The transformation of ocean ecosystems as a result of climate change and ocean acidification also poses significant challenges for RFMOs, which must now graft climate change impacts such as changes in species distribution and productivity onto the range of factors already to be considered in implementing the precautionary and ecosystem approaches to fisheries management (Rayfuse 2012, 147–74).

At its heart, the challenge for RFMOs relates to their ability to manage under conditions of uncertainty relating both to the fish under their management and to the broader marine ecosystem. Although initially designed for managing under more biologically stable conditions, in a climate-changed world RFMOs will need to ensure that their conservation and management measures are aimed at enhancing the climate resilience of the fisheries they are managing. Management practices will need to be both more conservative and flexible enough to account for changes in stock distribution and abundance. In other words, truly precautionary ecosystem-based approaches to management will be needed. Although both the precautionary and ecosystem approaches have now been adopted by virtually all RFMOs, experience shows that the application of these approaches is a complicated and difficult task at the best of times. Of course, RFMOs cannot solve the climate change problem. However, they can ensure that climate impacts are not exacerbated by inappropriate management practices by developing robust strategies and measures for responding to climate change–induced ecosystem changes. A review of RFMO reports and documents suggests, however, that they still have a long way to go in mainstreaming their adaptive capacity to anticipate climate stressors, to absorb the importance of these stressors into their decision-making processes, and to reshape their management regimes (Rayfuse 2018, 247–68).

In conclusion, RFMOs hold a privileged position as stewards of the world's high seas fisheries resources. In recent decades they have moved, with varying degrees of alacrity, to take seriously their mandates and to fulfill this role. However, as international organizations, RFMOs remain hamstrung by the limits of cooperation inherent in the very structures of international law. Thus, even if their status may not be open to question, the efficacy of their management may be. Moreover, they are creatures of a world viewed through essentially biologically stable conditions. Studies suggest that RFMOs are generally considered to be deficient in the essential capacities for adaptive, integrated governance and management that will be required to effectively support the resilience of marine ecosystems in an increasingly dynamic, climate change–challenged environment. Taken together, these issues represent the most fundamental and most critical challenges for contemporary international fisheries law and for the law of the sea as it relates to the conservation and management of the marine living resources of the high seas. They also represent fundamental and critical challenges to global food security. Fish in the high seas may be "out of sight"—but we consider them "out of mind" at our peril.

NOTE

1. The acronym *LOSC* is used in contradistinction to the acronym *UNCLOS*, which, although ubiquitous, is used in error to refer to the Law of the Sea Convention when it actually refers to the United Nations Conference(s) on the Law of the Sea, of which there were three: UNCLOS I (1956–1958), UNCLOS II (1960), and UN-

CLOS III (1973–1982). Use of the acronym *LOSC* is preferred by leading specialist international journals.

REFERENCES

1882 International Convention for the Regulation of the Police of the North Sea Fisheries Outside Territorial Waters, British and Foreign State Papers, LXXIII, 39.

1969 Vienna Convention on the Law of Treaties, 1155 UNTS 331, Article 34.

1982 United Nations Convention on the Law of the Sea, 1833 UNTS 397 (Law of the Sea Convention).

1993 Agreement to Promote Compliance with International Conservation and Management Measures by Fishing Vessels on the High Seas, 2221 UNTS 91.

1995 Agreement for the Implementation of the Provisions of the United Nations Convention on the Law of the Sea relating to the Conservation and Management of Straddling Fish Stocks and Highly Migratory Fish Stocks, 2167 UNTS 3.

1995 Code of Conduct for Responsible Fisheries. http://www.fao.org/fishery/code/en.

2009 Agreement on Port State Measures to Prevent, Deter and Eliminate Illegal, Unreported and Unregulated Fishing. http://www.fao.org/port-state-measures/resources/detail/en/c/1111616/.

2018 Agreement to Prevent Unregulated High Seas Fisheries in the Central Arctic Ocean, EC Document COM(2018) 454 final (Annex). https://eur-lex.europa.eu/legal-content/GA/TXT/?uri=CELEX:52018PC0454.

Bell, Justin D., Reg A. Watson, and Yimin Ye. 2012. "Global Fishing Capacity and Fishing Effort from 1950 to 2012." *Fish and Fisheries* 18, no. 3: 489–505.

Birdlife International. 2017. *Towards Seabird-Safe Fisheries: Global Efforts and Solutions.* https://www.birdlife.org/sites/default/files/bycatch_booklet_2017_w.pdf.

Food and Agricultural Organization (FAO). 1996. *The State of World Fisheries and Aquaculture.* Rome: FAO.

———. 1999a. *International Plan of Action for Reducing Incidental Catch of Seabirds in Longline Fisheries* (IPOA–Seabirds). http://www.fao.org/fishery/code/ipoa/en.

———. 1999b. *International Plan of Action for the Conservation and Management of Sharks* (IPOA–Sharks). http://www.fao.org/fishery/code/ipoa/en.

———. 1999c. *International Plan of Action for the Management of Fishing Capacity* (IPOA–Capacity). http://www.fao.org/fishery/code/ipoa/en.

———. 2000. *The State of World Fisheries and Aquaculture.* Rome: FAO.

———. 2001. *International Plan of Action to Prevent, Deter, and Eliminate Illegal, Unreported and Unregulated Fishing* (IPOA–IUU). http://www.fao.org/fishery/code/ipoa/en.

———. 2002. "Implementation of the International Plan of Action to Prevent, Deter and Eliminate Illegal, Unreported and Unregulated Fishing." *Technical Guidelines for Responsible Fisheries* No 9. Rome: FAO.

———. 2008. "Fisheries Management. 3. Managing Fishing Capacity." *FAO Technical Guidelines for Responsible Fisheries* No. 4, Suppl. 3. Rome: FAO.

———. 2018. *The State of World Fisheries and Aquaculture.* Rome: FAO.

Grotius, Hugo. [1608] 1916. *Mare Liberum,* or *The Freedom of the Sea or the Right which Belongs to the Dutch to take Part in the East Indian Trade,* translated by Ralph Van Deman Magoffin. Oxford: Oxford University Press.

Kurlansky, Mark. 1998. *Cod: A Biography of the Fish that Changed the World.* New York: Penguin.

Matz-Lück, Nele, and Johannes Fuchs. 2016. "Marine Living Resources." In *The Oxford Handbook of the Law of the Sea,* edited by Donald R. Rothwell, Alex G. Oude Elferink, Karen N. Scott, and Tim Stephens, 491–515. Oxford: Oxford University Press.

Pörtner, Hans-O., David M. Karl, Philip W. Boyd, William W. L. Cheung, Salvador E. Lluch-Cota, Yukihiro Nojiri, Daniela N. Schmidt, and Peter O. Zavialov. 2014. "Ocean Systems." In *Climate Change 2014: Impacts, Adaptation, and Vulnerability, Part A: Global and Sectoral Aspects. Contribution of Working Group II to the Fifth Assessment Report of the Intergovernmental Panel on Climate Change,* edited by Christopher B. Field, Vincente R. Barros, David Jon Dokken, Katharine J. Mach, Michael D. Mastrandrea, T. Eren Bilir, Monalisa Chatterjee, Kristie L. Ebi, Yuka Otsuki Estrada, Robert C. Genova, Betelhem Girma, Eric S. Kissel, Andrew N. Levy, Sandy MacCracken, Patricia R. Mastrandrea, and Leslie L. White, 411–84. Cambridge: Cambridge University Press.

Pringle, Heather. 1997. "Cabot, Cod and the Colonialists." *Canadian Geographic* 117, no. 4: 30–39. http://www.canadiangeographic.com/wildlife-nature/articles/pdfs/atlantic-cod-cabot-cod-and-the-colonists.pdf.

Rayfuse, Rosemary. 1999. "The Interrelationship between the Global Instruments of International Fisheries Law." In *Developments in International Fisheries Law,* edited by Ellen Hey, 107–58. Leiden: Kluwer.

———. 2004. *Non-Flag State Enforcement in High Seas Fisheries.* Leiden: Martinus Nijhoff.

———. 2010. "Moving Beyond the Tragedy of the Global Commons: The Grotian Legacy and the Future of Sustainable Management of the Biodiversity of the High Seas." In *The Future of International Environmental Law,* edited by David Leary and Balakrishna Pisupati, 201–24. Tokyo: United Nations University Press.

———. 2012. "Climate Change and the Law of the Sea." *International Law in the Era of Climate Change,* edited by Rosemary Rayfuse and Shirley Scott, 147–74. Cheltenham, UK: Edward Elgar.

———. 2016. "Regional Fisheries Management Organisations." *The Oxford Handbook of the Law of the Sea,* edited by Donald R. Rothwell, Alex G. Oude Elferink, Karen N. Scott, and Tim Stephens, 439–62. Oxford: Oxford University Press.

———. 2017. "Article 118: Cooperation of States in the Conservation and Management of Living Resources." In *United Nations Convention on the Law of the Sea: A Commentary,* edited by Alexander Proelss, 817–30. Munich: Beck, Hart, Nomos.

———. 2018. "Addressing Climate Change Impacts in Regional Fisheries Management Organisations." In *Strengthening International Fisheries Law in an Era of Changing Oceans,* edited by Richard Caddell and Erik J. Molenaar, 247–68. Oxford: Hart.

United Nations (UN). 1989. General Assembly Resolution 44/225 of 22 December 1989 on Large Scale Pelagic Driftnet Fishing and Its Impacts on the Living Marine Resources of the World's Oceans and Seas.

———. 1990. General Assembly Resolution 45/197 of 21 December 1990 on Large Scale Pelagic Driftnet Fishing and Its Impacts on the Living Marine Resources of the World's Oceans and Seas.

———. 1991. General Assembly Resolution 46/215 of 20 December 1991 on Large Scale Pelagic Driftnet Fishing and Its Impacts on the Living Marine Resources of the World's Oceans and Seas.

———. 2007. General Assembly Resolution 61/105 of 8 December 2007 on Sustainable Fisheries, Including through the 1995 Agreement for the Implementation of the Provisions of the United Nations Convention on the Law of the Sea of 10 December 1982 Relating to the Conservation and Management of Straddling Fish Stocks and Highly Migratory Fish Stocks, and Related Instruments.

———. 2009. General Assembly Resolution 64/72 of 4 December 2009 on Sustainable Fisheries, Including through the 1995 Agreement for the Implementation of the Provisions of the United Nations Convention on the Law of the Sea of 10 December 1982 Relating to the Conservation and Management of Straddling Fish Stocks and Highly Migratory Fish Stocks, and Related Instruments.

Wallace, Bryan P., Connie Y. Kot, Andrew D. DiMatteo, Tina Lee, Larry B. Crowder, and Rebecca L. Lewison. 2013. "Impacts of Fisheries Bycatch on Marine Turtle Populations Worldwide: Toward Conservation and Research Priorities." *Ecosphere* 4, no. 3: 1–49.

World Wildlife Foundation (WWF). 2017. Fish Bycatch—A Sad Topic. https://www.fishforward.eu/en/project/by-catch/.

———. 2019. Fishing Problems: Destructive Fishing Practices. https://wwf.panda.org/our_work/oceans/problems/destructive_fishing.

Chapter Eleven

Participatory Processes as Twenty-First-Century Social Knowledge Technology

Metaphors and Narratives at Work

Erena Le Heron, Richard Le Heron, June Logie,
Alison Greenaway, Will Allen, Paula Blackett,
Kate Davies, Bruce Glavovic, and Daniel Hikuroa

In this chapter we outline the metaphor and narrative work of participatory processes as they contribute to social knowledge in highly contested multi-use and multiuser marine spaces. By adopting a sociotheoretic approach, we can provide multiple layers of contextualization to establish the conditions in which participatory processes have emerged, explore their distinguishing characteristics as diverse and independent knowledge-generating entities, and consider their wider influences.

Common metaphors such as "ecosystem based management" or "consensus decision-making" are only two of the thought frameworks commonly used in the participatory processes discussed in this chapter. All metaphors both couch and frame the issue being engaged with and thus also affect the solutions that may be thought of, closing and opening possibilities. Likewise, narratives such as "fishing for abundance" and "no foreign ownership" story into place desired futures. "Story-ing" is using the power of narrative or storylines to elaborate on pathways and make them relatable, describing and framing understandings. Identifying metaphors and narratives at work and recognizing them as social knowledge technologies is therefore important to understanding the ways they shape knowledge making and new practices in participatory processes.

The research originates from a project in Aotearoa New Zealand's (Aotearoa NZ) Sustainable Seas National Science Challenge, in which the aim is to explore ways of reducing and transforming resource degradation while extending the use of natural resources. The Participatory Processes team (the authors of this chapter) was formed to critically examine the nature of contributions being made by recent participatory processes in contested heterogeneous marine spaces. Aotearoa NZ is an ideal setting to explore participatory processes. The participatory processes are independent, diverse in institutional, business, community, iwi[1] and public involvement, and are present in most sizable marine spaces.

In the chapter, we consider three questions derived from the geographic sociotheoretic emphasis on the collective and shared performative work of categories. The chapter's research emphasizes discourses, enactive and relational agency of humans and nonhumans, and power relations and politics, in a variety of settings. By taking this approach, we aim to illustrate how performative categories such as metaphors (that energize new directions of thinking) and narrative efforts (that story new desired directions) act as social knowledge technologies.

First, how have participatory processes achieved a switch toward forming collective imaginaries? This is their signature accomplishment. This entails probing what kinds of activities are wanted, and under what terms in the marine spaces. Second, by what means have participatory processes been able to create supportive analytic and political activities to advance their collectively inspired understandings? This question is prompted by the new knowledge demands of the novel and experimental narratives that define the nascent collective imagination of each participatory process. Third, in what ways have implementation efforts toward change been hampered by existing institutions and their legislative prescriptions? Or is it more realistic to take longer views when drawing conclusions about participatory processes and implementation in general? These questions reflect on exactly what the social presence of the participatory processes can be said to have achieved. The questions acknowledge the living character of participatory processes and their recent existence and uniqueness, and directs attention to the vitality embodied in the participatory processes, but also the multifaceted frustrations they assemble when formed.

Le Heron et al. (2018) discussed the implementation impasse met by participatory processes, but even if implementation occurs, the process or project may not hit the ground as it was imagined. Participatory processes are but one facet of wide engagement in marine spaces. Participatory processes do not always neatly move into implementation to desired outcomes even when everything goes as planned. Continual renewal and refreshing of public mandating and environmental awareness is required. Despite all this, the chapter argues that participatory processes use narratives inspired by attention to guiding metaphors to break new ground. The narratives are both disruptive and generative. Participatory processes *themselves* (regardless of implementation or outcome) are social technologies creating change.

In prioritizing participatory processes, it is important to retain a sense of proportion. The participatory initiatives selected for study, although widely acknowledged exemplars, are numerically small in number when compared with the more than 650 other (mostly single issue) marine participatory processes in Aotearoa NZ. The case study participatory processes ranged from relatively simple and time-bound to complex and ongoing. They were from a range of locations around Aotearoa NZ, and varied in size, type, style, object of interest, and marine space concerns. Some were institutional and others community driven. They all had in common that they were negotiating multiuse and multiuser marine spaces through collaborative processes, and were committed to incorporating Māori knowledge and practices. The five case studies were:

1. Integrated Kaipara Harbour Management Group 2005, ongoing
2. Sea Change Tai Timu Tai Pari Hauraki Gulf Marine Spatial Plan 2013–2016
3. Coalition against seabed mining 2005, ongoing
4. Gift Abel Tasman Beach (Awaroa) January–July 2016
5. Te Korowai o Te Tai o Marokura/Kaikoura Marine Guardians 2005, ongoing[2]

The research team conducted thirty-one in-depth interviews and undertook document analysis.

The distinctive character of the participatory process case studies as diverse and independent encouraged a customized approach to pick up on both relational agency and context from interviewing. A reflexive and coproduction conversational interview approach was adopted, in which we all added to a deep and productive dialogue. This meant contrary views could be gently teased out as momentum built over interviews, contentious power politics and conflicts were sometimes revealed, and all gained from the interaction. The methodology used conferred flexibility in accessing insights such as the role of metaphor and narrative in constructing participatory processes' visions and guiding practices (see also Blackett et al. forthcoming; Le Heron et al. 2018; Le Heron et al. 2019).

The chapter begins with a brief structural- and agency-centered review of the evolution of legislative and institutional developments flowing from reform momentum that has come to characterize the regulatory and marine environment. This is the context in which participatory processes are perceived as capable of exploring holistically conceived decision pathways to abate environmental problems. Following this, the chapter considers different sorts of collective knowledge formation in the case study marine spaces that have been made possible by participatory collectivities of diverse interests. The focus is on the role of narrative and metaphor in energizing and giving new dimensions to marinescape futures.

PARTICIPATORY PROCESSES IN CONTEXT
OF THE AOTEAROA NZ LEGISLATIVE ENVIRONMENT

The conditions before neoliberal-inspired reforms in the late 1980s and 1990s did not favor participation. Government departments were both policy and operational entities. The legislative regime was compartmentalized by discrete Acts; priority given to land-based sectors (e.g., forestry, sheep, and beef) for which indicative or illustrative planning was a standard tool; and interventions, when they occurred, were directed by departments. The research team surveyed the main legislation before the 1990s spanning marine activities and found limited consultation provisions in the Acts; when consultation was included, it was at the discretion of relevant departments.

Governmental economic and institutional reforms from the 1990s split policy and delivery amid a raft of new regulations. Two quite different legislative routes affected the marine realm in later decades. Responding especially to overfishing, a Fisheries Act (1996) introduced a quota management system–individual transfer quota (QMS-ITQ) system, with provision for formal consultation over annual quota setting. This framework of quota allocation conferred particular rights on iwi (the first Treaty settlement in 1992) and created a privileged class of quota holders. Although feted internationally, the often-overlooked aspect of the QMS-ITQ platform has been the intransigence of fisheries interests (Hale and Rude 2017) with other users in marine spaces.

In parallel, an omnibus Resource Management Act (RMA 1991) introduced an effects-based environmental management system of land and coast with obligations for ministries to consult and draw on participatory options. The reformist framework of RMA established distinctive pathways for "rights conscious" resource management choices about uses. Although rights are a complex field, they have not been widely researched. The arbitration of individual choices over use rights and by implication societal choices was a structure of consenting processes run by regional councils with appeals through the Environment Court. A coastal planning framework administered by regional councils emerged as part of the RMA.

A variant of the RMA framework, the Exclusive Economic Zone (EEZ) and Continental Shelf (CS) Act of 2012 features an Environmental Protection Authority to adjudicate in the marine realm. The intent was "to address gaps in the management of New Zealand's marine environment. . . . [P]rior to the EEZ Act, only some activities in the exclusive economic zone and continental shelf were regulated for their environmental effects" (Environment Foundation 2017). The EEZ Act covers activities that relate to disturbance of the seabed or water column, including petroleum and mineral exploration and production, aquaculture, marine energy generation, and carbon capture and storage. Significantly, in the EEZ Act "the activity status of marine activities is defined in regulations with limited public involvement compared to the RMA" (Environment Foundation 2017). Permitted, prohibited, or discretionary activities are decided by the Ministry of the Environment. In addition, there is no Treaty of Waitangi provision in EEZ and CS Act (cf. section 8 of RMA); instead, there is a Māori Advisory Committee and notification of consent applications to iwi authorities (Environment Foundation 2014).

In the 1990s, the nature of contestation in multiuse and multiuser coast and sea spaces began to exceed the regulatory and management reach of the RMA. In 2019, the research team found from surveying contemporary legislation defining the marine regulatory regime that there were disquieting trends. Some Acts had a reduction in public participation options, consultation and participation over successive Act amendments surpassed inclusion of these features, and provisions to exercise discretionary powers by regional councils especially over notifications of significant changes under Acts to the public and interested parties were being curtailed. The trends suggest consenting processes might be open to manipulation in favor of particular interests. Although a hollowing out of participatory options can be discerned, there is a strong counter trend involving increased opportunities for iwi in new Acts and Act amendments. Iwi have vigorously defended entitlements in legislation and exercised them to spearhead the agenda of Vision Matauranga.[3]

By the 2000s, it was realized that an effects-based system using the land-biased RMA was not designed to deal with complexities and complications of place-based, rights-fronted contestation in marine spaces (RMA applies up to the twelve-nautical-mile limit), especially when influences spilled beyond the bounds of the regulatory framework and immediate territory, whether on land, coast, or sea. The most obvious deficiency is the legislative blindness to the spatialities and temporalities of cumulative effects, a concept that draws attention to how user actions carried out anywhere could affect users in other places. In practical terms, this has reduced comprehension of, and means little provision for considering, territorial effects beyond the immediate location demarcated for consent. Indeed, mounting reservations about the RMA have led to an official review. Critics such as the Environmental Defence Society have called for a look at the whole legislative system, not just the RMA alone.

The section has sketched a legislative context that gives little attention to marine affairs and the limits for innovative participation. This shallow and narrow regulatory milieu is posited by the authors as forming conditions leading to a plethora of participatory processes—perhaps because building frustrations often find release in community action (Blackett et al. forthcoming).

PARTICIPATORY PROCESSES
AS SOCIAL KNOWLEDGE TECHNOLOGIES

The Science and Technology Studies literature broadly argues that science is a product of the culture and practices of those doing the science rather than a neutral and objective process (Hacket et al. 2008; Martin, Nightingale, and Yegros-Yegros 2012). As such, participatory processes that structure many engagements with the environment, institutions, and communities may be considered to be social knowledge technologies, which both produce knowledge and disseminate knowledge.

We argue that narrative and metaphor are some of the tools by which knowledge production is taking place in participatory processes. Metaphors are hugely powerful. Metaphors are known to entrench world views, whereas new metaphors may open up possibilities (Hanne, pers. comm., 2018). The participatory processes studied used collective metaphors to break out new categorizations and create new visions for the future of the marine space in question.

Closely interlinked with metaphor is narrative; they constantly reinforce and interrupt each other. Whereas metaphor categorizes the world and helps couch problems, narratives are stories that explain a series of events. They are from a particular view and voice, exist in multiple and are thus exclusive and partial. Significantly, the research showed that the guiding metaphors of how marine spaces can be best transformed were reinforced and interrupted by narratives of the specific futures participatory initiatives wished to create.

This chapter focuses on participatory processes as sites that stretch thinking away from existing institutional and investment patterns and behaviors, toward the intent of the new guiding metaphors. Supportive narrations of social reorganization accompany this, and are designed to leverage extensive change. To think of participatory processes as social or living laboratories *simultaneously* engaging in new relations of creativity involving present institutions, investors, and businesses, and where communities and iwi alter wider conditions and options in coastal and marine spaces, is a major shift from most understandings of participatory processes. The following section illustrates how metaphor and narrative were used to collectively imagine and configure collective knowledge and story specific futures.

COLLECTIVELY IMAGINING COLLECTIVE KNOWLEDGE

Metaphors Couch the Problem and Frame Solutions

The five case studies drew on multiple preexisting metaphors to conceptualize problems and response. Prominent examples include *ki uta ki tai* (mountains to sea),[4] *kaitiakitanga*,[5] marine spatial planning (MSP), ecosystem-based management (EBM), collaboration, consensus decision-making, science-based decision-making, mātauranga Māori[6]–western science integration, and *mauri*.[7] These are a mixture of overarching science and mātauranga Māori conceptual ideas; some are unique to Aotearoa NZ, and others are part of circulating global knowledge.

In each case study, these metaphors were employed as frameworks for understanding and conceptualizing the issues of interest. In some cases, such as "ki uta ki tai," "kaitiakitanga," and "mauri," these were powerfully new to some, radically reshaping world views, and part of customary practice to others. It is worth noting however that Māori metaphors are increasingly part of Aotearoa NZ's contemporary scene. Science metaphors such as *MSP* and *EBM* are created frameworks that have built up knowledge and practices around them, acting powerfully as thought-shaping metaphors in themselves. Yet other metaphors employed by the case studies encompass process, such as "consensus decision-making," "science-based decision-making," and "mātauranga Māori–western science integration." Each metaphor brings distinct framings of the world that both open and close possibilities—and as such they were often contested.

Although each case study cannot be simply tagged with its dominant metaphor as the reality is much more complex, messy, and entangled than this, we have attempted to highlight a selection of metaphors at play in each case study to aid reader interpretation.

INTEGRATED KAIPARA HARBOUR MANAGEMENT GROUP

Metaphors at Work in Shaping Thinking

Metaphors in this case study include ki uta ki tai (mountains to sea), kaitiakitanga, collaboration, mātauranga Māori–western science integration, and mauri.

Narrative: Healthy and Productive Estuary

Integrated Kaipara Harbour Management Group (IKHMG), an iwi-led participatory process, has shown leadership in coframing the ideas of kaitiakitanga and integrated catchment management. Historically, iwi's knowledge system of nurturing land-coast-sea interdependencies has taught that leadership was crucial to seeing the model of kaitiakitanga as *the* blueprint for restorative transitions.

The Kaipara Harbour and its catchment, the country's largest estuarine system, is bounded by European land-based legislation and institutions and iwi structures and has seen long-standing conflicts over the degradation and restoration of the ecosystem. Iwi initiatives over the health of the Kaipara catchment and Harbour deperipheralized the Kaipara in the eyes of traditional land-governing bodies. A holistic focus on water quality and use of science to identify the nature and sources of contaminants from the catchment, and in the Harbour, was simultaneously supported by iwi application and extension of kaitiakitanga principles. The IKHMG met resistance on the part of land users to acknowledge their role in the decline of the Kaipara's mauri.

When reviewing the achievements of the IKHMG, Makey and Awatere (2018) explained how it is important for Māori and stakeholders to build their own capacity according to kaitiakitanga principles. This, they stressed, is essential for the intergenerational journey to ensure kaitiakitanga becomes an ongoing culture. They acknowledged also the simultaneous engagement of research and practice, using diverse knowledge bases, and spatial planning.

New Knowledge and Practice Possibilities

Two innovative initiatives illustrate IKHMG efforts to showcase how their cogovernance model translates into new forms of respected agency and institutional and practice experimentation. These are framings that sponsor inhabitants' power and arrest the eroding effects of colonization on the land and estuary. The emphasis is on creating and narrating leverage points that are understandable to multiple interests.

Showcasing this philosophy in action has been achieved through the establishment of an IKHMG field program at flagship sites. Strategically placed within priority freshwater ecosystem catchments, IKHMG partnered with eight commercial farms (cattle, sheep, and dairy), two industrial sites (a lime quarry and a dairy milk processing factory), and a commercial flatfish fishery, to demonstrate the journey toward integrated EBM at a local scale. Informed by kaitiakitanga and EBM these microcosms of holistic on- and off-site activity are inspirational. However, several interviewees with different positioning in IKHMG were concerned at low public and farmer interest in field days. More recently, the Northern Wairoa Project began in 2017, drawing on experience of eight partners including Fonterra's Living Waters Partnership. This collaboration is unique in membership, working with landowners and *tangata whenua*[8] on sustainable land management practices informed by matauranga Māori.

SEA CHANGE TAI TIMU TAI PARI
HAURAKI GULF MARINE SPATIAL PLAN

Metaphors at Work in Shaping Thinking

Metaphors in this case study include MSP, EBM, collaboration, consensus decision-making, science-based decision-making, kaitiakitanga, ki uta ki tai.

Narrative: Voice of the Gulf

Sea Change is the most complex and ambitious participatory process in the country. Sea Change's contextual features are unusual—the Gulf has many iwi, activities and enterprises, territorial authorities, and a statutory monitoring entity in the form of the Hauraki Gulf Forum. Sea Change was a short-duration (three years), intensively resourced attempt to arrive at a marine spatial plan to reinstate the Gulf's mauri through giving voice to the Gulf. Mandating was the subject of protracted politics throughout as iwi contested their omission from key parts of the process. Iwi managed to influence later stages of the process.

The participatory initiative focused on navigating and negotiating a collective vision from and in spite of the competing science imaginaries about the ecosystem interdependencies, and government and management visions emanating from the large number of interests connected to the Gulf and beyond. Two goal-enriching metaphors, those of mauri and an MSP, energized the collective imagining and strategizing of Sea Change. The imaginative narrative "Voice of the Gulf" severed ties with mainstream thinking but exhorted listening to the aspirations and intentions of iwi seeking to connect and use the knowledge and principles kaitiakitanga alongside EBM. The stakeholder working group tried to embody the idea of "Voice of the Gulf," imagining themselves into this space and working and talking on behalf of the Gulf as an entity, rather than from usual representative positions. This is an uncomfortable zone for many. It credentialed both human and nonhuman agency by giving them conspicuous presence, materiality, and the status of active agency (all could speak). Further, it articulated new ways to gather knowledge. Voice could be combined into narratives of voices. In contrast, the technical metaphor of an MSP is less threatening as a byproduct of professional planning. The metaphor has promissory dimensions—a plan implies going somewhere; it can be implemented and is knowable. Much mainstream expertise can be called on.

New Knowledge and Practice Possibilities

Sea Change was ambitious and farsighted in its participatory design. A stakeholder working group (SWG) was formed in December 2013 via a workshop-based process involving more than 130 individuals from a range of community, business, and environmental groups. Ten community representatives from a pool of one thousand applicants were selected, using criteria of evidence of being able to work collaboratively and being able to articulate a collective vision for the Gulf. Together with four iwi representatives and an independent Chair, the SWG took charge of a participatory process designed to yield a marine spatial plan based on grounded knowledge of economy–environment relations. The SWG identified key issues; ran six issues-based roundtables; and used twenty-four listening posts, monthly public progress reports, and a framework of sixteen key themes to gather public input. This complemented formal scientific and management input from agencies and other stakeholders. The pathbreaking process did not satisfy all—challenges were made by those who felt SWG membership did not encompass their particular sectional interests, and by established agencies who claimed its procedures didn't meet "rules of consultation." However, the Office of Auditor General (2018) complimented the rigorous process of the MSP's development while acknowledging some flaws. While Flannery, Nealy, and Luna (2018) for the United Kingdom context contend that MSP is not facilitating paradigmatic shifts toward publicly engaged marine management, and that it may simply repackage power dynamics in the rhetoric of participation to legitimize the agendas of dominant actors, the testimony of Sea Change's innovativeness indicates more is possible than is commonly assumed. The Coalition government's commitment to appoint a ministerial advisory board to oversee the implementation of Sea Change suggests there is room for optimism.

COALITIONS AGAINST SEABED MINING

Metaphors at Work in Shaping Thinking

Metaphors in this case study include consensus decision-making, science-based decision-making, collaboration, kaitiakitanga, MSP, and legislation (Environmental Protection [EPA] Authority empowered by RMA Act 1991) as an appropriate framework for environmental decisions.

Narrative: Defend Environment through Participation in Process

The loose participatory process coalition mobilized around the Trans-Tasman Resources (TTR) application to seabed mine off the South Taranaki coast illustrates the contingent nature of participatory process composition in the face of changing context and circumstances of those involved and highlights difficulties in giving prominence to the scalar reach of seabed mining. The contest initially focused on using the consenting process of the EPA to mount opposition to TTR's proposal. This, however, gradually morphed into a challenge of the consenting process itself and its scope and effectiveness as a societal tool to decide on seabed mining applications. The different trajectories associated with the key protagonists over the life of three applications illuminates the creation of collaborative relations of convenience and negotiation on the part of major iwi (Ngati Ruanui),[9] Kiwis against Seabed Mining (KASM), Forest and Bird, and Department of Conservation (DOC). Some connections, such as between Taranaki Regional Council and TTR, were deemed inappropriate by both iwi and KASM. At stake in the applications lodged with the EPA were the nature and status of science, expertise, kaitiakitanga, and evidence that could be or could not be used. These concerns question the validity of the consenting process as a holistically sensitive procedure and an organizational guide. Bureaucratic logic, much from pre-Treaty settlement days, clashed with the assertion of place-based knowledge held by communities and demonstrated kaitiakitanga practices.

New Knowledge and Practice Possibilities

The EPA's consenting process since inception has been punctuated by "social" trials over the rights of procedure, conceptions and evidence that applicants could assume in the name of the use of resources and economic growth. The TTR application occasioned new levels of collective association and thinking that revealed that collective intentionality does not have a single subject. It showed that the "commons" of seabed mining is more than a demarcated mining location. Opponents in various ways articulated the ocean as three-dimensional; that is, it is inhabited water columns, extensive sea beds of biota, infrastructure stretching beneath the sea, and so on about which there is little knowledge. Generating and coordinating multiple lines of evidence to be introduced into the EPA process challenged the ingenuity and fortitude of parties. It also exposed the narrow knowledge base with which the EPA was working.

On a procedural level, the EPA process slowly became more centralized and less accessible to communities who found they had to travel to Wellington. Iwi in particular wanted a relocalization of the process and objected strongly when the EPA visited Taranaki but not South Taranaki, choosing instead to go New Plymouth, an affront to Ngati Ruanui.

GIFT ABEL TASMAN BEACH (AWAROA)

Metaphors at Work in Shaping Thinking

Metaphors in this case study include collaboration, consensus decision-making, mauri, kaitiakitanga, public ownership, and cogovernance.

Narrative: No Foreign Ownership, Aotearoa NZ Belongs to New Zealanders

The resolute commitment by the initiators to a full iwi partnership from the outset opened unprecedented possibilities. This participatory process is both pioneering in its genesis and in the way it illustrates a contemporary engagement with matters of access, ownership, and occupancy.

The participatory initiative was concerned with a "delicate" topic in Aotearoa NZ, namely private versus public ownership of, in this case, land (Awaroa beach) offered for sale adjacent to a national park. The intensity of decision making "on the hoof," "political lobbying and championing," and the reliance on crowdfunding to give substance and mana to a private citizen–driven participatory process, organized and funded in the name of the public, is salutary as an example of context and circumstance being reshaped by the relational actions of agency. This participatory process demonstrates that spaces of effective intervention can be created where it is deemed necessary by other actors. Other attempts to emulate the crowdfunding model used for the Awaroa purchase have been largely unsuccessful.

New Knowledge and Practice Possibilities

Although the campaign itself had a simple premise of preventing foreign ownership, commitment to full iwi partnership contributed to all-round satisfactory arrangements for access and ownership of the beach. The attitude of iwi was two stepped: first, keep the beach in Aotearoa NZ ownership and, second, address long-term thinking around future discussions of iwi ownership, management, or governance. Provisional transitions envisaged over the legal title and management of the beach being vested in iwi were renegotiated through *hui*[10] where the intergenerational interests of iwi and the public were debated with a view to finding an enduring framework. This involved detailing a new model of stewardship dependent on long-term trust by all parties and committed to an outward generosity when making choices. The solution set was for iwi to confer management on the DOC while retaining a cogovernance role. Awaroa was in the end a political statement that localities need not belong to private investors but could be recovered into the public estate.

TE KOROWAI O TE TAI O MAROKURA

Metaphors at Work in Shaping Thinking

Metaphors in this case study include kaitiakitanga, MSP, collaboration, consensus decision-making, science-based decision-making, mauri, EBM, and legislative product as outcome.

NARRATIVE: FISHING FOR ABUNDANCE

The narrative of "fishing for abundance" within the wider metaphors of MSP and kaitiakitanga has guided and energized the place-centered participatory model known as *Te Korowai*. The emergence of Te Korowai as a pioneering collaboration led by iwi is intimately tied to a long history of DOC leadership, Ngai Tahu influences and leadership, ministerial support of several governments, and interventions from major political parties. Review processes among those stakeholding at different times in the life of Te Korowai give considerable insight into the mutability of the participatory process and the challenges of producing a "product" suitable for legislative support by government and also translatable and workable in Kaikoura.

The longevity of Te Korowai is particularly salutary in three respects. First, Te Korowai has always asserted a responsibility to the area. Second, kaitiaki is about balanced holism, and not restricted to developing practices. Third, prominent spokespersons constantly reiterated the philosophy of only taking what you need to eat today and tomorrow. This multilayered set of values underscores the narrative of fishing for abundance. Users respectful of the health of different ecosystems (e.g., paua, whales, fur seals) have adopted protocols consistent with these principles and widely advertise their commitment to them.

New Knowledge and Practice Possibilities

Te Korowai was the enactment of special legislation, the Kaikoura (Te Tai-o-Marokura) Marine Management Act (2014), which fully empowered its pioneering co-evolved governance framework. The participatory process had its beginnings in collaboration between DOC and advocate ministers in the late 1990s. The Act prescribed the parameters within which Te Korowai could operate. It was understandably a fit-for-purpose Act dedicated to ensuring abundant biota. The Kaikoura earthquakes of 2016, however, exposed the lack of collective provisioning capacities in the community, local government, and iwi when a disruptive earthquake severely affected the infrastructure base on which Te Korowai's operation depended. Since the earthquake, two questions have preoccupied Te Korowai. First, positioning itself in the highly resource-intensive and demanding recovery process proved difficult. Demands on personnel for time, new kinds of decision making, and strategic planning on new matters took their toll. Second, as the community conduit, Te Korowai supported science providers in their efforts to evaluate the short- and long-term effects of the earthquake on the biophysical processes and biota upon which the area depended (Te Korowai 2018). Following the rebuild process, attention has turned to strategizing an international standard Education Research Hub based on marine biota and seeking ways to add to the accommodation base of the area through investment in hotels. This shift is fundamental, indicating a recognition that economy and environment need to be nurtured hand to hand, using Te Korowai's community and place-based practices to lay the ground for intergenerational security.

The chapter has investigated how the participatory processes have lived by metaphors of collective provocation. The backdrop of metaphors harnessed around marine spaces are collective in nature, although this is probably not widely recognized. EBM seeks to articulate managing the "whole ecosystem"; ki uta ki tai (mountains to sea) implies interdependencies across territory and over time; kaitiakitanga is grounded in connections of place and whakapapa over generations; MSP is a concept of organizing spaces in places; collaboration, consensus decision-making, and science-based decision-making are processes to engage others; matauranga Māori–western science integration constitutes sharing and respecting to forge a new knowledge frontier; and mauri is an expression of collective health. But what did the metaphors enable participatory processes to do?

   The findings indicate that the participatory processes engaged in storying the metaphors into existence by demonstrating steps necessary to change the *doing* of economic activities in conjunction with different environmental management techniques. There will always be doubt about whether each participatory process began their process with the intention of adopting or altering basic thinking categories, or whether their contributions have sprung from synergistic deliberations centered on resolving or at least constructively mediating contestation over investment options and externalities. Regardless, the localized narratives when viewed across the participatory processes have much in common. They are disruptive of entrenched narratives of others. They make visible by comparison much that has escaped attention or hidden. However, the extent of any disruptive influence has to be seen as partial and somewhere on a continuum of influence.

If we accept that actors are already changing the world, then we must ask forward-looking questions that engage more real-time. This is a crucial step, because it rotates thinking away from representing what has happened to representing catalyzing moments, remaining mindful of the past, and casting forward using knowledge in the present, to engage with the diversity of actors who are *already trying to enact different possible worlds*. Through the chapter's windows on the participatory processes, it becomes clear why actual dimensions and mechanisms of change are so difficult grasp. Metaphor and narrative provide means of couching problems and storying futures, and play a hitherto little recognized role in the way participatory processes engage with aspirations of change.

ACKNOWLEDGMENTS

Research was funded by the Sustainable Seas Ko nga moana whakauka National Science Challenge as part of Ministry of Innovation, Business and Employment contract C01X1515. Project: Testing EBM-supportive participatory processes for application in multi-use marine environments.

NOTES

1. *Iwi* refers to a Maori extended kinship group, tribe, nation, people, nationality, and race. It often refers to a large group of people descended from a common ancestor and associated with a distinct territory. (www.maoridictionary.co.nz)

2. For more information on each of the participatory processes highlighted in the chapter, see Le Heron and colleagues (2019) and the following websites for each initiative: Sea Change: www.seachange.org.nz for the finalized marine spatial plan and details of its emergence; IKHMG: www.kaiparaharbour.net.nz for a comprehensive and interesting insight into IKHMG and the Kaipara Harbour; Gift Abel Tasman Beach: www.facebook.com/saveatb for more on the campaign and further ongoing activity; Te Korowai: www.teamkorowai.org.nz for details on the Kaikoura (Te Tai-o-Marokura) Marine Management Bill as well as current activities; and KASM: www.kasm.org.nz for history and current involvement against seabed mining.

3. *Matauranga* refers to knowledge, wisdom, understanding, and skill (www.maoridictionary.co.nz). Vision Matauranga is a New Zealand government policy framework (Ministry of Research Science and Technology) "to unlock the innovation potential of Māori knowledge, resources and people to assist New Zealanders to create a better future." See https://www.mbie.govt.nz/assets/9916d28d7b/vision-matauranga-booklet.pdf.

4. *Ki uta ki tai* is a Māori concept meaning roughly "from mountains to sea" that complements and works with holistic, ecosystem-based thinking.

5. *Kaitiakitanga* means "guardianship," "stewardship," or "trustee."

6. *Matauranga* means "knowledge," "wisdom," "understanding," and "skill" (www.maoridictionary.co.nz). *Matauranga Maori* refers to Maori knowledge, wisdom, and so on.

7. *Mauri* refers to a life principle or life force. Also used for a physical object, individual, ecosystem or social group in which this essence is located (www.maoridictionary.co.nz).

8. *Tangata whenua* means "people of the land"; local people, hosts, indigenous people—people born of the whenua (i.e., of the placenta and of the land where the people's ancestors have lived and where their placenta are buried [www.maoridictionary.co.nz]).

9. Ngati Ruanui is an iwi group in the Taranaki region. See https://www.ruanui.co.nz/ for details of their multifaceted endeavours and aspirations.

10. A *hui* is a gathering or meeting, usually for indepth discussion.

REFERENCES

Environment Foundation. 2014. "Environment Guide: EEZ Māori and the EEZ Act." Accessed March 12, 2019. http://www.environmentguide.org.nz/eez/maori-and-the-eez-act/.

———. 2017. "Environment Guide: EEZ Introduction." Accessed March 12, 2019. http://www.environmentguide.org.nz/eez/.

Flannery, W., N. Nealy, and L. Luna. 2018. "Exclusion and Non-participation in Marine Spatial Planning." *Marine Policy* 88: 32–40.

Hacket, E., O. Amsterdamska, M. Lynch, and J. Wajeman, editors. 2008. *The Handbook of Science and Technology Studies*. Cambridge, MA: Massachusetts Institute of Technology Press.

Hale, L. Z., and J. Rude, editors. 2017. *Learning from New Zealand's 30 Years of Experience Managing Fisheries under a Quota Management System*. Arlington, VA: The Nature Conservancy.

Le Heron, R., P. Blackett, J. Logie, D. Hikuroa, E. Le Heron, A. Greenaway, B. Glavovic, K. Davies, W. Allen, and C. Lundquist. 2018. "Participatory Processes for Implementation in Aotearoa New Zealand's Multi-use/user Spaces? Unacknowledged and Unaddressed Issues." In *Towards Coastal Resilience and Sustainability*, Dynamics of Economic Spaces Series, edited by P. Heidkamp and J. Morrissey, 111–30. Milton Park, United Kingdom: Taylor and Francis.

Le Heron, E., J. Logie, W. Allen, R. Le Heron, P. Blackett, K. Davies, A. Greenaway, B. Glavovic, and D. Hikuroa. 2019. "Diversity, Contestation, Participation in Aotearoa New Zealand's Multi-use/user Marine Spaces." *Marine Policy*, forthcoming.

Makey, L., and S. Awatere. 2018. "He Mahere Pahekoheko Ma Kaipara Moana—Integrated Ecosystem-based Management for the Kaipara Harbour, Aotearoa New Zealand." *Society and Natural Resources*. DOI: 10.1080/08941920.2018.1484972.

Martin, B., P. Nightingale, and A. Yegros-Yegros. 2012. "Science and Technology Studies: Exploring the Knowledge Base." *Research Policy* 41, 7: 1182–1204.

Office of the Auditor-General. 2018. *Sea Change-Tai Timu Tai Pari: Creating a Marine Spatial Plan for the Hauraki Gulf.* Accessed May 3, 2019. https://oag.govt.nz/2018/hauraki.

Te Korowai. 2018. *Newsletter 19: June 2018.* Accessed March 12, 2019. http://www.teamkorowai.org.nz/docs/TK_NewsletterJune2018.pdf.

Chapter Twelve

When Penalizing Harm Propagates Harm

Rethinking Marine Resource Law Enforcement and Relations from South Africa

Marieke Norton

The notion of harm is central to the model of governance currently employed by the Fisheries Branch of South Africa's Department of Agriculture, Forestry and Fisheries. The reactive nature of the protocol that Marine Resource Compliance Inspectors are expected to follow means that the penalization of harm has superseded the *prevention* of harm as the de facto goal of the section, despite the language and intentions of the relevant Acts and policies. This approach does not, therefore, actively protect the marine resources and livelihoods in questions, but reactively judges and punishes. This does not offer incentives to comply, and so does little to preserve the marine fauna illegally harvested. Furthermore, it neglects to attend to issues of well-being, which are advocated by two central policies that South Africa has committed to implementing: the Ecosystems Approach to Fisheries (EAF) and the Small-Scale Fishers' Policy (SSFP). Despite progressive, protection-centered policies such as these, the actual relationship between resource users and fisheries compliance inspectors is framed by state-sanctioned violence. Indeed, as my research showed, the presence of physical and structural violence is so constricting that the agency of both resource users and inspectors are limited to the performance and enactment of harm.

Between 2011 and 2013, I did eighteen months of participant ethnographic fieldwork with the Marine Compliance Inspectors[1] based in nine different sites along the Western Cape Coast of South Africa, investigating what happens when policy becomes people through the act of policing. Our daily activities were all land-based, as land-based and sea-based compliance are split. We would inspect catches, permits, freezers, exports, gear, and coastlines for activity. We would then issue fines and make confiscations when required.

The majority of the photographs I was encouraged to take during fieldwork were of dead marine fauna. In particular, I often return to a photo I took in the

Fisheries Branch stores in Cape Town of confiscated abalone, worth several million South African rand, lying in a plastic-covered pile on the floor. The photo shows one corner of one freezer. There were three almost-full walk-in freezers packed with these see-through evidence bags.

I was taken to the storage facility, whose location is "undisclosed" but advertised on the doors outside, in order to be shown a visual representation of the success of the law enforcement measures against lobster and abalone poaching in the Western Cape. These packed stores were shown to me proudly as clear evidence that the men and women who risk their lives against violent poachers of shellfish and lobster were successfully executing their mandate to preserve and protect South Africa's marine resources.

Another example of marine resource law enforcement that met similar standards of so-called success occurred on a river bank in a small coastal town that is fairly isolated and surrounded by a farming region. An elderly, frail man had walked for hours to come fish, bringing one or two slices of bread, something to make a fire with, and some rudimentary fishing equipment. I was on the boat with two inspectors, and we almost missed him as we rode past, as he was fishing among some reeds. We stopped next to him. The inspectors searched his meager belongings and found one undersized white steenbras, which is illegal to catch. It was to be his lunch—hence the bread and fire-making tools. He had not meant to catch it—the fish had taken his bait and so he had pulled it up, not knowing what it was. It had been injured by the large hook, and he had let it die in order to eat it. He had not chosen to catch it, but had chosen to nourish himself with it once it was caught.

This garnered him a R750[2] fine—about half of what he was likely to earn a month as an elderly laborer. This elder began to cry and asked for leniency in a tone of voice that is horribly South African—one that is pleading with the master for some consideration and compassion. The only response he got was that if he returned on such and such a date to plead his case before the magistrate, he *might* have the fine reduced.

Not half an hour later, the inspectors fined a retired businessman on his boat R250 for intentionally taking out over three times the bag limit for bait. He laughed throughout the whole encounter, clearly not experiencing the same sense of submission as the elder in the previous encounter, and made it clear that the fine was simply regarded as an extra cost of his chosen leisure activity. The still-alive sand prawns were not returned to the mud. They were sealed in a plastic bag as evidence.

I was never with the inspectors when a crime was prevented; all the fauna that I saw was preserved only in a manner of speaking—frozen as evidence. Very little preservation, protection, or prevention was observed during my engagements. Every day, in every site, I engaged with the finding, evaluating, and penalizing of harm.

The authority responsible for enforcing South Africa's marine resource law and regulations is granted to the Compliance section of the Fisheries Branch of the Department of Agriculture, Forestry and Fisheries. Several laws prescribe their responsibilities and obligations, but the main piece of legislation concerned is the Marine Living Resources Act (#18 or 1998, referred to as *MLRA*). The mandate of the Fisheries Branch[3] is full of a language and ideal of caring: preservation, protection, promotion, prevention.

However, decades-long instability in the Western Cape fisheries, including increasing poaching activities and dire poverty, has led me to argue previously that these policy ideals are mostly missing from the actual implementation of marine resource law enforcement in the province (Norton 2014, 2015).

South Africa's fisheries management implementation is (despite many, many people's best intentions) primarily concerned with regulating the production of commodities, and not the maintenance of ecological or social well-being. This has made implementing both a national EAF and the domestic SSFP extremely problematic, with the actions of the Fisheries Branch often directly contradicting the principles of the policies (such as delaying the SSFP to the point that the biomass in question cannot support the economic well-being of the number of applicants). There are many obstacles to successful policy implementation in the South African fisheries; I argue that a particularly destructive one is the idea that well-being can be achieved by means of reactively penalizing harm.

The current model of law enforcement that is in operation almost solely focuses on reactively penalizing harm. The inspectors have no tasks that focus on or promote well-being. The protocols and operational parameters of their job effectively creates a tunnel vision that only recognizes harm as their incentive to act. Legally compliant behavior is largely unseen; there are no incentives to "be good," only incentives to not get caught.

Many of the inspectors, from all the different sites with each their own set of issues, are critical of this reactive protocol, and do not need prompting to say so. My discussions with them identified two dominant streams of interest expressed by inspectors that led to their current occupation: conservation and law enforcement. For those who are more interested in conservation work, the criticism of the reactive protocol was expressed in terms of a desire to "keep the fish in the sea," as one of my regular research participants termed it (personal communication between research participant and author, January 2012). They are aware that confiscating marine fauna does not protect those creatures, that the harm has already been done. For those more interested in law enforcement, the criticism of the reactive protocol was expressed by stating that crime *prevention*, not penalties, would be their ideal outcome of their work. Although the intention of some inspectors may be to prevent harm, the design of their job often denies them the opportunity to fulfill this. They are rendered incapable of enacting care by the very regulations that stem from the MLRA, despite its language of care.

The repeated articulation of this frustration, in being unable to prevent or protect, drew my attention to the notion of harm, and how the measurement thereof is a significant factor in how the effectiveness of compliance activities is judged.

For example, yearly evaluations, which determine promotions and bonuses, are largely determined by a comparison to the statistics of previous years. If more busts, confiscations, and arrests are made and more fines are issued than the previous year, that inspector or station is considered to be successfully implementing compliance regulations—an alternate reading of an increase in arrests or confiscations could be that the Compliance Section is particularly inefficient, that the levels of harm are increasing. An increase in "stats" (how statistics are referred to among inspectors) often means that the community experience of being well has significantly decreased, leading to more poaching activity. This contributes to damage to the ecosystem and therefore the cycle of harm amplifies as it creates a future of increased precarity for all.

However, in the current perspective, fewer arrests or confiscations means that the commitment and effectiveness of the station or inspector can be questioned, and bonuses deferred because of this uncertainty. Are they not doing their job properly because they are incompetent or because they are corrupt? This then leads to further precarity for the inspector in terms of reputation, job security, and even financial stability (as he or she will miss out on bonuses). This whole process runs counter to the logic of prevention, which asks for the potential for harm to be identified before it occurs. This calls for a different engagement with the idea of harm than is currently employed by the Fisheries Branch, a kind of engagement that regards a drop in the number of a station's cases of noncompliance, for the right reasons, as a measure of *success*.

The focus on harm, control, and penalization has created a space in which the tropes of war and battle, and even disease, dominate conceptual and tangible relations (Maldonado-Torres 2008; Norton 2015)—once again a language of care is replaced by activities that deal in harm. In poaching hotspots such as the towns of Paternoster, Hawston, and Gansbaai, the inspectors refer to both themselves and the divers who work for poaching syndicates as "foot soldiers," and the inspectors in Gansbaai often talk about "guerrilla tactics."

How can this still be the case in a context of progressive policies that foreground inclusivity and well-being, such as the EAF and SSFP? Both of these policies attempt to reimagine fisheries governance, and present differing but still ideal views on the necessity of a holistic view of the fisheries that takes well-being (human and ecological) into account in ways that promote sustainability and health—and in the case of the SSFP, social justice.

Although human well-being is considered an important dimension of EAF and a central tenet of the SSFP, it is largely considered in language (and in the case of EAF, indicators) entirely separate from that of governance (Norton and Jarre 2019). I argue that good governance also needs to be considered in terms of well-being, both for the ecosystem and for the entanglement of human relations that aggregates around it.

We can't recast our model of marine governance without rethinking our model of enforcement and engagement. While communities and inspectors are engaged in running battles over decisions made behind closed doors, there is no chance of well-being being promoted—for the inspectors, communities, or resources. The current model of marine resource law enforcement fails to be proactive, succumbing to what Sawatsky refers to as the "logic of problem responsiveness" (2007, 81). It is about assigning the blame for recognized harm, "based on the belief that if you gather the right nouns . . . you can have an adversarial fight between these nouns to determine whose noun label is correct: Guilty or Not Guilty" (Sawatsky 2007, 81). The emphasis on the use of nouns is pertinent here, as it speaks to the fact that what is often missing from reports or policies are people. It illustrates the thing-ification that Aimé Césaire notes as one of the central process of colonization (Césaire 1955)—the turning of relations into objects, in a way that reflects South Africa's painful colonial and segregationist relations of mastery, or *baasskap*[4] (Green 2017, 27). The inspectors are considered to be their tasks and duties (Inspector or Enforcer, Authority), and the resource users are categorized according to the legality of their resource activities (Fisher or Poacher, Submissive). The parties considered relevant by the applied law are invariably reduced to nouns and objects by the processes of jurisprudence. Behavior is seen to be determined by their category, and so their agency is subsumed by their perceived essence.

Although accountability is central to any collective endeavor that seeks to be just and equitable, there must also be an understanding of the externalities that create and are created by noncompliance in the marine sector. These externalities often do not originate in the fisheries sector, but impinge upon it. Therefore, marine resource law enforcement officials in South Africa are faced with dealing with the fallout from a number of underperforming or collapsing state institutions in a range of sectors: health care, education, public infrastructure, personal security, and employment. Costs are being produced, and those who can least afford it are being asked to pay. How are these costs produced, and what is their journey through these communities?

For example, many of our coastal communities do not have their own high school and so pupils must travel to the nearest one, often more than twenty kilometers away. Many cannot afford the transport costs. Children simply stop going to school, and in lieu of other appropriate activities, turn to fishing to keep themselves busy, join the trade of their fathers, and contribute to the household. As fish stocks dwindle and the rights allocation process drags on, often the only option is to join the local poaching network as a watcher, runner, or diver. The underperformance of the Department of Basic Education to ensure that every South African child can attend school until matriculation creates an externality that the Fisheries Branch must ultimately deal with, although they have no means of influencing the production of this externality. Many of these coastal settlements also almost entirely lack job-creation initiatives, such as is run by Public Works in other spaces. The failure to promote or provide alternative livelihoods for both men and women of all ages means that the choice of whether to comply or not is often not a moral or ethical one regarding the nature of the commons or the state of the resource—it is a practical one that considers hunger.

The reliance of coastal communities on the cash injection provided by poaching kingpins, in lieu of any significant government support, means that these figures gain authority that is often linked to, or contingent on, their affiliations to the black market and the gangs that control access to it. This relationship then again feeds drugs and violence back into the community. This draws down harsher measures of enforcement, which is often the tipping point to the kind of violence that exemplifies a breakdown in the social contract between the state and its citizens.

On the September 12, 2017, a protest action turned violent in Hangberg, a fishing community in the Cape Town suburb of Hout Bay. The protest was related to both the decrease in lobster quotas for the small-scale inshore sector (which is the sector that most of Hangberg's fishers fall into), and the lack of housing in the low-income community. This is a community that regards itself as, firstly, a community and, secondly, as one composed of fishers. Not only do the spatialized politics of their position play into the housing crisis by denying them the right to grow as a community, they are also constrained by the protectionist idea of nature that is manifested in the no-take zone of the Table Mountain National Park Marine Protected Area that encircles what they regard as their mountain. This is a spatialization of nature that excludes their historic resource use and the future growth of their community.

There is significant abalone and lobster poaching in the area, which has increased because of general unemployment but, importantly, there is also a reduction in number of permits and size of quotas. Fishers have been fighting not only for the right to fish but the right to quotas that are sufficient for sustaining a livelihood. Although they have been engaged in a decade of protest, representation, consultation, and bureaucracy during the development of the SSFP, their livelihoods and futures have remained insecure as the process has been delayed and dragged out. Much work has been done to repair the trust between the state and the small-scale fishers, but these efforts are continually eroded by what is experienced by the fishers as the state's indifference toward their well-being. In the meantime, it has become clear that the Nature that was being contested a decade ago—an inshore abundance that spelled economic development—is no longer there and exists only as a political object.[5]

The history of these protests has been one of violence and intolerance from both sides (protesters and law enforcement). This means that little provocation is needed to escalate a confrontation into a conflict. Perceived or expected harm is reacted to by inflicting harm, both by the fishers and the inspectors. This recent protest event in Hangberg saw stones and petrol bombs thrown at the police and nearby businesses, and rubber bullets fired by police at the bodies of protesters. A teenage boy sheltering under a picnic bench was shot twice in the mouth. It was this literal silencing of the child that drew public attention to the protest—some weeks such events are so frequent across the country, that only a fraction are reported on.

How harm is reacted to through policy and operational protocol propagates negative experience through the web of relations in ways that are not, perhaps cannot be, anticipated by our current model of marine conservation. These unacknowledged ramifications are *not* peripheral, but central to the functioning of the fisheries complex, and so must be identified and understood before accountability, collective or otherwise, can be determined. The act of determining accountability is not, however, done solely for the sake of blame (although there are times when this is necessary), but for the sake of determining which relations are most damaged, and how they could possibly be healed. An integrated approach is needed to address the harm that is done to these communities (which includes the inspectors) in ways that do not categorize it away through the blunt analytical tool of structural violence. The term *structural violence* refers to the ways in which the ingrained, often invisible, social structures, processes, and categorizations of people tend to consistently disadvantage those who are either people of color or economically vulnerable—the way that society is organized is biased against certain groups without the bias necessarily being consciously intended. Although there is indeed structural violence, simply attributing harm to this hold-all category can obscure more than it illuminates, and so does little work in developing alternatives that are tailored to specifics. If we replace the abstract notion of structure with people, we instead look to what enables the levels of relational violence experienced in the fisheries sector—here I mean the ways in which certain people, not the abstracted idea of society, do harm to each other through how they treat each other in daily or long-term interactions.

To understand relations between people is to apprehend experiences of personhood. Recognizing the relevant perspectives on, negotiations between, and embodiment of personhood is necessary both to tracking harm and to identifying alternatives to measuring success. Personhood is, among other things and other ways of describing it, the convergence of relations on and in a body. Current law and policy do not acknowledge this: the person of the inspector is absent from marine resource law enforcement policy and proto-col, and this absence (or blindness) endangers their bodies and marginalizes them as people (Norton 2014, 2015). The same can be said for resource users, although this has been acknowledged through means of the Constitutional Court judgment of 2004 and the deliberate inclusion of human rights into the SSFP (although there is still a gap between the acknowledgment of this in the policy and the accommodation and implementation in practice).

My research covered several coastal settlements and therefore several types of fisheries; the common feature was that marine inspectors have a significant presence in the everyday life of resource users and coastal com-munities. This is unacknowledged by the policies and protocols that regulate both the inspectors' and the resource users' actions, by not considering the practicalities of things such as where the inspectors live, eat, worship, or send their children to school. As distinct as these two groups may be in some respects, they nonetheless tend to belong to the same geographic community.

Whereas the inspector-on-paper is an individual imbued with the authority of the state to reach in and control resource users, the inspector-on-the-ground is in fact already *in* and often must find a way to express authority while be-ing in a position that is reliant on those around him or her, some of whom the inspector may be policing. By not acknowledging this, marine governance authorities not only do not anticipate the harm that may be visited on the inspectors, but do not have the means by which to register these harms. This is true also for many of the harms that may be visited on the fishers and their communities. However, although the institutions of marine governance do not have the means by which to register these harms, as people on the ground talking to other people, the inspectors do.

Inspectors-as-people use their presence in networks of relations to work against their own objectification as inspectors-as-policy. In among it all, I saw many moments of empathy and compassion between resource users and inspectors, where an awareness of context and well-being informed an inter-action that sought to minimize harm. In my doctoral thesis, I speak of it as knowing that "the dogs are biting at home."[6] The idiom refers to a situation of such hunger that the pet dogs have turned on their people. It acknowledges the home, economic need, and multispecies relationships. It speaks to an acknowledgment that compliance is not simply a question of individual choice, but a leveraging of bodily need and social constraint, of responsibility to not only self but importantly also to kin, however constituted. Importantly, the acknowledgment of this by inspectors is an act that places themselves in relation to the fisher, as people and not relative nouns. For inspectors, too, sometimes choices are not made according to duty but according to the circumstances of their own marginalization, which sees them removed from the process that determines the nature of their work and lives (Norton 2014).

Quoting Donna Haraway discussing Isabelle Stengers: "[D]ecisions must take place in the presence of those who will bear the consequence" (Haraway 2016, 12). This is not news to the fishers or inspectors. Fishers, whose personhoods and livelihoods bear the consequences of marine resource decisions, have been fighting (often, literally) to have their presence permitted and acknowledged by resource management and decision-makers.[7]

The inspectors too need to be present for and involved in the decisions taken, as they, too, bear the physical and relational consequences of the same decisions (Norton 2014). In the face of crises, Haraway argues for "intensely inhabiting specific bodies and places as the means to cultivate worldly urgencies with each other" (2016, 12). This, too, is an idea with which some inspectors and fishers are familiar. It speaks to the tension that the inspectors are placed under when required to produce evidence. Although the inspector must witness the crime to penalize it, his or her human presence or experience must be connected to a technology that records the act in specific ways for it to be accepted by the various stages of jurisprudence: for a person to submit evidence, he or she needs to be a cyborg. In other words, inspectors must augment their senses or bodies with technologies, become more-than-human, to be deemed sufficiently expert. If they are insufficiently hybridized with technology, their testimony cannot be explicitly trusted, even if their embodied knowledge and skill based on years of occupying specific spaces and processes allows them certainty.

Quoting Donna Haraway paraphrasing Marylin Strathern: "It matters what matters we use to think other matters with" (Haraway 2016, 12). This requires "staying with the trouble" (Haraway 2016, 12) and, in this case, following the fish (Probyn 2016). The trouble here is harm-as-dead-fish. Of course, it is not the object of the dead fish itself that is policed, but the manner of its death. What is policed is how harm was inflicted—essentially, was it legal or illegal? Legal harm is of course the managed and regulated exploitation (death) of marine fauna according to scientifically backed and capital-friendly policies. Illegal harm is when that death occurs outside of or in contravention of the parameters of those policies. So, in the legal sense, not all dead fish have been harmed. Furthermore, harm in this perspective is an act or event that is contained by the body of the fish and is entangled with the type of human act, and does not relate beyond that to the world other than in general terms of concern for the abstracted "ecosystem."

This results in a kind of tunnel vision that allows peripheral harm to go unchecked, despite its effect on well-being. Harm moves beyond acts or events, and is certainly not contained by moments of human agency. If we accept the expanded notion of what it is to be and to become a part of a relational web of life that is forever expanding and contracting as its nodes draw vital breath, then we need to expand our notion of harm as an energetic consequence that, if unchecked, propagates along this web to reach far beyond the moment and the human act.

The contestations in the South African fisheries are, as articulated here, about so much more than just total allowable catches and permits. Entangled in these examples are a host of subjects that make collective life together: inspectors, fishers, poachers, gangsters, the State and its various departments, the fish, the shellfish, those who rely on those that fish, observers, property developers, the ocean, those who process the fish, those who sell the fish, those who buy the fish, the climate, the historically disadvantaged, the currently disadvantaged, privilege, acts of dispossession, guns, rocks, pieces of paper, human rights, and legal categories.

The problem arises when this entanglement is ignored or simplified by focusing on the categories of objects as discrete. For example, one can be both a fisher and a poacher, as well as someone who relies on that protein as part of their diet. One can be an inspector, as well as a marginalized member of a fishing community. One can be a lobster, as well as a valuable commodity, and a dead piece of evidence that will never be eaten. Pieces of paper can ignore human rights while still expounding ideas of well-being.

The problem of the fisheries on South Africa is, I argue, at least partially the result of the focus on harm in lieu of care, on the protocol that demands reactions instead of responses. It needs to be addressed as not only a resource management problem, but as a breakdown of the social contract.

ACKNOWLEDGMENTS

Financial assistance for this research was provided by the Department of Science and Technology/National Research Foundation South African Research Chairs Initiative through the Research Chair in Marine Ecology and Fisheries and the National Research Foundation's SeaChange Fund, via the School of African and Gender Studies, Anthropology and Linguistics (University of Cape Town).

NOTES

1. Marine Compliance Inspectors work for the Compliance Directorate of the Department of Agriculture, Forestry and Fisheries, Fisheries Branch: Monitoring, Compliance and Surveillance Chief Directorate.

2. About US$60 at 2014 rates.

3. "The aim of the branch will be to contribute to maintaining and restoring the productive capacity and biodiversity of the marine environment, ensuring the *protection* of human health, as well as *promoting* the conservation and sustainable use of marine living resources. The branch further aims to ensure that the degradation of the marine environment from land-based activities is *prevented* by facilitating the realization of the duty of [the Department of Agriculture, Forestry and Fisheries] to *preserve and protect* the marine environment through the application of the respective policies, priorities and resources [my emphasis]." www.daff.gov.za.

4. In Afrikaans, a *baas* is literally a boss, but also, historically, a master. It is this second meaning, in the context of a history of slavery, colonialism, and apartheid, that is being referred to here—your *baas* is not only your employer, but also your constructed social superior.

5. For example, the 2018 court case between World Wildlife Fund South Africa (WWF–SA) and the Department of Agriculture, Forestry and Fisheries, in which the WWF–SA was challenging the west coast rock lobster total allowable catch decision by the Department: http://awsassets.wwf.org.za/downloads/wchc_judgment_wcrl. pdf. It was a case that was quickly framed (by many in the Department but by some media also) as a choice between the interests of ecology and the interests of the historically disadvantaged: https://www.sentinelnews.co.za/news/rocky-road-to-fair-fishing-16146907.

6. In Afrikaans, "die honde byt by die huis."

7. This has been a call from fisher-representation groups since the advent of democracy in 1994, and, although there has been much improvement in the access and space granted to these groups in various consultation processes and fora (which has led to the current SSFP, in particular after an order by the Constitutional Court mandated its development), the slow implementation of the SSFP and the disorganization of DAFF has meant that consultation and good intentions have not always resulted in actual improvements in the sector.

REFERENCES

Césaire, Aimé. [1955] 1972. "Discourse on Colonialism." Translated by Joan Pinkham. *Monthly Review Press.* Originally published as "Discours sur le colonialisme" by *Editions Presence Africaine*, 1955.

Green, Lesley. 2017. "Thinking Decoloniality with Perlemoen." *Catalyst: Feminism, Theory, Technoscience* 3, no. 1: 25–31.

Haraway, Donna Jeanne. 2016. *Staying with the Trouble: Making Kin in the Cthulucene.* Durham, NC: Duke University Press.

Maldonado-Torres, Nelson. 2008. *Against War: Views from the Underside of Modernity*. Durham, NC: Duke University Press.

Norton, Marieke. 2014. *At the Interface: Marine Compliance at Work in the Western Cape*. PhD Thesis. Western Cape, South Africa: University of Cape Town.

——. 2015. "The Militarisation of Marine Resource Conservation and Law Enforcement in the Western Cape, South Africa." *Marine Policy* 60 (C): 338–44.

Norton, Marieke, and Astrid Jarre. 2019. Being Well Governed: Including inspectors in a systems approach to fisheries management. *AMBIO*. https://doi.org/10.1007/s13280-019-01237-3

Probyn, Elspeth. 2016. *Eating the Ocean*. Durham, NC: Duke University Press.

Sawatsky, Jarem. 2007. "Rethinking Restorative Justice: When Geographies of Crime and of Healing Justice Matter." *Peace Research*, 29, no. 1/2: 75–93.

Chapter Thirteen

The Protection of Small-Scale Fisheries in Global Policymaking through Food Sovereignty

Alana Mann

Small-scale fisheries—defined as fisheries "where fishers operating from the shore or small fishing vessels use simple methods to catch fish from inland or coastal waters" (WorldFish 2017)—support the livelihoods of 90 percent of all people dependent on capture fisheries. They serve as vital "economic and social engine[s], providing food and nutrition security, employment and other multiplier effects to local economies" (Food and Agriculture Organization [FAO] 2015, v). Yet fisheries governance systems have largely been developed with a primary focus on the more profitable large-scale commercial fishing sector, as well as powerful interests from the tourism and oil and gas sectors. Further, efforts to mitigate climate change have created "no fishing" zones that further marginalize food-insecure communities. According to Dr. Pip Cohen, leader of the WorldFish program on Resilient Small-Scale Fisheries, these approaches "concentrate power and generate wealth for a relative few, fail to recognize the diverse rights and flexibility of small-scale fishers, and interrupt mechanisms that previously distributed economic and food security benefits broadly—particularly to those most in need" (WorldFish 2017). In other words, the people who are most affected by policies are overlooked.

These blind spots characterize economic globalization as "an increasing disconnect between those who make decisions that generate ecological harms and risks (such as states, investors, producers, consumers), and those for have expert knowledge of them (scientists), the victims who are exposed to them (typically the most marginal and least represented) and those who must take formal political responsibility for them (political representatives)" (Eckersley 2019, 6). They are the product of a neoliberal ideology that "operates with a view of economic life that does not value voice and imposes that view of economic life onto politics, via a reductive view of politics as the implement-ing of market function" (Couldry 2010, 2). Encouraged to thrive by vested interests, the market "evacuates entirely" the role of the social in political regulation of economics, creating "a contemporary crisis of voice" across not only political and economic but also social and cultural domains. Accord-ingly, articulating one's voice is a form of resistance against an economic system that strives to silence voices, especially those who lack "opportunity to compete as a commodity" (Couldry 2010, 9). These contests are escalat-ing in an increasingly corporatized food system in which the voices of those most affected by hunger and malnutrition are silenced in a market dominated by powerful elites.

Alternative narratives that reclaim productive resources as a commons, in-cluding food sovereignty, are about equal and proper valuing of the multiple dimensions of food and reclaiming the voice of food producers and eaters in food politics (Mann 2019). The first internationally agreed instrument dedi-cated to the small-scale fisheries sector, the *Voluntary Guidelines for Secur-ing Sustainable Small-scale Fisheries in the Context of Food Security and Poverty Alleviation* (referred to here as the "SSF Guidelines"; FAO 2015), represents a human rights–based approach to small-scale fisheries manage-ment that reflects several key principles of food sovereignty, as defined by the peasant farmers' movement *La Vía Campesina*. The SSF Guidelines, which emerged from a "bottom-up participatory process" involving more than four thousand representatives of governments, small-scale fishers, fish works, labor organizations, researchers, and nongovernmental organizations (NGOs) from 120 countries (FAO 2015, v), include acknowledgment of the cultural and social significance of fisheries, respect for First Nations' ancestral knowledge, and the empowerment of both men and women to par-ticipate in the management of freshwater and marine ecosystems. I evaluate the potentials for the realization of these principles through the incorporation of fishers' voices and knowledge in participatory, ecosystem-based manage-ment, drawing on fisheries research and cases in the Global South where new concepts such as "aquaecology" (Mills 2018) open space for the voice of the fisher.

OCEAN-GRABBING IN A CHANGING CLIMATE: THE STATE OF SMALL-SCALE FISHERIES

The privatization of the oceans is the tragedy of the commons writ large. In comparison to land, the privatization of the oceans and the resources it contains came late, for two reasons: firstly, fish are "fugitive resources" of which it is difficult to claim ownership and, secondly, the effect of the fishing industry on growth and fecundity of fish stocks was, until recently, perceived as negligible (Hannesson 2008). We "eat the ocean . . . every day, often without knowing it" and until recently "we thought we could eat it with impunity" (Probyn 2016). In 1883, eminent English biologist Thomas Huxley rallied against legislative restrictions on fishing at the International Fisheries Exhibition in London, declaring "it may be affirmed with confidence that, in relation to our present modes of fishing, a number of the most important sea fisheries, such as the cod fishery, the herring fishery, and the mackerel fishery, are inexhaustible" (Huxley 1883). It was the advancement of fishing technology from the 1800s and until the end of World War II that triggered the first demands from coastal states to exclusive rights to the fish stocks off their shores. A major boost to these claims was provided by the Truman Proclamation of 1945, claiming all resources on and underneath the seabed on the continental shelf of the United States were the property of the federal government. Countries who relied heavily on fish took this a step further and claimed exclusive rights to the fish stocks above the continental shelf. The three states on the west coast of South America, which have a narrow continental shelf and fish stocks close to the surface, claimed two hundred nautical miles without any reference to the continental shelf whatsoever. This two-hundred-mile norm was then recognized at the Third United Nations (UN) Law of the Sea Conference in the early 1970s. Following this, countries established a two-hundred-mile exclusive economic zone, marking the establishment of state property rights to ocean resources (Hannesson 2008).

These processes of centralization of fisheries management, technological developments, and demographic changes have replaced, in many regions, customary practices of resource allocation and sharing that had been in place for generations. One consequence is the severe depletion of fishery stocks. In 2012, Olivier de Schutter, the former Special Rapporteur for the Right to Food, stated that "without rapid action to claw back waters from unsustainable practices, fisheries will no longer be able to play a critical role in securing the right to food of millions," noting that "with agricultural systems under increasing pressure, many people are now looking to rivers, lakes and oceans to provide an increasing share of our dietary protein" (de Schutter 2012). Since 1961, the annual average global growth in fish consumption has been twice as high as population growth; consumption of seafood per capita has skyrocketed from less than three kilograms in the 1950s to approximately twenty kilograms in 2015 (Mills 2018, 1273). This contributes to claims that the fisheries sector is "crucial in meeting FAO's goal of a world without hunger and malnutrition" (FAO 2018, vii). Global fish production peaked at about 171 million tons in 2016 (88 percent for human consumption) "thanks to relatively stable capture fisheries production, reduced wastage and continued aquaculture growth" (FAO 2018, vii). In 2016, fisheries and aquaculture production was valued at US$362 billion, US$232 billion from aquaculture production (2). In 2015, fish accounted for about 17 percent of animal protein consumed by the global population and provided about 3.2 billion people with almost 20 percent of their average per capita intake of animal protein (FAO 2018, vii). Official statistics indicate that in 2016, 59.6 million people were engaged (on a full-time, part-time, or occasional basis) in the primary sector of capture fisheries and aquaculture (5). Nearly all those engaged in the industry were based in the Global South: 85 percent in Asia, 10 percent in Africa, and 4 percent in Latin America and the Caribbean (5). Alarmingly, the marine fishery resources, based on the UN Food and Agricultural Organization's (FAO) monitoring of assessed marine fish stocks, has continued to decline. The percentage of stocks fished at biologically unsustainable levels increased from 10 percent in 1974 to 33.1 percent in 2015, with the highest percentages of assessed stocks fished at unsustainable levels occurring in the Mediterranean and Black Seas, the southeast Pacific, and the southwest Atlantic (6). On top of this, 59.7 percent of fish stocks are "maximally sustainably" fished, and only 7 percent of fish stocks are underfished. The persistence of overfished stocks is an "area of great concern. . . . [I]t seems unlikely that the world's fisheries can rebuild the 33.1 percent of stocks that are currently overfished in the very near future" (6).

The 2017 report *Food from the Oceans* (European Commission 2017) posed the question "How can more food and biomass be obtained from the oceans in a way that does not deprive future generations of their benefits?" Humans eat at higher trophic levels in seafood than agricultural plants and animals which comprise 98 percent of our food. So even though seafood is only 1.6 percent (in weight) of human food, we are generally eating top predators—salmon, tuna, cod, haddock—and much less of the filter feeders and algae that is economically viable and nutritious (Bonhommeau et al. 2013; Duarte et al. 2009; Duarte, Marbá, and Holmer 2007; Olsen 2015). The food chain concept is misleading in that ecosystems are not a linear series of simple steps; the notion of the food web is much more accurate in illustrating that many organisms feed on the prey of different trophic levels in the ocean, and trophic levels must be estimated from the trophic levels of all the prey consumed (FAO 2019). Our challenge is to define the role ocean-derived protein should play to fulfill the Sustainable Development Goals (SDGs), "leaving no one behind" (United Nations [UN] 2018) while also respecting planetary boundaries as described in the UN Framework Convention on Climate Change. In 2012, fishing vessels contributed about 0.5 percent to total global carbon dioxide emissions (Barange et al. 2018, vi). Consuming seafood from lower trophic levels could contribute to achieving several SDGs—"life below water" (SDG 14), "sustainable cities" (SDG 11), "quality education" (SDG 4), and "climate action" (SDG 13).

Security of tenure rights to coastal and waterfront land are critical to those who need access to fisheries and the attendant processing and marketing activities and infrastructure that supports them, including housing and services (FAO 2015). *Ocean-grabbing* describes a set of contentious industrial fishing tactics that transform community fishing grounds into private spaces and allow foreign vessels into waters previously reserved for domestic use (Mills 2018). Practices include the creation of opaque access agreements that marginalize small-scale fishers, and unreported and illegal catches. Like land-grabbing, ocean-grabbing diverts resources away from local populations (de Schutter 2012). Consistent with related instruments that support rights to tenure and food, including the *Voluntary Guidelines on the Responsible Governance of Tenure of Land, Fisheries and Forests in the Context of National Food Security* (FAO 2012) and the *Voluntary Guidelines to Support the Progressive Realisation of the Right to Adequate Food in the Context of National Food Security* (FAO 2004) the "Guiding Principles" of the SSF Guidelines include respect of cultures; gender quality and equity; consultation and participation; transparency; accountability; and economic, social, and environmental sustainability in the achievement of the right to food and productive resources (FAO 2015, 2–3). All these documents are tools supporting the UN and FAO visions to "eradicate hunger and promote sustainable development." Being voluntary, they rely on the implementation of Member States, which must "ascertain which activities and operators are considered small-scale, and to identify vulnerable and marginalised groups needing greater attention" (FAO 2015, 2).

The civil-society companion document to the Guidelines, the *People's Manual on the Guidelines on Governance of Land, Fisheries and Forests* (International Planning Committee 2016), identifies specific challenges to the application of the SSF Guidelines and fisheries in general. Although fishing is the "greatest threat" to future global fish production, causing changes in the distribution, demography, and stock structure of individual species and direct or indirect changes in fish communities and marine ecosystems, climate change is having profound effects on fish physiology, behavior, growth, development, reproductive capacity, mortality, and distribution, and also alters the productivity, structure, and composition of the ecosystems on which fish depend for food and shelter (Brander 2007, 19710). It can also enable both competitive species, such as the Pacific oyster, and pathogenic species to spread to new areas, which can lead to mass mortalities of many aquatic species (19710). Many inland fisheries are also threatened by alterations to water regimes that can lead to entire rivers and lakes, such as Lake Chad in Africa, disappearing. In sixty years Lake Chad, once one of the largest in the continent, has decreased in area by 90 percent (from 26,000 square kilometers in 1963 to less than 1,500 in 2018) as a result of extended drought, inefficient use, and the effects of climate change (UN Environment 2018). These effects include reduced precipitation and greater evaporation, and more water being used for irrigation to offset lower rainfall. All have profound consequences for inland fisheries and those who depend on them. Climate change will lead to significant changes in the availability and trade of fish products, with potentially important geopolitical and economic consequences, especially for those countries most dependent on the sector. Regions most affected include the tropics, mostly in the South Pacific, and inland fisheries in Pakistan, Iraq, Morocco, and Spain (Barange et al. 2018). Aquaculture will experience short-term climate change effects, including losses of production and infrastructure arising from extreme events such as floods, increased risks of diseases, parasites, and harmful algal blooms.

The real value of seafood is little understood, poorly protected, and somewhat marginally integrated into global food security and nutritional policy considerations (Béné et al. 2015). It is not just a problem of nutrient uptake but one of "fisheries justice" in the context of "overlapping exclusions" (Mills 2018). The first category is industrialization and privatization, existing processes enabled by large-scale mechanized fishing operations and increasing demand for seafood, and the loss of traditional fishing grounds through processes such as ocean-grabbing (1273). The newer processes of exclusion include climate change and mitigation agendas that exacerbate the first exclusion. Although there is little research on the effects of mitigation projects on fishers' livelihoods and access to foods, Elyse Mills identifies how, in attempts to develop fossil fuel alternatives and harness sustainable energy sources, companies are targeting inland freshwater rivers for dam projects, causing widespread loss of livelihoods, displacement, and contamination, and destroying fish habitats. Further, "blue carbon" initiatives whereby governments and corporations buy credits by investing in the protection of coastal areas to offset their carbon emissions impinge on small-scale fishers' access to waterways where they live and work. Although this is "done in the name of conservation, it displaces local populations and seriously disrupts local livelihoods" (1275). Here it is important to note how adaptation discourses themselves tend to limit and depoliticize the forms of power that are produced and reproduced depending on one's ability "to influence, profit from and find security" (Taylor 2015, 8). Adaptation is less a "valid analytical tool" than a "politically constructed concept" that erases difference in highly unequal contexts. Although climate change needs to be addressed on a global scale through coordinated and cooperative efforts to limit emissions and mitigate effects, vulnerability will continue to be produced and reproduced in specific ways in different places unless we, as a species, acknowledge the relationships between local vulnerabilities and established structures of power and privilege.

Fishers' movements and organizations contesting these structures are offering alternatives to the dominant approaches currently applied to govern fisheries. Mills describes the emergence of a "fisheries justice movement" built on existing conceptualizations of agrarian, environmental, and climate justice: "a fishers' collective struggles for inclusion, equal rights, and the democratization of access, ownership, and control of natural resources and fishing territories" (Mills 2018, 3). At the heart of their struggle is what she coins the "aquaecological" approach to fishing.

AQUAECOLOGY: FOOD SOVEREIGNTY ON THE SEAS

Aquaecology is a natural approach to fisheries, based on species-specific equipment and techniques, following life-cycle and breeding patterns, protecting coastal ecosystems, and adhering closely to catch limitations. Sharing with the *agroecological* approach to farming a politically mobilizing frame focused on the empowerment and agency of those directly engaged in the practice (Altieri and Nichols 2017; Altieri and Todelo 2011; Mann 2019), aquaecology promotes an alternative "to the mitigation strategies used by most governments worldwide, which allow control over resources to be redirected away from communities and into the hands of state and private actors" (Mills 2018, 1278). Like agroecology, aquaecology actively supports food sovereignty, "the right of peoples to healthy and culturally appropriate food produced through ecologically sound and sustainable methods, and their right to define their own food and agriculture systems" (Declaration of Nyéléni 2007). This concept originated in Mexico in the 1980s, before the North American Free Trade Agreement led to the decimation of that country's agrarian economy, traditional diets, and, subsequently, public health (Gálvez 2018; Mann 2019). Popularized, and politicized, by the international peasant movement *La Vía Campesina* at the World Food Summit (Wittman, Desmarais, and Wiebe 2010), food sovereignty deploys the notion of a *peoples' sovereignty* "to resuscitate small-scale farming [and fisheries] as a public resource for food security and nutrition" (McMichael 2018, 208). This demands the democratization of community and regional food systems, which includes respecting the rights of small-scale farms and fishers to produce food, embracing gender equity and rights for workers, and determining territorial rights for First Nations people.

In the international food sovereignty movement, there has been a strong focus on land-based resource grabs, their effects on peasants and small-scale farmers, and the resistance movements that have emerged from these struggles. Accordingly, "the struggles of small-scale fisher movements have until recently been overlooked in the primarily 'land-centric' global campaigns" (La Vía Campesina 2017, 2). To rectify this, fisher movements and allies committed to research published a 2015 report titled *The Global Ocean Grab* (Transnational Institute 2015). It elaborated the definition of *ocean-grabbing* to include "the capturing of control by powerful economic actors of crucial decision-making . . . including the power to decide how and for what purposes marine resources are used, conserved and managed." This definition has been adapted by the two global fishers' movements: the World Forum of Fish Harvesters and Fish Works and the World Forum of Fisher Peoples (2015) to include "the exclusion of small-scale fishers from access to fisheries and other natural resources."

This exclusion encompasses strategies of cooptation including "blue growth," which includes the aforementioned blue carbon schemes. Claiming to protect wetland areas to mitigate global carbon emissions through their natural abilities to capture and store carbon, these instruments place an economic value on these areas based on how much carbon they can capture and store, and they are then introduced as yet another commodity in the carbon markets to be financialized and speculated on by investors and multinational corporations. Through such investments, these actors can offset their carbon emissions or simply trade with them for profit. The World Forums of Fish Harvesters and Fish Workers and the World Forum of Fisher Peoples (2015) deem this "a false solution" that "legitimises the continued emissions and plunder of natural resources in one place as long as wetlands that store and absorb carbon somewhere else are protected." Blue growth projects frequently result in the displacement of the people who need the productive resources in these areas for their survival; they involve expelling communities, reducing customary or community access rights, and fundamentally changing communities' relationships with the resources in a narrow framing of "conservation" that prioritizes profit-driven eco-tourism and marine protected areas over human rights. La Vía Campesina (2017, 3) deems these practices "ocean-grabbing under the cloak of 'sustainability.'" In opposition to these "elite solutions" that insist on further privatizing and marketizing fisheries and ocean resources, fishers' movements are engaged in building counterpower with other mass-based movements in the pursuit of climate justice and food sovereignty. The goal of this convergence is "to link up the struggles resisting against land, water, ocean and green grabbing—all of which indeed intersect" (3; see also Mills 2018). The movements insist real solutions to the climate crises should be based on First Nations and local traditional knowledge, and on the simple principle that human rights come before profits and the rights of the corporate sector. As the "true guardians of the fisheries resources" (La Vá Campesina 2017), small-scale fishers should be given back control of their waterways.

Intrinsic to this is gender equality in fisheries. Although women often play critical roles in providing, managing, and safeguarding water resources, their participation in decision-making over water resources remains limited. The empowerment of women in fisheries governance is highlighted throughout the SSF Guidelines on the basis of their prominent roles in the postharvest and processing subsectors and their often-limited access to tenure and capital (FAO 2015). Solving today's most pressing water issues depends on not just the participation, but the leadership, experience, and guidance of women, as highlighted in the following statement, developed at the inaugural Women and Rivers Congress in Nepal in March 2019:

> We have come together in solidarity to recognise and celebrate women as leaders and custodians of rivers and freshwater ecosystems. We are here to elevate women's voices, build alliances, and to learn from and strengthen each other's struggles and initiatives. Water is life and rivers are our global commons yet . . . [e]verywhere, rivers, springs, and freshwater ecosystems are being destroyed, diverted and degraded by dams, pollution, plantations, and over-extraction[,] . . . threatened by large-scale infrastructure and profit-driven development that ignores and obscures local and indigenous knowledge, culture, and ways of life. (Women and Rivers 2019a)

The Women and Rivers Congress, which included one hundred women from more than thirty-two countries, followed the first Women Fishers' Forum in Dili, Timor-Leste, in 2018. This event aimed to develop consensus statements to support women to participate in policy discussions as a critical step in realizing gender equality in small-scale fisheries in which women are consistently underrepresented. These patterns of exclusion reduce the effectiveness of management actions, perpetuate inequities in the opportunities small-scale fisheries offer, and hinder the achievements of food and nutrition security outcomes (CGIAR 2019). The imperative to engage in more meaningful and relevant gender analysis to improve socioecological approaches to fisheries research and management is supported by a review of case studies of small-scale fisheries from the last twenty years that detail the participation of women in fishing (Kleiber, Harris, and Vincent 2014). These studies reveal a quantitative data gap in the characterization of gender in small-scale fisheries and highlight how commonly used research methods may perpetuate biased sampling that overlooks women's fishing. Applying the gender lens provides a more complete view of the issues facing fishers across the entire supply chain and avoids a narrow focus on fish production and stocks, and those who work in those areas (Williams 2008).

LEARNING FROM THE GLOBAL SOUTH

The Global South offers numerous examples of how fishing communities can bring about change by "mobilising and directly challenging government policies for neglecting the rights of small-scale fishers" (Mills 2018, 1275). In many cases, local communities collaborate with capacity-building partners to determine appropriate governance and technical solutions. For example, since 2011 the international nonprofit research body WorldFish has worked locally with fourteen villages in Cambodia to empower them to comanage fisheries in the Stung Treng Ramsar stretch of the Upper Mekong River, a wetland of international importance. This approach includes community patrolling, the creation of five conservation zones, and a new knowledge-sharing network. It has led to reports of increasing fish stocks in the area, more equitable access to fishing for local stakeholders, and a greater sense of shared responsibility between all stakeholders. In the Solomon Islands, WorldFish has worked directly with at least thirty different villages to facilitate discussions on shared fishery goals, to exchange contemporary and local knowledge on solutions, and to collectively design fisheries management measures. It has also worked with the Ministry of Fisheries and Marine Resources to design a fishing strategy and related legislation that formalizes the role of communities in the management of coastal fisheries.

Small-scale fishers and traders are also informing regional policies that support more equitable economic, social, and ecological outcomes in the Caribbean, through the Western Central Atlantic Fishery Commission. Strengthening regional fisheries governance in collaboration with the Caribbean Natural Resources Institute and the Caribbean Network of Fisherfolk Organisation (2019), the Commission has produced the *Statement Calling for Attention to Enhancing and Developing Mechanisms for Effective Engagement of Fisherfolk and Other Stakeholders in Regional Fisheries Governance.* This Statement recognizes and commends existing commitments, policies, and practices of Member States for effective engagement of fisherfolk and other stakeholders in fisheries governance and management, including international commitments such as the 2030 Sustainable Development Agenda and the SSF Guidelines. Notably, it also expresses concern that "the farther away fishers are from decisions made, the greater the negative impact[s] are on small scale fisheries and livelihoods of fishers and their communities." It therefore recommends "specific attention to enhancement and development of mechanisms for effective engagement of fisherfolk and other stakeholders," including capacity building to facilitate "participation" and "effective engagement and voice" to facilitate effective input into the regional decision-making processes.

The vital need for fishers to regain voice and agency in their own futures has also been recognized in South Africa, largely thanks to the efforts of mass-based community organizations such as Coastal Links, established in 2003 with the mandate of enabling small-scale fishers to claim their human rights to secure livelihoods. *The Policy for the Small-Scale Fisheries Sector in South Africa* was completed in 2012 in consultation with small-scale fishers through the capacity-building efforts of Coastal Links, which by 2014 had grown to a two-thousand-member-strong organization supported by the NGO Masifundise Development Trust. The policy introduces new strategies for managing the sector; highlights human rights, gender, and development as key issues; and aims to secure rights and controlled access to marine protected areas. It marks an important victory for South African fishers, demonstrating "their capacity for mobilisation, and their organisational power to enact changes at the national policy level" (Mills 2018, 1277; see also Green, chapter 1, and Norton, chapter 12, this volume).

The historic, yet perverse, disconnection of fishers from the "practices, processes and relationships that are the focus of fisheries science, policy, and management" (Heyman and Granados-Dieseldorff 2012, 144) is being overcome in research projects that focus on fishing communities and fishers themselves as rich sources of data. In the Gulf of Honduras, the "Voice of the Fishermen" process acknowledges and draws on traditional fisher knowledge, increasing fishers' involvement in the fisheries management process and promoting sustainable management through collective action. Demographic and socioeconomic profiles of fishers and their gear, harvest data expressed as estimated landings and values by fishing gear and location, fishers' perception of the state of the resource and suggestions for improved management, and fishers' degree of interest in economic alternatives to commercial fishing contribute to an elaborate understanding of "the seasonal geography of fishing." Several new marine protected areas have been implemented with a level of cross-national collaboration that, the authors say, would have been virtually impossible "without the trust built between fishers and managers throughout the region via Voice of the Fishermen process" (Heyman and Granados-Dieseldorff 2012, 145).

Although fishing communities have always been important actors within public debates over fisheries resources, they have rarely been considered worthy as sources of data in fisheries science and management. In other words, fishers have not traditionally had "voice." They have been "invisible in the space of stock calculations and consequently, their management has been inherently biased towards the actions and desires of larger commercial fishers and their short-term interests" (Heyman and Dieseldorff 2012, 130).

This chapter has explored how the SSF Guidelines, as a framework for responsible and inclusive fisheries management, foreground the voices of those directly involved in practices of food production, harvesting, and processing. The Guidelines reflect many of the broad principles of food sovereignty including a focus on responsible governance of tenure, ensuring the rights of women and First Nations people and recognizing their unique contributions to the protection and management of aquatic ecosystems. The onus is on the State to create enabling conditions for the full participation of fishers in policy making, however. Given the constraints of the neoliberal, market-driven economy, the SSF Guidelines can only be realized with the grassroots energy and impetus of peoples' movements and allies, including capacity-building NGOs and researchers, in the struggle for food sovereignty. Food sovereignty "pushes back against capitalism" by insisting that food is treated "not as a commodity but as a common good" (Vivero-Pol 2017, 17). In creating a "self-controlled and self-managed resource base" that enables a form of "co-production of man and living nature that interacts with the market," small-scale fishers, like their counterparts the peasant farmers, reduce their dependency on the corporations that seek to control them (van der Ploeg 2008, 23). Countering the "accumulation of capital with an accumulation of the commons," their approach recognizes the significance of public spaces where people share, communicate, exchange, and live together, liberated from the "character of exclusion from the rights of both use and decision-making, instituting instead schema of open, shared use and democratic governance" (Hardt and Negri 2016, 100; see also De Angelis 2017). Reinvoking the commons strategies and interventions such as those presented in this chapter encourages us to "read for difference" (Gibson-Graham 1996). They demand the development of "alternative standpoints" that "acknowledge attempts to imagine and enact a food politics that achieves different socio-environmental justice outcomes to those of conventional food systems" (Harris 2009, 60).

The aquaecology concept strengthens the food sovereignty movement in several ways. It broadens our understanding of food sovereignty beyond land and agriculture, to more purposefully and effectively integrate fishers and aquatic resources and spaces. It also communicates the relational nature of land and water, extending discussions on resource production, access, and related conflicts, and drawing on the lived experience and often intergenerational knowledge of those most affected by changes. Further, it highlights the convergence between fishers' and other resource justice movements, and advocacy campaigns such as Slow Fish (2015). Importantly, it promotes First Nations' valuable contributions to effective governance of food production and climate change mitigation agendas. This is particularly vital where land title claims continue to be unresolved, and are often actively obstructed by patronage of commercial fishing interests in countries across the Global North and South, including Australia (Allam 2019). As researchers we have a responsibility to support these efforts through interdisciplinary and intersectional approaches to small-scale fisheries research. In this way we can support the emerging aquaecology movement in generating knowledge and influence, mobilizing resources, strengthening the movement, and framing a narrative that unites "water stewards and protectors" (Women and Rivers 2019b) such as women, First Nations people and other small-scale fishers in common struggle.

ACKNOWLEDGMENT

The author thanks Drew Rooke for his assistance with the research reported in this chapter.

REFERENCES

Allam, Lorena. 2019. "Aboriginal Land Rights Claims Unresolved Despite All-clear from Independent Review." *The Guardian*, March 29. https://www.theguardian.com/australia-news/2019/mar/29/aboriginal-land-rights-claims-unresolved-despite-all-clear-from-independent-review.

Altieri, M. A., and C. I. Nicholls. 2017. "Agroecology: A Brief Account of Its Origins and Currents of Thought in Latin America." *Agriculture and Sustainable Food Systems*, 3–4: 231–37.

Altieri, M. A., and V. M. Toledo. 2011. "The Agroecological Revolution in Latin America: Rescuing Nature, Ensuring Food Sovereignty and Empowering Peasants." *The Journal of Peasant Studies* 38, no. 3: 587–612.

Barange, Manuel, Tarûb Bahri, Malcolm C. M. Beveridge, Kevern L. Cochrane, Simon Funge-Smith, and Florence Poulain, editors. 2018. "Impacts of Climate Change on Fisheries and Aquaculture: Synthesis of Current Knowledge, Adaptation and Mitigation Options." *FAO Fisheries and Aquaculture Technical Paper* No. 627. Rome, FAO. 628 pp.

Béné, C., M. Barange, R. Subasinghe, P. Pinstrup-Andersen, G. Merino, G. I. Hemre, and M. Williams. 2015. "Feeding 9 Billion by 2050: Putting Fish Back on the Menu." *Food Security* 7, no. 2, 261–74. http://doi.org/10.1007/s12571-015-0427-z.

Bonhommeau, S., L. Dubroca, O. Le Pape, J. Barde, D. M. Kaplan, E. Chassot, and A. E. Nieblas. 2013. Eating Up the World's Food Web and the Human Trophic Level. *Proceedings of the National Academy of Sciences* 110, no. 51: 20617–20.

Brander, K. 2007. "Global Fish Production and Climate Change." *Proceedings of the National Academy of Sciences of the United States of America* 104, no. 50: 19709–14. DOI: 10.1073/pnas.0702059104.

Caribbean Natural Resources Institute and Caribbean Network of Fisherfolk Organisations. 2019. "Statement Calling for Attention to Enhancing and Developing Mechanisms for Effective Engagement of Fisherfolk and Other Stakeholders in Regional Fisheries Governance." *Special Session of the Western Central Atlantic Fishery Commission*, Barbados, March 26. http://www.canari.org/wp-content/uploads/2017/08/wecafc-statement-by-cnfo-and-canari.pdf.

CGIAR. 2019. "Gaining a Voice: First Women Fishers' Forum Held in Timor-Leste." January 23, 2019. https://fish.cgiar.org/impact/stories-of-change/gaining-voice-first-women-fishers%E2%80%99-forum-held-timor-leste.

Couldry, N. 2010. *Why Voice Matters: Culture and Politics after Neoliberalism.* London: Sage.

De Angelis, M. 2017. *Omnia Sunt Communia: On the Commons and the Transformation to Postcapitalism.* Chicago: The University of Chicago Press.

Declaration of Nyéléni. 2007. Accessed on February 2, 2019. https://nyeleni.org/spip.php?article290.

De Schutter, O. 2012. "Ocean Grabbing as Serious a Threat as Land Grabbing." New York: United Nations. Accessed March 5, 2019. http://www.srfood.org/en/ocean-grabbing-as-serious-a-threat-as-land-grabbing-un-food-expert.

Duarte, C. M., M. Holmer, Y. Olsen, D. Soto, N. Marbà, J. Guiu, and I. Karakassis. 2009. "Will the Oceans Help Feed Humanity?" *BioScience* 59, no. 11: 967–76. doi:10.1525/bio.2009.59.11.8 48 49.

Duarte, C. M., N. Marbá, and M. Holmer. 2007. "Rapid Domestication of Marine Species." *Science* 316: 382–83.

Eckersley, Robyn. 2019. "Ecological Democracy and the Rise and Decline of Liberal Democracy: Looking Back, Looking Forward." *Environmental Politics.* DOI: 10.1080/09644016.2019.1594536.

European Commission. 2017. *Food from the Oceans: How Can More Food and Biomass be Obtained from the Oceans in a Way That Does Not Deprive Future Generations of Their Benefits?* Accessed March 10, 2019. https://ec.europa.eu/research/sam/pdf/sam_food-from-oceans_report.pdf.

Food and Agriculture Organization (FAO). 2004. *Voluntary Guidelines to support the Progressive Realisation of the Right to Adequate Food in the Context of National Food Security.* Accessed January 5, 2019. http://www.fao.org/3/a-y7937e.pdf.

———. 2012. *Voluntary Guidelines on the Responsible Governance of Tenure of Land, Fisheries and Forests in the Context of National Food Security.* Accessed January 5, 2019. http://www.fao.org/3/a-i2801e.pdf.

———. 2015. *Voluntary Guidelines for Securing Sustainable Small-Scale Fisheries.* Accessed October 10, 2018. http://www.fao.org/voluntary-guidelines-small-scale-fisheries/en/.

———. 2018. *The State of World Fisheries and Aquaculture 2018—Meeting the Sustainable Development Goals*. Rome. Accessed March 25, 2019. http://www.fao.org/3/I9540EN/i9540en.pdf.

———. 2019. Trophic Levels. Accessed March 25, 2019. http://www.fao.org/fishery/topic/4210/en.

Gálvez, Alyshia. 2018. *Eating NAFTA: Trade, Food Policies, and the Destruction of Mexico*. Oakland: University of California Press.

Gibson-Graham, J. 1996. *The End of Capitalism (As We Know It): A Feminist Critique of Political Economy*. Minneapolis: University of Minnesota Press.

Hannesson, R. 2008. *Privatization of the Oceans: Handbook of Marine Fisheries Conservation and Management*. Oxford University Press, New York, 666–74.

Hardt, M., and A. Negri. 2017. *Assembly*. Oxford: Oxford University Press.

Harris, E. 2009. "Neoliberal Subjectivities or a Politics of the Possible? Reading for Difference in Alternative Food Networks." *Area* 41, no. 1: 55–63.

Heyman, W. D., and P. Granados-Dieseldorff. 2012. "The Voice of the Fishermen of the Gulf of Honduras: Improving Regional Fisheries Management through Fisher Participation." *Fisheries Research* 125–26: 129–48.

Huxley, H. T. 1883. "Inaugural Meeting of the Fishery Congress: Address by Professor Huxley." Accessed March 25, 2019. https://mathcs.clarku.edu/huxley/SM5/fish.html.

International Planning Committee for Food Sovereignty. 2016. *People's Manual on the Guidelines on Governance of Land, Fisheries and Forests: A Guide for Promotion, Implementation, Monitoring and Evaluation*. Accessed January 8, 2019. http://www.foodsovereignty.org/wp-content/uploads/2016/06/peoplesmanual.pdf.

Kleiber, Danika, Leila M. Harris, and Amanda C. J. Vincent. 2014. "Gender and Small-Scale Fisheries: A Case for Counting Women and Beyond." *Fish and Fisheries* 16, no. 4: 547–62. DOI: 10.1111/faf.12075.

La Via Campesina. 2017. *Nyéléni Newsletter: Oceans and Water*. September 2017, 31.

Mann, Alana. 2019. *Voice and Participation in Global Food Politics*. Abingdon: Routledge.

McMichael, Philip. 2018. "Going Public with Notes on Close Cousins, Food Sovereignty, and Dignity." *Journal of Agriculture, Food Systems, and Community Development* 8, no. A, 207–12. https://doi.org/10.5304/jafscd.2018.08A.015.

Mills, Elyse N. 2018. "Implicating 'Fisheries Justice' Movements in Food and Climate Politics." *Third World Quarterly*. DOI: 10.1080/01436597.2017.1416288.

Olsen, Y. 2015. "How Can Mariculture Better Help Feed Humanity?" *Frontiers in Marine Science* 2, no. 46: 92. DOI: 10.3389/fmars.2015.00046.

Probyn, Elspeth. 2016. *Eating the Ocean*. Durham, NC: Duke University Press.

Slow Fish. (2015). "Mangroves, Education and Health in Tamil Nadu." Accessed March 25, 2019. https://www.slowfood.com/slowfish/pagine/eng/blog/dettaglio-id_edit=667.lasso.html.

Taylor, Marcus. 2015. *The Political Ecology of Climate Change Adaptation: Livelihoods, Agrarian Change and the Conflicts of Development*. London: Routledge/Earthscan.

Transnational Institute. 2015. *The Global Ocean Grab*. Accessed March 26, 2019. http://worldfishers.org/wp-content/uploads/2015/01/The_Global_Ocean_Grab-EN.pdf.

United Nations. 2018. Leaving No One Behind. Accessed March 16, 2019. https://sustainabledevelopment.un.org/content/documents/2754713_July_PM_2._Leaving_no_one_behind_Summary_from_UN_Committee_for_Development_Policy.pdf.

United Nations Environment. 2018. The Tale of a Disappearing Lake. Accessed June 4, 2019. https://www.unenvironment.org/news-and-stories/story/tale-disappearing-lake.

Van der Ploeg, J. 2008. *The New Peasantries: Struggles for Autonomy and Sustainability in an Era of Empire and Globalisation*. Oxford: Earthscan.

Vivero-Pol, J. 2017. "Food as Commons or Commodity? Exploring the Links between Normative Valuations and Agency in Food Transition." *Sustainability* 9: 442.

Williams, M. J. 2008. "Why Look at Fisheries through a Gender Lens?" *Development* 51: 180–85.

Wittman, Desmarais, and Wiebe. 2010. *Food Sovereignty: Reconnecting Food, Nature and Community*. Black Point, Nova Scotia: Fernwood Publishing, Food First, Pambuzuka Press.

Women and Rivers. 2019a. "Nagarkot Statement." https://www.womenandrivers.com/statement.

———. 2019b. "Women and Rivers." https://www.womenandrivers.com/about.

WorldFish. (2017). "From Local to Global: How Research Enables Resilient and Sustainable Small-scale Fisheries." https://www.worldfishcenter.org/pages/from-local-to-global-how-research-enables-resilient-sustainable-small-scale-fisheries/.

World Forum of Fisher Peoples' and World Forum of Fish Harvesters and Fish Workers. 2015. *No to Blue Carbon, Yes to Food Sovereignty and Climate Justice!* http://worldfishers.org/wp-content/uploads/2015/12/Blue_Carbon_Statement.pdf.

Section 4

EMBODYING THE MARINE

Chapter Fourteen

The Sea and the Breathing

Astrida Neimanis (words)
and Janet Laurence (artwork)

HOLD YOUR BREATH (1)

"When the seas dried," philosophers Deleuze and Guattari remind us, "the primitive Fish left its associated milieu to explore land, forced to 'stand on its own legs,' now carrying water on the inside" (1987, 55). Just as these ancestral kin had to relearn movement, gravity, and communication, they also had to learn breathing anew.

Yet dry land was not, for everyone, all it was cracked up to be. When Darwin was writing his theories of evolution, the origin stories of whales were something of a mystery. Darwin nonetheless speculated that a "race of bears" could have conceivably evolved into whales. While his speculations were for the most part laughed right out of subsequent printings of *On the Origin of Species*, we know now that his musings were not as ludicrous as they seemed at the time. Over fifty-five million years ago, some of those terrestrial bodies—early whales, dolphins, porpoises—mutated into a cetaceous state. They took a long, deep breath, and headed back to sea.

Beneath the waves, the early cetaceans would have met others who had never left that watery milieu. Perhaps they encountered N'ba N'ga, a character in an old story by Italian fabulist Italo Calvino. N'ba N'ga is the paternal uncle of the protagonist Qfwfq, the time-traveling narrator of this tale that describes an historical epoch where "the water period was coming to an end." While for most aquatic species, this terrestrialization was cause for excitement, N'ba N'ga (a lobefish, from the extinct species of *Coelacanthus*) refuses to let go of the ocean. As Qfwfq puts it, "it just wasn't possible to make him accept a reality different from his own" (Calvino 2014, 71). The mortification Qfwfq experiences toward his uncle, though, is not only that N'ba N'ga is out of step with the terrestrial zeitgeist; worse than that, the

Figure 14.1. Janet Laurence, *Deep Breathing: Resuscitation for the Reef* (detail), 2015–16 / 2019, installation view, Janet Laurence: After Nature, Museum of Contemporary Art Australia, Sydney, 2019, wet, coral, coral drill core and dredge specimens, resin castings, 3D-printed skeletons, fish bones, tubing, laboratory glass, silicon, mirror, pigment, thread, acrylic, sand, glass, shells, clay, video projections, coral block CT scan from the Geocoastal Research Group, School of Geosciences, Marine Studies Institute, University of Sydney, selected specimens from the Collection of the Australian Museum, Sydney, selected specimens from the Geocoastal Research Group, School of Geosciences, Marine Studies Institute, University of Sydney, collection of the artist, image courtesy the Museum of Contemporary Art Australia © the artist.
Photo by Jacquie Manning.

aquatic uncle lures Qfwfq's fiancée, Lll, back into the water: N'ba N'ga and Lll fall in love, and Lll relearns to breathe below the surface.

If we were to examine the descendants of N'ba N'ga and Lll today—*coelacanths* who, unlike most that died out around the time of the dinosaurs, can still be found off the southeastern coast of Africa and in a pocket of Indonesia—we would find only barely perceptible traces of their brief flirtation with land and the open skies: vestigial lungs present only in an embryonic state.

We also now know that nineteenth-century naturalist, biologist, and philosopher Ernst Haeckel got it wrong when he suggested that ontogeny recapitulates phylogeny; evolution does not progress up through increasingly sophisticated species in the same way that a human begins as zygote, then becomes amphibian, and eventually a fully fledged, air-breathing bipedal Human. Citing fish as a particularly telling example, phenomenologist Merleau-

Ponty reminds us that "we do not find either less numerous or simpler types by going back in the history of the earth" (2003, 260). So when whales leave life on land to return to the sea, and when Lll rejects Qfwfq in order to join N'ba N'ga in his aquatic habitat, this is not a "regression" as in progress thwarted, or a reversal of evolutionary teleology. Neither Lll nor the whale regress. Rather, they select another possibility.

This all might be another way of asking: what might the whales, holding their breath for hours, maybe days, know that we are just starting to learn?

LEARNING TO BREATHE (1)

When I was pregnant, I learned that newborns, expelled from the seas of their amniotic home, learn to breathe air by following the rhythms of their mother's breath.

I worried a lot about my breathing.

In her important essay on situated knowledges, feminist science studies and environmental scholar Donna Haraway (1988) teaches us that all vision is prosthetic. When we see the world, she argues, human binocular vision is the prosthetic that determines how and what we see. No god's eye view here. Seeing the world—or knowing it in any other way—is only ever via an embodiment that is always already prosthetic: seeing/knowing *this* way, not *that*; seeing/knowing through *this* lens, not *that*. One's human bodily affordances (binocular vision, bipedal movement, a capacity for empathy or rational judgment), but also one's sex, one's age, one's language, one's habits, one's exposures, one's imaginaries, one's historical and geographic location in time and place: these things are uniquely entangled to compose a specific body's prosthetics of seeing and knowing.

Yet we rarely consider that all breathing is prosthetic, too. What and how we breathe is also a question of specific capacities and affordances: lungs versus gills, skin versus mouth, a capacity to extract oxygen from water, versus the terrifying prospect of a windpipe filling with wet.

While we might readily understand breathing, then, as a species-specific endeavor, like Haraway, we must also add that not all humans are situated in the same way: breathing is never neutral. Difficulty breathing, for example, has been analyzed under the rubrics of "settler atmospherics" (Simmons 2017, n.p.) and "socio-atmospherics of power" (Choy and Zee 2015, 211), whereby racism and colonialism are material phenomena the effects of which are activated when they meet black and brown bodies. We are reminded that air quality is a question of environmental justice, and that the weaponization of air and breath targets certain populations differentially within settler capitalist white supremacies. As Timothy Choy writes, "atmospheres do not equalize, and . . . breathing together rarely means breathing the same" (2016, n.p.). Feminist black studies scholar Christina Sharpe argues that breathless-ness as weather has always accompanied black life, from Frantz Fanon's understanding of the connection between revolt and an inability to breathe to Eric Garner's final words on a New York sidewalk: "I can't breathe" (Sharpe 2016, 108–10).

White feminist philosopher Luce Irigaray (1999) worries that we've forgot-ten about air (another enabler of the white male philosopher that he too easily takes for granted), but that is not the only forgetting that should concern us: while some breathing prosthetics are obvious (gills, lungs, but also an inhaler, a scuba mask, or an oxygen tank), in our current climate we have naturalized and forgotten all the other prosthetics—unstressed lungs, high ground, white collar job, white skin—that allow some bodies to breathe without noticing, for this "aerial matter" to "remain unthought" (Irigaray 1999, 12).

Hold your Breath (2)

(It had taken the rising sea for us to realize that not all breathing came so eas-ily. Where were our prostheses then? our white-skinned bodies would come to wonder.)

BAD WEATHER AND THE BREATHLESS SEA (1)

As terrestrial weathers heat up, they are spared even more intense effects of rising temperatures as the seas relieve them from a large portion of this bur-den. The watery world is a convenient resting place for all that heat, and the oceans are getting warmer. But rising ocean temperatures mean that the sea is also slowly suffocating. Whereas in the last fifty years overall oxygen levels in the world's oceans have dropped by an average of 2 percent, in coastal zones, hypoxic conditions have increased far more dramatically. Dead zones in the ocean, where oxygen is almost or totally depleted, have quadrupled since the 1950s (Carrington 2018; Isensee and Valdes 2015).

Even small changes in sea temperatures dramatically affect the behavior, reproductive patterns, and movements of ocean creatures, which has rever-berations up and down ocean food webs, affecting oxygen levels in turn. For phytoplankton, hotter water temperatures mean moving to different habitats, becoming smaller—affecting their ability to soak up the carbon we are pump-ing out. As the oceans grow warmer, the capacity of the deep to draw down carbon also decreases (International Union 2017). Simply put: in multiple and reinforcing ways, all of our greenhouse-gassed and fossil-fueled desires are starting to take up all of the oxygen in that watery room. We are weathermakers, even in the places our bodies do not and cannot belong.

This might be ironic if not so devastating: even as we may still imagine the oceans an unfathomable elsewhere—the convenient dump for out-of-sight, out-of-mind discards, and the ultimate away-washer of all of our sins—our weathermaking returns to us in multiple ways. Ocean photosynthesizers gift to humans every second breath we take, meaning that human air-breathers such as ourselves will also be profoundly affected by the suffocation of the seas (Morsink 2017). We are making the ocean breathless, and in turn deoxygenating ourselves.

Hold Your Breath (3)

Similarly, old N'ba N'ga, first conjured from Italo Calvino's pen in 1965, doesn't predict the poisonous gifts that we humans would bestow upon our oceans in the coming decades. "Down there," he explains to his nephew and his bride-to-be, Lll, "changes would be very few, space and provender were unlimited, the temperature would always be steady; in short, life would be maintained as it had gone on til then" (Calvino 1965, 79). Even if N'ba N'ga "didn't conceal the problems, even serious ones," his version of the ocean world is hardly what we swim in today.

What have we done, that we have forgotten to tell the whales?

BAD WEATHER AND THE BREATHLESS SEA (2)

The breathless sea reminds us that bad weather is a problem for all kinds of bodies—not only human ones, but for life below the waves, as well. Climate justice is radically extended to a multispecies justice of the deep, where the liveliness of the ocean is itself at stake.

But must we not also ask: what is lost when the words "I can't breathe" slip, almost unnoticed, off the body of Eric Garner to be washed into the ocean, now eddying around the bodies of phytoplankton, brachyuran crab, or a bivalve mollusk? Grappling with the weather underwater cannot be the simple inflation of categories, such as weather, or climate, or justice. We must remain alert to how differences morph, connect, and come to matter, differently.

Sinking to the sea floor, we discover not only Spam cans and car tires, chemical traces and carbon takeovers, but also the persistence of the same forces that take breath from some bodies more than others—the *undersea* of human systems of mastery, dominance, and violence. In the benthos, do we not also find the rotting wrecks of the Middle Passage, and the residue of slave bodies forced into the sea? Do we not also find the terror of 268 Syrian refugees, drowned sixty miles south of Lampedusa whose distress signals went unheeded by Italian authorities for hours, or of hundreds of asylum seekers who, in spite of the savagely inhospitable Australian border defense "turn back the boats" policy, nonetheless risked a journey of six thousand miles, only to crash upon the rocks of Christmas Island? Do we not also find the deep connection of Pacific Islanders to their lands slowly sucked up by the rising tides, while continental coal extraction on Indigenous country continues apace, and deep-sea mining waits, breath bated, in the wings?

Understanding bad weather as making a breathless sea also demands reckoning with the ways in which the Anthropocene (or the Sixth Extinction, or the apocalyptic collapse of our planetary systems "as we know them") comes as no great surprise to black and colonized people whose worlds have long been given over to upheaval and ruptured relation (Davis and Todd 2017; Yusoff 2018).

We know these things—human violence, and the breathless sea—are not unrelated, but we must also refuse their facile equivalence. We must refuse analogy ("marine hypoxia is *like* racist death") and instead connect these bad weather systems in their common alibis of force, breathlessness, and power. We must trace their flows into and out of one another, while remaining attentive to scales of mattering whose objective measures of comparison have also been dissolved in the strange material admixtures of the sea.

LEARNING TO BREATHE (2)

In her book-length prose poem, *M Archive: After the End of the World* (2018), Afrofuturist lesbian feminist poet Alexis Pauline Gumbs also brings us into the breathless sea. A "speculative documentary" that evokes the inextricability of environmental devastation from social justice, particularly in terms of gender and race, the poem features a narrator-archaeologist in the future who sifts through the rubble of our contemporary disasters. She is trying to understand what precipitated (and survived) these end times.

Diving below the waves, the poem's narrator discovers many things: traces of surface violences and undersea ecologies heaving under the weight of all we have thrown overboard. But most importantly, the narrator finds possibility: multitudes of life that still persist in the ocean because of and in spite of the bad weather and the breathless sea.

Among these possibilities, the narrator also discovers her own transformed self. Her bodily transformations of possibility include relearning to breathe:

"she considered that maybe something was happening. . . . she had a sense of herself one day soon, being able to breathe through her whole body" (113). The narrator continues:

> *what is the menstrual cycle of a whale?* she found herself wondering as she forced herself to breathe deeper and quiet the rising complaints. she resumed the posture, forehead to the ground to remember the training, to remember the earth was holding her when it felt like she was holding the earth. she cued the sound of ocean waves and cried and cried and cried until she could breathe again. did a whale ever feel like it was swimming in its own tears? she lay there and dreamt of whales, or what she thought were whales, wailing at each other until they could surface again. (Gumbs 2018, 120)

Violence and control, and survivance (Vizenor 2009) by whatever means in the face of them, are not to be romanticized. But especially in the wake of breathlessness, nor is poetry a luxury, as Audre Lorde (1984, 36) insisted. When asked in an interview about her breathing practice, Gumbs asks the interviewer back: "What are the forms of breathing that make survival possible?" (Gumbs, Brown, and Brown n.d.).

Or we might ask: what prostheses might be developed to survive the end of the world? More importantly, how will those of us who have precipitated breathlessness—through domination, complicity, or carelessness—undergo our own transformations, to undo the version of the "human" that demanded these violences and extracted those breaths?

Nor is sea level rise a metaphor. Nor is the melting ice that has given up on its terrestrial existence, nor the very real thermal expansion of the oceans that is evermore inundated with our panting hot breath, to take from the sea her own.

Human and more-than-human. Whale and more-than-whale. The sea as increasingly breathless, but also so much more. Gumbs's text is about finding ways of relating, of crossing over, not to an elsewhere that could pretend these disasters do not matter. She instead finds fugitivity within the incalculable inexhaustibility of black feminist poetics and the meaning of the sea. "How are we related," Gumbs asks, "beyond what we even understand our species to be right now? How are we related beyond what we understand to be our dominant relationship to the planet?" (Gumbs, Brown, and Brown n.d.). Could it be that in the damage of the breathless sea, there might also be a possibility for a new relation?

> here the oxygen in water is not screaming. here the o in h2o is not a sob. here the roundest molecules are breathing quite peacefully, like they chose the job.
> here the water is not waiting to waste you. here the sun is not stripping your skin. this is the dark water of renewal. offering only one message:
> begin. (Gumbs 2018, 123)

REFERENCES

Calvino, Italo. 2014 (1969). *The Complete Cosmicomics*. Translated by M. McLaughlin, T. Parks, and W. Weaver. Boston: Houghton Mifflin Harcourt.

Carrington, Damian. 2018. "Oceans Suffocating as Huge Dead Zones Quadruple since 1950, Scientists Warn." *The Guardian*, January 5, 2018. https://www.theguardian.com/environment/2018/jan/04/oceans-suffocating-dead-zones-oxygen-starved.

Choy, Timothy. 2016. "Distribution." *Theorizing the Contemporary, Cultural Anthropology*. January 21, 2016. https://culanth.org/fieldsights/distribution.

Choy, Timothy, and Jerry Zee. 2015. "Condition—Suspension." *Cultural Anthropology* 30, no. 2: 210–23.

Davis, Heather, and Zoe Todd. 2017. "On the Importance of a Date, Or, Decolonizing the Anthropocene." *ACME: An International Journal for Critical Geographies* 16, no. 4: 761–80.

Deleuze, G., and F. Guattari. 1987. *A Thousand Plateaus*. Translated by B. Massumi. Minneapolis: University of Minnesota Press.

Gumbs, Alexis Pauline. 2018. *M Archive: After the End of the World*. Durham, NC: Duke University Press.

Gumbs, Alexis, Autumn Brown, and Adrienne Marie Brown. n.d. "A Breathing Chorus with Alexis Pauline Gumbs." *How to Survive the End of the World*. https://soundcloud.com/endoftheworldshow/a-breathing-chorus-with-alexis-pauline-gumbs.

Haraway, Donna. 1988. "Situated Knowledges: The Science Question in Feminism and the Privilege of Partial Perspective." *Feminist Studies* 14, no. 3: 575–99.

Irigaray, Luce. 1999. *The Forgetting of Air in Martin Heidegger*. Translated by M. Nader. Austin: University of Texas Press.

Isensee, Kirsten, and Luis Valdes. 2015. "The Ocean Is Losing Its Breath." *GSDR 2015 Brief*. IOC-UNESCO. https://sustainabledevelopment.un.org/content/documents/5849The%20Ocean%20is%20Losing%20its%20Breath.pdf.

International Union for the Conservation of Nature. 2017. *Issues Brief: Ocean Warming*. https://www.iucn.org/sites/dev/files/ocean_warming_issues_brief_final.pdf.

Lorde, Audre. 1984. *Sister Outsider: Essays and Speeches*. Freedom, CA: The Crossing Press.

Merleau-Ponty, Maurice. 2003. *Nature: Course Notes from the College de France*. Translated by R. Vallier. Evanston, IL: Northwestern University Press.

Morsink, Katja. 2017. "With Every Breath You Take, Thank the Ocean." *Ocean: Find Your Blue*. The Smithsonian. https://ocean.si.edu/ocean-life/plankton/every-breath-you-take-thank-ocean.

Sharpe, Christina. 2016. *In the Wake: On Blackness and Being*. Durham, NC: Duke University Press.

Simmons, Krista. 2017. "Settler Atmospherics." Dispatches, *Cultural Anthropology*. November 20. https://culanth.org/fieldsights/1221-settler-atmospherics.

Vizenor, Gerald. 2009. *Native Liberty: Natural Reason and Cultural Survivance*. Lincoln: University of Nebraska Press.

Yusoff, Kathryn. 2018. A *Billion Black Anthropocenes or None*. Minnesota: University of Minneapolis Press.

Chapter Fifteen

We Drain East to the Pacific

Or, a Sydney-Centric Theoretical Description of Anthropocene Stormwater Drainage

Jennifer Mae Hamilton

The compounds *rainwater*, *freshwater*, *groundwater*, and *stormwater* are linguistic effects of a material situation. The words categorize water because freshwater is essential; stormwater is water that becomes polluted and has to be managed and both start out as rainwater or groundwater. Rainwater becomes either freshwater or stormwater through a relationship with land. When rain falls on impervious city streets it transmogrifies materially *and* semiotically into stormwater. Rain that has not yet fallen on land can still be consumed, but water does not remain potable once it has hit the ground (unless it receives treatment). The specific geographical setting of a place determines where the stormwater goes. In a coastal city, much of that stormwater ends up in the sea. In fact, most of the world's stormwater will drain toward the sea, if it does not soak into the ground or evaporate en route. Sydney Water estimates that on average, a Sydney Harbour load of stormwater—about five hundred billion liters—drains off the city each year into the sea. Before colonization, water on these courses would have made it into a tributary without undergoing this particular change, or could have been harvested for drinking along the way. Swimming and fishing could have taken place after a storm along the coast. But now, because of stormwater run-off from land, swimming after storms is a health risk to humans. The point of all this is that, when thinking about how to sustain the seas, we need to interrogate what happens on land.

In relation both to water and drainage, the interrogation of what happens on land on behalf of the sea is not just a question of more effective stormwater filtration infrastructure, but also how that infrastructure supports certain individual, social, historical, and political formations better than others. Building on Astrida Neimanis's description of existence as both occurring in common and in difference on account of a watery "posthuman gestationality"—noting

211

Figure 15.1. Image by Cucombre Libre from Flickr Creative Commons https://cre-ativecommons.org/licenses/by/2.0/. "Sydney 2592" Image https://creativecommons.org/licenses/by/2.0/.

that *"we are all"* both *"bodies of water"* *and* expressions of "watery differ-ence" (2017, 66, emphasis in original)—what kind of worlds, subjectivities, imaginaries, relationalities, aspirations, domesticities, bodies do urban storm-water drains best support? The work of this chapter is not to answer directly, but to know that the discussions herein are buoyed by a transformative inclusive feminist ethos: that for the water to be cleaner—for the seas to be sustained—our social formations have to be made otherwise. Stormwater is a byproduct of the current status quo; to change the quality of the stormwater requires changing the status quo.

Rainwater transubstantiates into stormwater when encountering a surface be-cause the water's quality literally changes too. And stormwater is gross. It is a mixture of water and oil, petrol, fertilizer, pesticide or herbicide, chalk, food, food waste, plastic bottles, soil, sand, gravel, anything, anything at all—old shoes, tissues, unopened mail—that can be washed away. Although modern industrial cities are iconically toxic bodies (think of nineteenth-century Lon-don, for example), it is not until 1996 and an article by Chester L. Arnold Jr. and C. James Gibbons that the mainstream planning and policy discourse on impervious surfaces begins to shift. "Impervious surfaces," they write, "can be defined as any material that prevents infiltration of water into the soil. While roads and rooftops are the most prevalent and easily identifiable types of impervious surface, other types include sidewalks, patios, bedrock outcrops, and compacted soil" (244). Importantly, however, "impervious surfaces are not only indicators of urbanization, but are also major contributors to the [implicitly negative] environmental impacts of urbanization" (244). Although there are small-scale projects seeking to reverse-engineer the imperviousness, like rain gardens, most city surfaces around the globe are becoming increasingly less porous. For a striking and apocalyptic example, between 1984 and 2016, 16 percent of the world's spongy, aquatic tidal flats were lost to *both* development *and* sea level rise (Murray et al. 2019). Although we now know about the downstream effects of stormwater, we are doing little to curb the kind of development that literally produces and widely distributes this kind of water.

From space, contemporary Sydney is a brown blotch on the lower third of the otherwise green belt that is the eastern seaboard of Australia. It is one of the largest cities on the island continent. When living in or visiting this place of multiple landmarks—a harbor, a bridge, an opera house, and several iconic beaches—the arrangement of life can feel like an inevitable effect of progress, if not (perversely) the best of all possible worlds. Sydney, with its cafes, malls, roads, and fancy houses, models the high standard of living that sets international development targets and orients individual aspirations. Aspects of the city have been there for a long time: the estuary known as *Sydney Harbour*, for example, formed at the end of the last ice age, as did the beaches (Flannery 2015). The same estuary supported the life of Aboriginal people for millennia, too. The city itself—buildings, bitumen roads, nice cafes, expensive houses—is a colonial artifact, edifying at once the violent dispossession of Aboriginal people and social structures and the great acceleration of global capitalist industrialization. Although Sydney is both locally much more and cosmologically much less than this short description allows, one thing is key: given all the water (the rivers, creeks, lakes, estuaries, bays, and harbors), the network of stormwater drains is a cornerstone of the city's contemporary organization. Despite the occasional nuisance flood, as one the drains are hydrologically functional. That is, they are functional insofar as they enable both the ongoing dispossession of First Nations people and the acceleration of capital. What will it take to both imagine them otherwise and literally change their functionality? As Astrida Neimanis reminds us, "no imaginary can be installed simply with a triumphant flourish" (2017, 185). This is definitely true for an imaginary against concrete drains. Under the pressure of a triumphant flourish, the concrete, steel, plastic, brick, and mortar conglomerate of drainage systems seems to solidify even further.

Figure 15.2.   Harbourside Drain in Milson Park Kirribilli.
Photo by Jennifer Mae Hamilton.

Drains facilitate a particular kind of modern life in any place marked by modern development; what makes a place particular is its hydrology. Sydney has some local particularities in this regard. Sydney has multiple, large bodies of water; is coastal and tidal; and is wild and undulating, bearing the scars of a long-term relationship with all that liquid. The city is hydrologically and topographically complex. Combined with decent average annual rainfall patterns and strong tides from the world's largest ocean to the east, Sydney is an aquatic city. The aquatic life of the city, however, is tightly controlled for the benefit of the city. This management (water provision and drainage) enables commodification (harbor or ocean views are expensive) and work (people can get to and from work quickly and efficiently on roads that are designed to move water quickly).

Although many of the smaller waterways are now underground, life in relation to these streams was once very different. Take as prime and iconic example the Tank Stream, a small, fresh watercourse that runs from Hyde Park under the contemporary central business district and deposits its waters into the harbor at Circular Quay. Used by the Cadigal for a long time before colonization for drinking, cooking, and bathing, the Tank Stream is also the key "reason" for colonial Sydney's geolocation in Port Jackson: it was the first freshwater course stolen from First Nations and commandeered by the British. Less than fifty years after settlement—by the 1820s (I still cannot get over how *quickly* we completely trashed it)—the Tank Stream was effectively an open sewer and capped to avoid the spread of disease (Sydney Water, n.d.). So, as well as being the first drinking water stream, the Tank Stream is also the colony's first stormwater drain. Colonialists had to look further afield for potable water and today the city's freshwater is piped in from reservoirs that sit in the mountains to the west of the city, while waters native to the city's lands are managed away by stormwater drains and flushed out to sea.

Stormwater drains are the infrastructural equivalent of the pharmakon: the stormwater drain is neither cure nor poison. These drains both deliver health—sanitation and shelter during wild weather—but at the same time have both ecological and social consequences. In the context of thinking stormwater in cities, the concepts *drains* and *drainage* likely seem to most people benign or even essential aspects of urban nature, divorced from history and politics. In particular, the stormwater drain is remote from the more extractive and morally loaded valence of the verb form "to drain": to diminish, to kill. But they are neither separate from these negative valences, nor are they outside of history or politically neutral. The historical observation is not new: many drainage systems in Sydney have heritage status. But how to think about the political world built by drains in relation to an environmentalist project is a bigger project.

These systems have made themselves indispensable in cities, as embedded in the high-consumption, high-waste, high-carbon functioning of everyday life in late liberalism as a supermarket or motorway. Drains deliver sanitation at the same time as being agents of violent colonization and dispossession, extractive capital and climate change. The Tank Stream was only the beginning of what is now a massive network. These systems, although often ugly (think: the concretized drains of urban creek lands and estuaries) seem essential, but they are only essential in relation to a particular trajectory of development, a particular kind of economy and a particular system of land tenure. They are only essential in relation to the practices that radically pollute water; they are only essential because the Tank Stream became a sewer and drain because of colonial activities.

In Sydney, these drains have made life and work in cities possible whatever the weather, they facilitate sanitation and have opened up once-swampy or even tidal zones for real estate development. In Sydney, the most visible index of how development and drainage are in lock-step is the danger posed to human health caused by swimming in the city's beaches after a rainstorm. Beaches close because they are toxic, but pollution does not begin in the sea.

Figure 15.3. Sydenham Stormwater Drainage Pit and Pumping Station.
Photo by Jennifer Mae Hamilton.

What is taken from the seas to serve land-based animals is one concern, and what drains into the sea is another: when thinking about how to sustain the seas, we need to give equal (if not greater) import to what happens on land. If not-draining into the sea is related to sustaining the seas, then the questions stemming from stormwater drains in this regard engage the political situation on land on behalf of the sea: *do stormwater drains mirror the dominant political ideology? Can stormwater drains that developed under liberal colonial capitalism form the foundation for a radically different social order? If not, what kind of water infrastructure is required to undergird a reimagined social order? How to reengineer the current system to move toward the latter, for the good of the sea?* Drainage is a useful concept for thinking with water in the Anthropocene, precisely because it describes formations of life that urgently need to change. In fact, I'd wager that if the structure of life, work, economy, and politics in Sydney could change *enough* so the harbor and ocean beaches were swimmable after a rainstorm, then *drainage* would no longer be a useful term to describe the dominant relationship with water. In fact, I'd wager that if that were the case, the classification of certain kinds of water as stormwater would be meaningless too.

The generation of knowledge in the broad field of the environmental humanities is, in part, thus motivated by an assumption that we need to better know the structure of our relations with the more-than-human aspects of the world to change them. Some scholars are beginning to integrate an age-old critical question into the field's foundational assertion: who is this "our"? Who is the "we"? Who needs to change what, and how? As the "we" of the field is conceptually qualified and problematized in theory (Hamilton and Gunaratnam 2018; Hamilton and Neimanis 2018), a particular and anthropocentric conception of a collective "we" remains solidly materialized in the infrastructure that governs daily life. This vision of a "we" is held together by concrete and steel, ready for fossil-fueled lives to roll over its vast impervious surfaces and enabling the continuation of life whatever the weather. The stormwater drain contains a vision of a collective we; a we that, at the same time, needs to be deconstructed, thought otherwise, and remade. What will it take to enable us to swim in the city's estuaries and sea beaches after a storm? Who is the "us" doing the swimming?

Stormwater is the drainage of *weather*. Recent theorizations of the weather by Christina Sharpe (2016) and Quill Chrissie-Peters (2018) posit the weather as both metaphor and metonymy for the whole earth system of racialized oppression in which we live. What Sharpe calls "the weather" is an antiblackness in American society that "is pervasive *as* climate" (Chrissie-Peters 2018, 106, emphasis in original). It is a total atmosphere. Similarly, Anishinaabe artist Chrissie-Peters asserts that "much like the weather, settler colonialism is immersive, generating an environment that I must necessarily interact with" (2018, n.p.). The weather both is and is like the unjust system within which they, within which we, live; on one hand, the weather is the invisible system of power that produces different, violent, unjust earthly experiences— following Neimanis (2017), these are differentiated in a range of particular ways, as much as they are in common in the watery world. On the other, however, the curious thing about thinking with weather is that for both Sharpe and Chrissie-Peters, weather presents the potential for the development of an otherwise. For Sharpe, the weather also "necessitates changeability and improvisation; it is the atmospheric condition of time and place . . . those in the wake also produce out of the weather their own ecologies" (2016, 106). For Chrissie-Peters, despite the immersiveness of the weather, Anishinaabe have also "created shelters that shield our loved ones from harsh rains, have learnt to live and love in bodies harmed and hated, have softly tended to new worlds in the stillness of the night" (2018, n.p.). Which brings me back to swimming in Sydney Harbour after the storm, right up near where the Tank Stream deposits itself into present-day Circular Quay. If a sustainable drainage system were to be reengineered and the city remade in this wake, what brilliant group of people would come together to bathe in the moonlight?

REFERENCES

Arnold, Chester L., and C. James Gibbons. 1996. "Impervious Surface Coverage: The Emergence of a Key Environmental Indicator." *Journal of the American Planning Association* 62, no. 2: 243–58. https://doi.org/10.1080/01944369608975688.

Christie-Peters, Quill. 2018. "Kwe Becomes the Moon, Touches Herself so She Can Feel Full Again." *GUTS*, March 26, 2018. http://gutsmagazine.ca/kwe-becomes-the-moon/.

Flannery, Tim. 2015. *The Birth of Sydney*. New York: Grove.

Hamilton, Carrie, and Yasmin Gunaratnam. 2018. "Environment." *Feminist Review* 118, no. 1: 1–6. https://doi.org/10.1057/s41305-018-0096-9.

Hamilton, Jennifer Mae, and Astrida Neimanis. 2018. "Composting Feminisms and Environmental Humanities." *Environmental Humanities* 10, no. 2: 501–27. https://doi.org/10.1215/22011919-7156859.

Karskens, Grace. 2010. *The Colony: A History of Early Sydney*. Crows Nest, NSW: Allen and Unwin.

Murray, Nicholas J., Stuart R. Phinn, Michael DeWitt, Renata Ferrari, Renee Johnston, Mitchell B. Lyons, Nicholas Clinton, David Thau, and Richard A. Fuller. 2019. "The Global Distribution and Trajectory of Tidal Flats." *Nature* 565 (7738): 222. https://doi.org/10.1038/s41586-018-0805-8.

Neimanis, Astrida. 2017. *Bodies of Water: Posthuman Feminist Phenomenology*. London: Bloomsbury Academic.

Sharpe, Christina. 2016. *In the Wake: On Blackness and Being*. Durham, NC: Duke University Press.

Sydney Water. n.d. "Tank Stream Stormwater Channel No. 29E." Sydney Water Corporation. Accessed April 7, 2019. https://www.sydneywater.com.au/SW/water-the-environment/what-we-re-doing/Heritage-search/heritage-detail/index.htm?heritageid=4573709&FromPage=searchresults.

Chapter Sixteen

Toxic Bloom

Remaking William Hunter's Obstetric Illustration to Represent the Epigenetic Toxification of Bodies

Clare Nicholson

Installed on a reclaimed industrial chute from an old Sydney warehouse, and confined within a concrete "sarcophagus" womb, an orange ballistics gel full-term fetus recoils in a complicated breech position. The baby's body has congenital deformities, and its skin is pitted with holes. A hidden light-emitting diode illuminates the concrete uterine vault via the cervix causing the infant to glow. *Toxic Bloom* (Figure 16.1) connects gestational catastrophe from the ecocide Agent Orange with the affluent Sydney peninsula of Rhodes in Homebush Bay.

Predominately associated with the Vietnam War, Agent Orange was spawned from amoral political military-technoscience. The ecocide chemicals were manufactured by Union Carbide at Rhodes "leaving an ongoing dioxin legacy across Sydney Harbour" (McGrath 2012, paras. 1–33). *Toxic Bloom* addresses this insidious industrial past, but also suggests epigenetic ecocidal entanglements may be invisibly active not only in disparate lands and distant shores, but also much closer to home.

Epigenetics is a comparatively new line of molecular enquiry creating a radical scientific shift. It is now understood that environmental influences have the ability to impress upon the body, altering gene expression without altering the sequencing of DNA (Landecker and Panofsky 2013, 333–57). Body-environment relations are revealed as interdependent and permeable, rather than discrete entities, with acquired genetic alteration considered heritable across several generations (Heijmans et al. 2009, 526–31). Consequently, epigeneticists are particularly interested in maternal-fetal programming—maternal effects causing transgenerational fetal adversity leading to childhood- and adult-onset chronic disease caused by harmful exterior influences (Lock 2018, 449–74). As a result, there is an increasing focus on

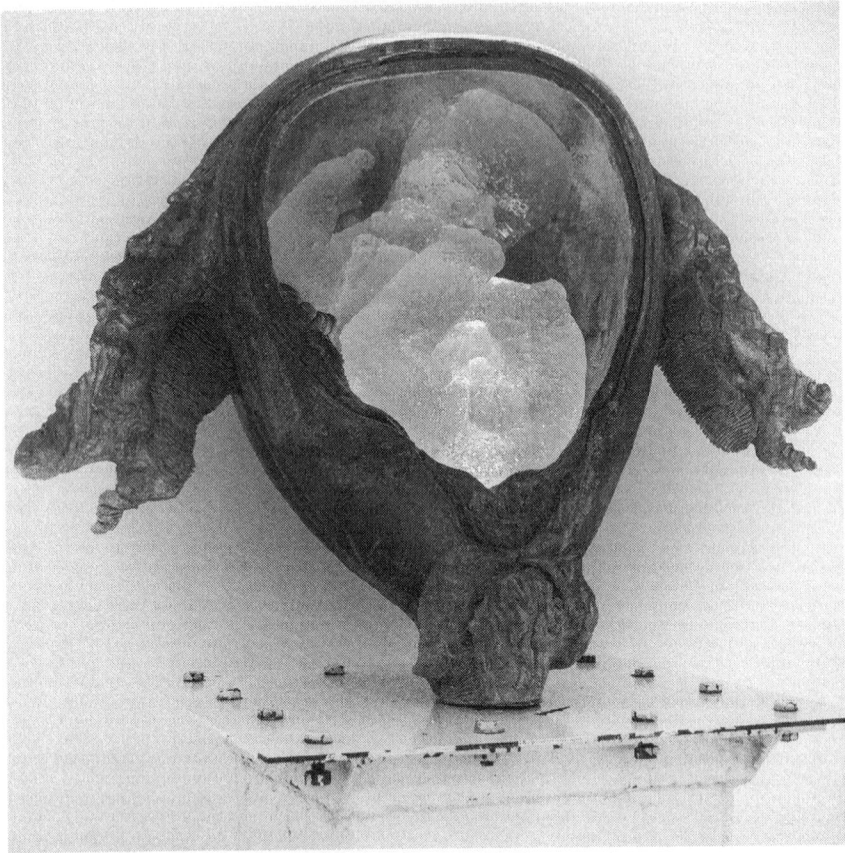

Figure 16.1. Toxic Bloom, (detail), 2017. Hydrostone cement, ballistic gel, pigment, steel, paint, LED. H171 × W57 × D52cm.
Image by Clare Nicholson.

maternal behaviors, with an onus of responsibility placed on mothers to manage and control etiologically detrimental environmental exposures (Yehuda and Lehrner 2018, 243–57; Meaney 2001, 1161–92). Epigenetic illustrations within clinical texts and across popular media repeatedly position gestating mothers in conflict with fetal health because of the mother's poor maternal choices. These images are designed to be hard-hitting, evoking a range of unspoken associations regarding mothers and maternal care.

Blaming mothers for offspring adversity has occurred since antiquity (Park and Daston 1981, 20–54) including within the domain of medical science (Caplan 2013, 99–103). Although epigenetic investigation promises improved health, simplistic misinterpretations of epigenetic discourses may inadvertently repeat unsettling sociohistorical mother-blaming tropes. *Toxic Bloom* shifts the focus toward the toxification of female bodies outside of maternal sovereignty, to address dominating sociopolitical structures and power networks as causal triggers instead.

Traditional pedagogical medical illustrations, models, and phantom prosthesis as substitutes for living bodies are supposed to impart an objective, empirical, universal truth about the standardized and fixed body, but are of course powerful cultural mediators toward how we see and understand bodies (Haraway 1988, 585–87). Taking a materialist approach to remake historical models with contemporary materials and entangled with epigenetic findings suggests the "mattering" effect (Barad 2003, 801–31) of exterior worlds materializing as interior worlds (Landecker 2019).

Toxic Bloom conveys "chemical ecologies, where species, bodies, and environments shape one another through absorption, ingestion, respiration, and sensation" (Kenney 2019, 9). In particular, this obstetric model represents the agential fusion of manufactured chemical militarism with gestational corporeality to convey biopolitical toxic burdens that are entirely outside of the management, responsibility, or decision-making of the individual mother.

Once a large industrial site, the densely populated Rhodes peninsula is situated twelve kilometers northwest of Sydney's central business district. Emmett and colleagues (2004) explain that highly toxic byproducts, including dioxin, were "dumped into the wetlands and mangroves, discharged into Homebush Bay and used [as landfill] to reclaim land along the Rhodes foreshore, leaving a legacy of significant dioxin contamination" (1). Dioxin is considered "the most toxic substance discovered by mankind to date" (Tuyet and Johansson 2001, 156) and is bioaccumulative, prevailing through water, sediment, vegetation, food, fatty tissue, and breast milk (Micheletti, Critto, and Marcomini 2007, 45–55; Rifkin and LaKind 1991, 103–12). Furthermore, gestational maternal-to-fetal dioxin transfer persists through amniotic fluid, placenta, maternal blood, cord blood, and the waxy protective layer covering neonates, *Vernix caseosa* (Morokuma et al. 2017, 1–4). Dioxin induces estrogen-like gene expression, disrupting physiological reproductive health and embryonic development, making exposure particularly lethal for girls and women (Silliman et al. 2016, 139).

Dangerously high dioxin levels in seafood initiated the 2006 ongoing ban of commercial fishing within Sydney Harbour four years after the government suspected dioxin levels were exceeding international safety standards (Needham 2006, para. 1). Remediating the ocean is impossible, so the ocean and river systems around Rhodes rely on clean sedimentary layer formation to cover up the toxic sediment (Davies 2010, para. 2).

Remediating Rhodes and the Homebush Bay area was the largest dioxin remediation project undertaken in the world using thermal treatment technology, and included burying waste on site (Troxler, Hunt, Taylor, and McNiven 2010, 613–14). Rhodes was marketed as an exclusive waterside suburb before the remediation project was complete (Creagh 2007, para. 1) and progressively blanketed in concrete (Figure 16.2). Along with the built urban environment, ocean containment walls, and burial vaults, this anthropogenic rock is frequently used to encase and isolate hazardous materials, including radioactive waste. Concrete also appears to be highly effective toward promoting public amnesia.

Conceptually, concrete felt like an obvious material to remake the womb from William Hunter's illustration *Plate XXVII*, in *The Anatomy of the Human Gravid Uterus exhibited in figures* (1774). I am suggesting the political concealment and erasure of our homeland's morally bankrupt involvement in chemical militarism and the blanketing of Rhodes industrial contamination. By carving a botanical motif onto the exterior wall of the uterus (Figure

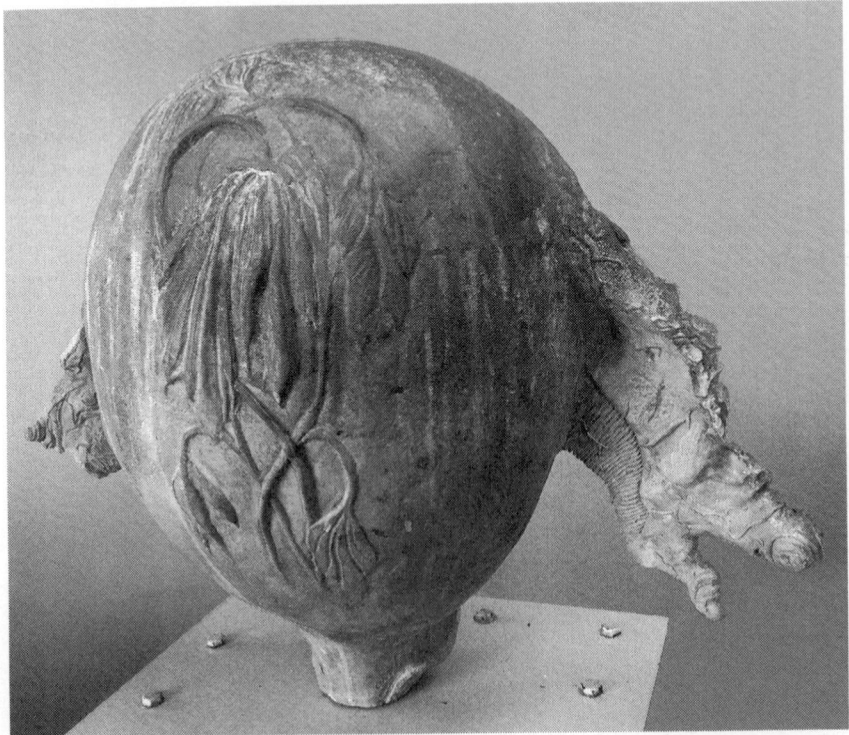

Figure 16.2.   Toxic Bloom, (detail), 2017. Hydrostone cement, ballistic gel, pigment, steel, paint, LED. H171 x W57 x D52cm.
Image by Clare Nicholson.

Figure 16.3.   Rhodes. Photograph.
Image by Mark Merton, 2012. With kind permission of photographer.

16.3), I am suggesting ornamental architectural stone carvings that are long preserved as artifacts of tangible cultural heritage. However, carving a wilted Vietnamese spinach, a staple food in Vietnam, as the motif links the devastating ecocidal maternal effects that remain out of sight and mostly forgotten in other geographic locations today. Using an industrial chute reclaimed from a derelict Sydney warehouse as a medical display stand suggests Sydney as a site historically implicated in biopolitical profiteering from a developing country. Despite the insipid magnolia paint, the chute represents the unscrupulous maneuvers behind commercialization of chemical mass production.

Ballistics gel is a synthetic flesh designed to test military projectile ammunition, but is increasingly used for phantom bodies in medicine to teach sonographic, X-ray, and surgical procedures. Antenatal ultrasound is a routine clinical intervention to examine fetal development and to reveal birth anomalies, but as Barad (2007) explains, "the sonogram does not simply map the terrain of the body; it maps geopolitical, economic, and historical factors, as well" (194). To represent the agency of Agent Orange, I stained the ballistics gel orange to cast the fetus. Agent Orange was color-coded because it was dispatched in drums identified with an orange band (Tuyet and Johansson 2001, 156) but it occurred to me this color also signifies the defoliant's ability to reduce vibrant green autotrophic vegetation into hues of orange before complete necrosis. By representing an orange ballistics gel fetus in a concrete womb devoid of amniotic fluid, I am implying the hazardous agency between dioxin-drenched oceans and gestational life; the toxic water-marking fails to nourish, hydrate, or sustain fecundity. Merging these material elements in *Toxic Bloom* signifies the convergence of cultural, historical, and biological agencies affecting the developing fetus: the toxic weaponization of disparate places imprinting upon maternal bodies, that might cause the sort of gestational catastrophe ultrasounds were developed to reveal and diagnose.

Epigenetic enquiry brings a new understanding toward the subtle poisoning of ecosystems as molecular activity plays out through bodies in minute doses, unpredictably unfolding corporeal, spatial, and temporal boundary violation. Guthman and Mansfield (2012) explain previous chemical concerns vacillated between "acute toxicity and carcinogenesis [but] at the forefront now are subtle reproductive, neurological, and morphological effects" (486). Epigenetic ecotoxicology delivers new concerns, making government statements regarding the complete remediation of landscapes such as Rhodes highly questionable. We are assured treated soil buried at Rhodes has been rendered inert, but what does *inertia* mean when the very nature of nature has been found anything but inert? And what of water tables—rising or not? Will dioxin leach from its subterranean tomb to combine with the uncontrollable, nomadic, saline-bound dioxin which already laps and undercuts the artificial concrete Rhodes foreshore?

The temporality of epigenetic environmental contamination is unpredictable, because, "[m]atter doesn't move in time, matter doesn't evolve in time, matter materialises and enfolds different temporalities" (Barad 2014). So far, the Rhodes community appears to have had conventional time on its side. But will impending generations suffer the materialized mattering caused by accumulative, epigenetic dioxins?

ACKNOWLEDGMENTS

This research has been undertaken with the generous support of the Australian Government Research Training Program Scholarship.

REFERENCES

Barad, Karen. 2003. "Posthumanist Performativity: Toward an Understanding of How Matter Comes to Matter." *Signs,* 28, no. 4: 801–31. https://doi.org/10.1086/345321.
———. 2007. *Meeting the Universe Halfway: Quantum Physics and Entanglement of Matter and Meaning.* Durham, NC: Duke University Press.

———. 2014. "Re-membering the Future, Re(con)figuring the Past—Temporality, Materiality, and Justice-to-come." Annual Feminist Theory Workshop filmed at Duke University. https://www.artandeducation.net/classroom/video/66314/karen-barad-re-membering-the-future-re-con-figuring-the-past-temporality-materiality-and-justice-to-come.

Caplan, Paula. 2013. "Don't Blame Mother: Then and Now." In *Gender and Women's Studies in Canada: Critical Terrain*, edited by Margaret Helen Hobbs and Carla Rice, 99–103. Toronto: Women's Press.

Creagh, Sunanda. 2007. "Approved: Waterfront Homes Next to Toxic Site." *Sydney Morning Herald*, December 28, 2007. http://www.smh.com.au/news/national/approved-waterfront-homes-next-to-toxic-site/2007/12/27/1198345162026.html.

Davies, Anne. 2010. "The Poison That Got Away." *Sydney Morning Herald*, October 30, 2010. http://www.smh.com.au/environment/water-issues/the-poison-that-got-away-20101029-177i0.html.

Emmett, Katy, Trish Mannes, Alan Willmore, Jill Kaldor, Vicky Sheppeard, and Tim Churches. 2004. "Rhodes Peninsula Small Area Cancer Incidence and Mortality Study." *New South Wales Government Health Report*, 1–21. https://www.health.nsw.gov.au/environment/Publications/rhodescancer.pdf.

Guthman, Julie, and Becky Mansfield. 2012. "The Implications of Environmental Epigenetics: A New Direction for Geographic Inquiry on Health, Space, and Nature-Society Relations." *Progress in Human Geography* 37, no. 4: 486–504.

Haraway, Donna. 1988. "Situated Knowledges: The Science Question in Feminism and the Privilege of Partial Perspective." *Feminist Studies* 14, no. 3: 575–99.

Heijmans, Bastiaan, Elmar W. Tobi, L. H. Lumey, and P. Eline Slagboom. 2009. "The Epigenome: Archive of the Prenatal Environment." *Epigenetics* 4, no. 8: 526–531. doi: 10.4161/epi.4.8.10265.

Hunter, William. 1774. *Anatomia Uteri Humani Gravidi Tabulis Illustrata.* [Etched plate]. https://wellcomecollection.org/works/ykdr24rz.

Kenney, Martha. 2019. "Fables of Response-ability: Feminist Science Studies as Didactic Literature." *Catalyst: Feminism, Theory, Technoscience* 5, no. 1: 1–39. https://doi.org/10.28968/cftt.v5i1.29582.

Landecker, H. 2019, January 25. *Outside In: Microbiomes, Epigenomes, Visceral Sensing, and Metabolic Ethics.* [Video file]. https://www.youtube.com/watch?v=ZioN-wabNfg.

Landecker, Hannah, and Aaron Panofsky. 2013. "From Social Structure to Gene Regulation, and Back: A Critical Introduction to Environmental Epigenetics for Sociology." *Annual Review of Sociology* 39, no. 1: 333–57. https://doi.org/10.1146/annurev-soc-071312-145707.

Lock, Margaret. 2018. Mutable environments and permeable human bodies. *Journal of the Royal Anthropological Institute*, 24, no. 3: 449–74. https://doi.org/10.1111/1467-9655.12855.

McGrath, Chris. 2012. "Sydney Harbour's Toxic Legacy Shows Value of Green Safety Net." *The Conversation.* December 17, 2012. https://theconversation.com/sydney-harbours-toxic-legacy-shows-value-of-green-safety-net-11197.

Meaney, Michael, J. 2001. "Maternal Care, Gene Expression, and the Transmission of Individual Differences in Stress Reactivity across Generations." *Annual Review of Neuroscience* 241, no. 1: 1161–92.

Micheletti, Christian, Andrea Critto, and Antonio Marcomini. 2007. "Assessment of Ecological Risk from Bioaccumulation of PCDD/Fs and Dioxin-like PCBs in a Coastal Lagoon." *Environment International* 33, no. 1: 45–55. https://doi.org/10.1016/j.envint.2006.06.023.

Morokuma, Seiichi, Kiyomi Tsukimori, Tsuguhide Hori, Kiyoko Kato, and Masutaka Furue. 2017. "The Vernix Caseosa Is the Main Site of Dioxin Excretion in the Human Foetus." *Scientific Reports* 7: 1–4. https://doi.org/10.1038/s41598-017-00863-9.

Needham, Kirsty. 2006. "Toxic Fish Advice Kept Secret." *Sydney Morning Herald.* February 2, 2006. https://www.smh.com.au/national/toxic-fish-advice-kept-secret-20060202-gdmw3u.html.

Park, Katherine, and Lorraine J. Daston. 1981. "Unnatural Conceptions: The Study of Monsters in Sixteenth- and Seventeenth-century France and England." *The Past and Present Society* 92 (August): 20–54. http://www.jstor.org/stable/650748.

Rifkin, Erik, and Judy LaKind. 1991. "Dioxin Bioaccumulation: Key to Sound Risk Assessment Methodology." *Journal of Toxicology Environmental Health* 33, no. 1: 103–12. https://doi.org/10.1080/15287399109531509.

Silliman, J., M. G. Gerber Fried, L. Ross, and E. Gutierrez. 2016. *Undivided Rights: Women of Color Organizing for Reproductive Justice.* Chicago: Haymarket Books.

Troxler, William, John W. Hunt, Justin Taylor, and Campbell McNiven. 2010. "Thermal desorption treatment of dioxin-contaminated soil at the former allied feeds site, Sydney, Australia." *Environmental Engineering Science,* 27, no. 7: 613–22. https://doi.org/10.1089/ees.2009.0423.

Tuyet, Le Thi Nham, and Annika Johansson. 2001. "Impact of Chemical Warfare with Agent Orange on Women's Reproductive Lives in Vietnam: A Pilot Study." *Reproductive Health Matters* 9, no. 18: 156–64. doi: 10.1016/S0968-8080(01)90102-8.

Yehuda, R., and Lehrner, A. 2018. Intergenerational Transmission of Trauma Effects: Putative Role of Epigenetic Mechanisms. *World Psychiatry* 17, no. 3: 243–57. https://doi.org/10.1002/wps.20568.

Chapter Seventeen

Looking for Skin, Finding Kin

Kassandra Bossell

BREATHING

Imagine if we couldn't take every second breath. Half the oxygen in the at-mosphere is breathed out by microscopic marine plants called *phytoplankton*, which are adapting and changing because of the acidification and warming of the oceans (Hoegh-Guldberg and Bruno 2010, 1523–28). They are primary producers, forming the base of all aquatic food webs and thereby supporting all ocean, lake, and river life. They obtain energy through photosynthesis, fixing carbon dioxide through the growth of corporeal structures. When they shed "skin" or die, they generate a continuous rain of calcium carbonate to the deep ocean, thereby sequestering carbon into what is known as a "carbon sink."

Most are smaller than the eye can see, so phytoplankton are easy for humans to disregard; yet the oxygen they make is essential to our ongoing survival. Our interconnected network of life-forms relies on these and many other nonhuman species to survive. One could say we are in a species assem-blage with these microscopic marine plants. I created the series *S'Kin* (Figure 17.1) as a reflection on our interdependence with nature. It offers a critique on recent human-driven trophic cascades that have led to the widespread extinc-tion of species in what may be leading to the sixth mass extinction (Barnosky et al. 2011, 51).

As skin is an organ shared by most living beings, I used its material pres-ence to introduce empathetic ties and bonds between species such as those encountered in kinship relationships within a species. I sculpturally enlarged the fine corporeal structures of phytoplankton in clay to huggable human scale using images from scanning electron microscopy (SEM). Using this shedding process as a reference, I used a molding process to produce rubber membranes. These textural sheaths appear like discarded animal casings, yet their scale, color, and flesh carry an uncanny resemblance to human skin. Essentially cast in tree sap, they suggest a delicate and transient material relation between tree blood and animal blood. This visual bridge offers a vibrant material connection between the human and nonhuman; like human skin, these membranes degrade over time. In this process of ongoing decomposition, the phytoplankton membranes betray an ephemeral life cycle of transformation connected to our own.

I cast numerous skins from the phytoplankton forms and exhibited them as specimens in the style of a natural history museum. Yet these skins are flaccid, merely suggesting evolutionary form, now bereft of agency. Peeled or flayed? Sheared or scaled? The viewer is asked if these are remnant skins from natural remains or hides prepared for market? These tangents return the viewer to a capitalist view of nature as resource or marketable commodity.

This idea of nature as resource reflects the assumptions made by a traditional Western humanistic framework that sees nature and culture to be held in binary opposition. The traditional European idea of the human, symbolized by Leonardo da Vinci's *Vitruvian Man*, the "measure of all things," assumes the individualistic unity of the subject, an "ideal of bodily perfection, which doubles as a set of mental, discursive and spiritual values. Together, they hold up a view of what is 'human' about humanity" (Ashrafian 2011, 593). This subjectivity sets a standard that is not merely individualistic but also cultural.

Figure 17.1. *S'Kin*, 2018, latex rubber, dimensions variable, AD Space, UNSW.
Image by Kassandra Bossell.

It is equated with the tenets of self-reflexive reason: consciousness, rationality, and self-regulating ethical behavior.

It is the delinking of the human from these fundamental premises of the Enlightenment that may change the modus operandi in the relationship between human and nature. Cultural theorist Rosi Braidotti, among others, has inspired me in her rejection of these conventions by proposing a "self-organizing or auto-poetic force of living matter." The unit of reference for the human is transformed by what Braidotti conceives as the posthuman; an opening up of the cultural standard for what is human offers new rights to those freshly included. She replaces the apparent universal human supposedly in charge of history with "a more complex and relational subject, framed by embodiment, sexuality, affectivity, empathy, and desire as core qualities" (Braidotti 2006, 26).

Posthumanisms may have opened the field of human rights to more of humanity, but what about the rights of all other biota, the more-than-human, or the landscape that nurtures them? From an evolutionary perspective, the human status as apex predator has allowed us to adopt a directorial role, ignoring the essential interdependence between all biota. It is only by forgetting our interdependent status that we can see nature as solely a material resource. By creating an artwork that frees the viewer from the anthropocentric framework, I aim to offer a memory of coevolution and interdependence with other species.

In the face of contemporary Western dystopian fantasies that focus on colonial survival, *S'Kin* takes on a view of evolutionary flow, including and beyond the human. Through art, a viewer may use imaginative leaps to enter another being's world. As *S'Kin* suggests a visual and metaphoric link between the human and nonhuman, the ensuing idea of multispecies plurality associates the work with Braidotti's regrounding of the subject in a posthuman age. Philosopher Donna Haraway (2016) goes beyond this human focus to call for the dismantling of any supposed interspecies chain of command and to view *us all* as a system of interrelated kin. In her conception, we are responsible to "oddkin," multispecies kin, including humans and other-than-humans. She rejects individualism or *autopoiesis* as a failed endeavor, proposing instead *sympoiesis*, meaning collective creation or organization. She uses science, anthropology, and creative storytelling to imagine the "Chthulucene," which describes the past, present, and future of multispecies creative collaborations. Although this sounds just like evolution, her point is the frame of reference and the conscious use of it.

Natural history museums traditionally represent organisms as activated individuals and often leave out the essential interdependent relationships on which they depend to survive. By depicting living beings as separate from their web of life, they seem to be isolated, independent heroes, deprived of the coagency they enact within a coevolutionary context. My exhibition at AD Space, an art gallery in Sydney, mimicked such displays by showing them evenly spaced in a line, side by side and well lit. Without even the fanfare of an antihero, these depleted transitory specimens mirrored a fleshy, human degradation. Along with ephemerality, the shed skins invoke regeneration or extinction. In materiality and scale, these sheaths operate as ghosts of sensible beings, connecting them with humans in a visual and textural bridge of kinship. This quality of kinship is aimed to induce an identification in the viewer, associating the demise of the phytoplankton with the potential demise of humans.

The slightest visual or tactile affinity with *S'Kin* may call into question the viewers' subjectivity with a complex and open-ended set of relations. Now open to the nuances of biotic networks, human identity is rewilded as it perceives itself within a field of agency. The concept of shared agency generates multispecies plurality within human identity. It is in this place of interconnection with other biotic agents that I find a defiance at my own anthropocentric self-importance. In this place I can challenge my tendency to conceive myself within an individual biological frame of reference that would block any imagination of coagency.

CHOKING

Considerations of *us* as a species assemblage led to my wall work *Choker* (Figure 17.2). I undertook experiments of enmeshing life casts from my own body within impressions taken from found sea shells, corals, nuts, and other unnameable organisms I found while exploring a tiny, unnamed island on the Great Barrier Reef. I cast impressions of my body and other organisms together in concrete to create sculptural forms that appear like future fossil remains. From these, I also cast shared skins in tree sap rubber to show our connection in living and in dying. Humans are among numerous organisms who will disappear without oxygen.

This disparate collection of concrete fossils and sheathed skins seemed to speak to me of the preciousness of ephemeral materiality and struck me as authentic jewels, perfect elements for a giant necklace. Titled *Choker*, this work becomes simultaneously a cartography of interdependence with surrounding organisms and a parody of human materialism in the face of the priceless value of ephemeral life. Touching on issues of agency, species interrelations, and evolution, this oversized adornment could be a vestige of evolutionary form from the future.

Figure 17.2. *Choker*, 2018, concrete and latex rubber, 200 × 150 × 35cm, Sustaining the Seas Conference, University of Sydney.
Image by Kassandra Bossell.

This affective viewpoint offers an expansion of relations within material life. Once we see ourselves inside the ecosystem, we can ask: who is this *we*? In the face of expanded relations, the proverbial *us* and *them* disintegrate. The meaning of the word *we* opens to include all living organisms, and the materials and conditions that make up their local environments. A pivotal, situated understanding of *we* in this context is borne out by feminist philosopher Astrida Neimanis. Her formulation of ontology is likened to a body of water, "connected, indebted, dispersed, relational. [It is] not only about correcting a phallogocentric understanding of bodies, but about developing imaginaries that might allow us to relate differently" (Neimanis 2017, 11). She conceives of a more-than-human hydrocommons in which we are, share, and connect through water, which engenders and sustains the more-than-human *us*. In this sense, embodiment is extended beyond our human selves to incorporate other bodies, and water itself.

Choker connects to this notion in a shared conception of *us*. As simple and complex organisms, *we* are and share water and through interdependence, *we* species are and share the planet, our different bodies constituting the situated distinction between us. Neimanis's inclusion of water in this formulation extends a "commoning" not only to other species, but to matter itself, as it flows in and out of life. The inclusion of matter into an earthly ontology is developed by political theorist Jane Bennett in a notion of vital materiality that is allied with the conceptual development underlying this installation. Agency is, in Bennett's conception, always a work group involving humans and nonhumans. She suggests that all things—living and not—are, in Gilles Deleuze's words, "ontologically one, formally diverse" (Deleuze 1993, 66). So it is not merely a matter of rendering nonhumans as important as humans, but of understanding the political relations between them (Bennett 2009, 4). My use of materials in this work aims to include multispecies beings and their elemental constituents into an embodied relationship with the viewer.

S'Kin and *Choker* use materiality, scale, and metaphor to question our assumptions about other organisms. They offer an assimilation of the idea of interconnectedness by helping to thaw the frozen veil of an anthropocentric gaze and reveal our kinship within nature. These artworks call the viewer to investigate perspective, evaluate difference, and find new relations.

ACKNOWLEDGMENTS

This publication was incorporated into my thesis for my master of fine art (research) at the University of New South Wales Art and Design program in August 2018, titled Humans inside Nature: Co-agency in Multispecies Art. During the research period, I received the Research Training Program scholarship from the Australian government.

REFERENCES

Ashrafian, Hutan. 2011. "Leonardo da Vinci's *Vitruvian Man*: A Renaissance for Inguinal Hernias." *Hernia* 15, no. 5: 593–94.

Barnosky, Anthony D., Nicholas Matzke, Susumu Tomiya, Guinevere O. U. Wogan, Brian Swartz, Tiago B. Quental, Charles Marshall, et al. 2011. "Has the Earth's Sixth Mass Extinction Already Arrived?" *Nature* 471, no. 7336: 51–57.

Bennett, Jane. 2009. *Vibrant Matter: A Political Ecology of Things*. Durham, NC: Duke University Press.

Braidotti, Rosi. 2006. "Posthuman, All Too Human: Towards a New Process Ontology." *Theory, Culture and Society* 23, no. 7–8: 197–208.

Deleuze, Gilles. 1993. *The Fold: Leibniz and the Baroque*. Minneapolis: University of Minnesota Press.

Haraway, Donna. 2016. *Staying with the Trouble: Making Kin in the Chthulucene.* Durham, NC: Duke University Press.

Hoegh-Guldberg, Ove, and John F. Bruno. 2010. "The Impact of Climate Change on the World's Marine Ecosystems." *Science* 328, no. 5985: 1523–28.

Neimanis, Astrida. 2017. *Bodies of Water: Posthuman Feminist Phenomenology.* New York: Bloomsbury Publishing.

Section 5

LIVING HUMAN
AND MARINE ECOSYSTEMS

Chapter Eighteen

Operation Crayweed

Merging Art and Science
to Restore Underwater Forests

Adriana Vergés, Michaelie Crawford, Lana Kajlich,
Ezequiel M. Marzinelli, Alexandra Söderlund,
Peter D. Steinberg, Jennifer Turpin,
Georgina Wood, and Alexandra H. Campbell

Imagine that you are in a vast, living forest.

You are at the base of a large tree.

You look up and see branches rustling up in the canopy.

You look a little closer: among the branches, you see creatures with ten legs, or three hearts, beaks, suckers, fluorescent colors. And some have just one foot.

More than five hundred animals cling to the trees' swaying branches. But they don't seem to weigh it down.

Why? Because the forest you are in . . . is underwater.

Around the world, underwater forests created by large brown seaweeds like the one we have just described are in decline. In fact, forests like this have not existed in Sydney for about forty years. This disappearance is part of a global trend of habitat degradation that has already affected biodiversity, food security, economies, and ecosystem functions such as carbon capture around the world. Scientists have focused on quantifying this type of decline, with increasing accuracy. However, because these underwater forests are "out of sight, out of mind," their disappearance has largely gone unnoticed by the public.

But things are changing. In this chapter, we describe our work as Operation Crayweed, where we combine solutions-focused science, community engagement, and art to raise awareness about the importance of underwater forests through the restoration of crayweed, a dominant, canopy-forming seaweed on the east Australian coast. This species disappeared from a seventy-kilometer stretch of coastline along the Sydney metropolitan region during the 1980s

Figure 18.1.   The Art Above, The Science Below.
Photo: John Turnbull.

(Coleman et al. 2008), and our team has been effectively restoring it since the early 2010s (Campbell et al. 2014).

We explain how we have used a good-news story about reversing the local extinction of a seaweed species to engage local communities and boost public awareness about the local marine environment in Sydney, Australia's largest city. This was possible through an art-meets-science collaboration that was part of the 2016 Sculpture by the Sea annual exhibition in Bondi Beach, in Sydney, Australia.

A five-hundred-meter art installation was erected along the part of the coastline where crayweed restoration was taking place underwater, thus making the invisible ecological work happening beneath the waves visible to all. This artwork attracted over 450,000 visitors and highlighted the restoration of a crayweed site that we planted especially for the occasion. More than one hundred children participated in the artwork from local public schools.

By championing participatory outreach, our project aimed to increase public awareness of marine habitat degradation while highlighting how science, art, and the community can come together to facilitate environmental recovery.

UNDERWATER FORESTS AS THE BIOLOGICAL ENGINES OF THE GREAT SOUTHERN REEF

Everyone the world over has heard about the Great Barrier Reef. But Australia has another, much larger, wonder-filled reef. It's called the Great Southern Reef and it spans more than eight thousand kilometers along Australia's southern coastline from Brisbane, Queensland, in the east to Kalbarri, West Australia (Bennett et al. 2016). The Great Southern Reef is a major hotspot of biodiversity and provides a home for many species that are found nowhere else on Earth, such as the giant cuttlefish (*Sepia apama*) and the weedy seadragon (*Phyllopteryx taeniolatus*).

Although this system of interconnected reefs contributes at least $10 billion per year to the Australian economy and most Australians (70 percent of our population) live, work, and play directly alongside the Great Southern Reef, most people have never heard of it, much less understand its critical importance to our nation (Bennett et al. 2016). And with lower awareness comes less research investment and inadequate protection.

Just like trees on the land, seaweed forests on the Great Southern Reef are home to a highly diverse community of marine organisms, including fish, shellfish, and many other invertebrates. In fact, examining all the tiny creatures (collectively known as *mobile epifauna*) inhabiting just one individual seaweed provides an impressive insight into the vast diversity and abundance of life in the sea. These critters include some of Australia's most valuable fishery species, like the rock lobster (*Sagmariasus verreauxi, Jasus edwardsii*) and abalone (*Haliotis sp.*).

As well as creating habitat, underwater forests also provide many underwater vegetarians their primary source of nutrition. Herbivorous fish and invertebrates consume seaweed, and these herbivores are themselves food for predatory animals, who are then consumed by larger predators and so on, throughout the food web. Thus, seaweeds are in a crucial, basal position that underpins marine food webs and supports biodiversity along the coast.

Beyond providing food and habitat for marine organisms, seaweeds perform other essential functions for all life on Earth by photosynthesizing, producing oxygen, and capturing carbon dioxide. They are also natural water filters: they can remove excess nutrients, heavy metals, and other contaminants from the water column, making the water cleaner and safer for all. Improvements in water quality not only enhance the diversity and functioning of marine habitats, they also make the seafood produced there safer to eat. In these multiple, important ways, seaweed forests are essential for healthy marine ecosystems and, more broadly, to all life on our planet.

DECLINING UNDERWATER FORESTS
AND THE DISAPPEARANCE OF CRAYWEED

Given the immense importance of seaweeds, the global patterns of declines of these underwater forests are alarming. In most places around the globe where we know seaweed forests exist, reports of their declines have emerged (Wernberg et al. 2019). These declines are often linked to warming ocean waters, poor water quality caused by nutrient run-off and other types of pollution, outbreaks of disease, or the removal of predators at the top of food webs by overfishing (Steneck and Johnson 2014).

For example, in Western Australia, hundreds of kilometers of the Great Southern Reef were affected by a marine heat wave in 2011, when temperatures were well above average for more than three months. This heat wave led to the contraction of the distribution of several key seaweed species and a resulting reorganization of marine ecosystems over hundreds of kilometers of reef (Smale and Wernberg 2013; Wernberg et al. 2016). Simultaneously, on the eastern side of the Great Southern Reef, increases in herbivore abundance caused by both warming waters and the removal of large predatory fish by fishing, has led to the loss of kelp populations across several hundred kilometers (Ling et al. 2009; Vergés et al. 2016). In Tasmania, the intensification of the East Australian Current, which brings warm, tropical waters south along the east coast, has created conditions that are decimating giant kelp (*Macrocystis pyrifera*) populations. Today, the area covered by giant kelp is less than 5 percent what it was twenty-five years ago (Johnson et al. 2011). These warming waters have also allowed populations of black sea urchin (*Centrostephanus rogersii*) to increase dramatically in recent decades, leading to devastating effects on the golden kelp (*Ecklonia radiata*).

All around the Great Southern Reef and elsewhere in the world, climate-mediated diseases are affecting seaweed forests, with warmer waters simultaneously eroding seaweed defenses against pathogens and inducing many opportunistic pathogens to become virulent (Case et al. 2011). Some seaweed diseases are lethal, whereas others have dramatic effects on seaweed reproduction, photosynthesis, and health.

Although the loss of large forests from the land would be conspicuous and inspire an impassioned response from local communities, seaweed forests are out of sight for most people and, as a consequence, their declines and disappearances go largely unnoticed, often for many decades. This was well illustrated by the disappearance of crayweed (*Phyllospora comosa*) from the metropolitan coastline of Sydney. Crayweed was once common along Sydney's coast, but disappeared completely around the 1980s. Despite this dramatic disappearance occurring along the most densely populated coastline in Australia, it was not noticed until 2008, when marine scientists observed it was dominant in the upper subtidal both north and south of Sydney, but absent from the metropolitan coastline (Coleman et al. 2008).

Although we do not know exactly why crayweed disappeared from Sydney, the timing coincided with a peak in the sewage pollution problem that Sydney had until the construction of deep-water outfalls and treatment upgrades. At the time, beaches were frequently closed because of public health concerns and locals knew to avoid certain surf breaks on a particular wind or tide, as sewage was dumped straight onto our shoreline. Despite major improvements to the water quality around Sydney since the disappearance of crayweed (whales now frequently visit Sydney Harbour; the beaches are almost always open and the diving is exceptional), the immense crayweed forests that once thrived have never returned.

HABITAT RESTORATION:
FROM THE TERRESTRIAL TO THE OCEANIC DOMAIN

Stories of environmental degradation and destruction may be prolific the world over, but we are also beginning to have some positive effects on the state of the environment. Over the past two to three decades, restoration activities around the globe have surged (Dobson, Bradshaw, and Baker 1997), in some cases, changing and even rebooting ecosystem trajectories. Global habitat restoration efforts have already enhanced biodiversity and ecosystem services in previously degraded areas by an average of 44 percent and 25 percent, although these numbers differ wildly between environments (Benayas et al. 2009) and represent a small proportion of the global area of each environment.

The successes of restoration efforts globally, combined with improvements in Sydney's coastal water quality and an understanding of the unique and important biodiversity that crayweed forests support, inspired us to investigate the feasibility of restoring crayweed forests to the Sydney coastline. We knew the odds were stacked up against us, as the record of restoration success in seaweed systems is considerably lower than for habitats on land. This is due to the logistical challenges associated with working in open and dynamic marine systems that are at the mercy of currents and waves, but is also due to the relative invisibility of what happens beneath the waves.

In 2011, we began initial trials to restore crayweed forests in Sydney (Campbell et al. 2014). We transplanted small numbers of fertile adult crayweed from donor populations outside of the city, with the initial aim of seeing whether they might now survive given water quality improvements over the last couple of decades. We found that the seaweeds not only survived but transplanted crayweed also reproduced at several sites, and their offspring (so-called craybies) survived at rates comparable to or higher than control sites (Campbell et al. 2014). After a year, the craybies themselves matured into fertile adults and established self-sustaining populations that have now spread several hundred meters from the original restoration patch over multiple generations, with no ongoing maintenance required. Beyond re-establishing crayweed itself, we also now have evidence that the crayweed forests are leading to enhanced biodiversity at restored sites by providing vital habitat that's been missing for decades (Marzinelli et al. 2016).

The success of our initial trials inspired us to restore crayweed forests at the scale of the initial degradation, that is, along the entire seventy-kilometer coastline adjacent to the Sydney metropolis where crayweed disappeared. We are accompanying upscaled restoration efforts with ongoing scientific research to evaluate the factors that mediate restoration success, including crayweed genetic diversity and the effect of herbivores like sea urchins, which in some sites have prevented the reestablishment of crayweed transplants.

IT IS NOT JUST ABOUT SCIENCE

The success of Operation Crayweed as a marine restoration project is underpinned by rigorous science, with data ranging from detailed information on the genetics and population biology of crayweed to trials of different methods of restoration (Campbell et al. 2015; Marzinelli et al. 2014; Marzinelli et al. 2018). But all this science also leads to the question: who cares? And how do you respond if the initial answer to that question is "no one"?

These questions motivated our efforts to broaden Operation Crayweed beyond science. The answer to "who cares?" comes in a number of forms and reflects the changes in community attitudes to science and the environment over the last decades. If there is a message from major environmental issues such as climate change it is that scientific evidence in and of itself does not convince the broader public of the need for action or even for caring about the environment (Moser 2010). We need to be more effective at communicating science and understanding our audience. But what does that mean?

The traditional method of simply conveying scientific information to an audience—also known as the *knowledge-deficit system*—has been proven time and time again to not work (Groffman et al. 2010). This conclusion has been challenging to scientists, who strive to live in an evidence-based world and use words like *objectivity, falsification*, and *data* a lot, and for whom the relevance of an obscure but ecologically important species of beetle is often glaringly apparent. So why don't people care about the beetle when you tell them why they need to?

The reason is that humans have evolved to learn information through storytelling. Moreover, stories need to resonate with people's feelings and values—social, cultural, economic as well as environmental, to have an effect. As the saying goes, "people will forget what you said, people will forget what you did, but people will never forget how you made them feel." We thus need a broader approach to environmental issues. This is reflected these days in new accounting frameworks, such as the triple bottom line, which broadens the standard financial bottom line focus of businesses by also including social and environmental considerations, or in community well-being approaches to environmental issues (Voyer et al. 2017). We also need a broader approach to communication of environmental issues.

Operation Crayweed is not just about science and restoring one species of seaweed; it is about the story of Sydney's coastline over the years and the value of looking after the place that you love. Sydneysiders and many Australians generally hold the marine environment close to their hearts, even if they know relatively little about the specifics of ecosystems beneath the waves. We chose to connect people to the science of Operation Crayweed and the story of their coasts through art (see Figures 18.1 and 18.2) and gave them the opportunity to interact and engage with the project in a variety of ways using this art as a centerpiece. By combining science with opportunities for our audience to engage in a way that resonated with them, the story of crayweed disappearance and restoration became a stronger, more compelling story.

OPERATION CRAYWEED ART-WORK-SITE: PARTICIPATORY ART IN NATURE

Art highlights the creative spirit. It has the potential to attune us to making unexpected, revelatory connections from which emerge new ways of seeing, hearing, and thinking. Art can inspire such connections between ourselves and the world around us, the natural world, the cultural world, and the world of the mind.

Art can also mediate our lost relations with the world of nature, re-creating associations with plants and other species. It can invoke the nature in all of us and give expression to the yearning for a closeness to nature felt by many alienated in contemporary urban landscapes like Sydney.

Operation Crayweed Art-Work-Site, part of the Bondi Sculpture by the Sea exhibition in 2016, was a participatory artwork comprising a number of active parts. Through direct scientific learning and joyous creative engagement, it aimed for heightened conscious and subconscious connections between humans and nature. Hundreds from the community were, and will continue to be, involved in a project that seeks to link land with sea, humans with plants, and humans with plants and other marine animals. It is a creative exploration of the concept of biodiversity.

A Walk along the Shoreline at Bondi Bay

Running half a kilometer along the famous coastal walk from Marks Park to the South Bondi headland, a bright yellow artful safety fence was installed, snaking its way around Bondi Bay. This yellow fence followed the meandering shoreline and encircled an underwater worksite, where a new crayweed restoration site was being installed.

We took the concept of the worksite and applied it to the underwater domain and labor of scientists. The land-based worksite is well known as a place of intense and temporary human construction. It is demarcated with strong and easily recognizable visual cues, including high-visibility safety fences, lines of bunting, signs, tapes, traffic cones, and hard hats. But a marine worksite is not such a known concept, and we wanted to draw attention to the substantial physical work done by marine ecologists and the site of their underwater work.

We combined the well-known land-based work site safety materials with materials used in marking marine sites, including floating marker buoys and long lines of crab floats and thus created a half-kilometer-long art worksite fence. All was bright yellow, the color used in the marine and boating world to mark a special project in maritime safety language. Together, the hybrid of materials formed a strong visual language highlighting and making visible the underwater, and otherwise hidden, worksite.

This art fence also sought to transform everyday materials. By layering the materials, we created visual and kinesthetic effects. As the sun moved across the sky each day, the light cast patterns onto the path through the barrier mesh, creating a moiré of ever-changing shadows, the bunting rippling in the wind and the tactility of crab floats attracting the touch of passers-by. This multilayered fence demarcated the relationship between landscape and seascape.

Out in the Water

A floating yellow buoy was anchored above the exact site of the reforestation. It rolled with the motion of the waves bobbing variously in the changing swell. Inscribed on the buoy were messages for daring ocean swimmers to discover, such as "You are swimming in the Bondi Bay Crayweed Nursery," "Babies expected late summer," and "We are reforesting Sydney's Underwater Coastline from Cronulla to Palm Beach."

Encircling the buoy's floatation compartment were the names of the sea creatures who rely on crayweed for habitat and who are expected to return once the crayweed reestablishes itself, such as weedy sea dragon, eastern crayfish, Sydney spiny crayfish, common stingray, abalone, and many more.

Ocean Swimmers

During the exhibition, a team of ocean swimmers wearing yellow bathing caps and flippers swam around the buoy and dived beneath the waves in a synchronized display that drew further attention to the crayweed underwater reforestation site.

Up on the Cliff

The South Bondi headland was a perfect vantage point for viewing the site of the underwater work. Here we installed three large Viewscope artworks, made of the same combination of land-based work site and marine marker materials. The three yellow buoys with inverted traffic cones at their apex were covered in *Operation Crayweed* signs and positioned to point to the underwater reforestation site out in the bay. Like road signs, they were yellow and diamond in shape with black text. Standing on the buoy's platform and looking through the narrow end of the cone, visitors saw a large cutout "X" at the end of the cone. This superimposed itself on the ocean's surface, marking the spot of the crayweed restoration underwater.

Like the buoys in the water, the signage explained their meaning in the context of the greater project with lines such as "Scientists working below the waves," "Crayweed reforestation in progress," "Seaweeds are the biological engine of the Great Southern Reef," "Seaweeds are the trees of the underwater world," and "Habitat creates diversity." Again, the lower platforms of the

Viewscopes were ringed with names of creatures known to inhabit crayweed, such as Port Jackson shark, smooth ray, eastern kelpfish, and many more.

Embodying Another Creature

Meanwhile, rolls and boxes of those yellow worksite and marine demarcation materials were inventively cut up, folded, bent, and reshaped in the creation of sea-creature costumes fashioned by over one hundred local school children.

The children first learned about the life history of crayweed and the biodiversity it fosters in community outreach workshops led by the scientists. Then, in art workshops led by the artists with assistance from parent volunteers, they learned to make wearable sculptures. Whether octopus or sea urchin, sting ray, weedy sea dragon, or sea snail, the children studied the look and physical attributes of their chosen creature. They also focused on the way their marine creature propelled itself in the water and how their costume would reflect these particular dynamic characteristics. Making the costumes involved activities such as attaching black cable ties to yellow hard hats for a sea urchin's head, spreading sheets of barrier mesh to create a sting ray's body, or joining segmented pieces to articulate a crayfish body or coiling up foam logs for a snail's body and attaching large eyes on flexible stalks.

Figure 18.2.   Sea Urchins on Parade.
Photo by Ian Hobbs.

While wearing their costumes, the children imagined being a marine creature living in a crayweed forest. They were encouraged to think beyond the human frame and psyche and see themselves inhabiting an underwater world.

We All Live in the Yellow Crayweed

The project reached its climax in a lively launch featuring a performative event and parade. A local children's orchestra and choir performed a song about crayweed to the tune of "We All Live in a Yellow Submarine." The scientists and artists spoke, the mayor cut the ribbon, and the blowing of a conch shell heralded the beginning of the children's parade.

Wearing their all-yellow creature costumes, and in species groups, they blew yellow whistles while happily walking the meandering coastal path. They enlivened the yellow fence as they followed alongside to the enjoyment of the passing walkers.

As a visual and aural activation of the land artwork, the parade brought the whole project together as a lively performance.

ART CONNECTS US TO THE ENVIRONMENT

Operation Crayweed Art-Work-Site was a temporal and site-specific project whose meaning lies in the interplay between art, science, nature, the urban environment, and the local community. It reminded us of nature's beauty, its powerful energy, and its fragility. It highlighted humanity's role in protecting the marine environment.

Operation Crayweed, with its art and science collaboration, honors and fosters the intuitively felt relationship between ourselves and other species of the natural world. We know this relationship—this reassociation—is at the heart of what will help us heal some of the damage we have wrought on nature.

Bondi Bay was the first in a series of crayweed reforestation sites combining art and science along Sydney's eastern seaboard stretching from Cronulla to Palm Beach. As the scientists move up and down the coast, so will the artists, working with other local communities in the creative work of reconnecting with their coastal environments.

COMMUNICATING RESTORATION SCIENCE
THROUGH STORYTELLING: TALES FROM THE FIELD

Operation Crayweed Art-Work-Site, shown at the 2016 Sculpture by the Sea exhibition, was seen by over 450,000 people. This provided an exceptional opportunity for the scientists and artists involved in *Operation Crayweed* to connect directly with the general public (see Figures 18.1 and 18.2).

We seized this opportunity and accompanied the artwork with a collection of three-minute podcasts and a film that brought the science behind the project to life. These media explained the ecological value of seaweeds and described the science behind our restoration project. In some podcasts, the science was explained using narrative and storytelling, whereas in others it was communicated in a factual, textbook style.

After either listening to one of the short podcasts or watching the film, we asked participants questions about the local Sydney marine environment, their connection to this urbanized coast and the restoration project. We also had a so-called control treatment, during which we asked participants the same questions but without exposing them to our science communication.

Almost five hundred people offered their time and voice for our study on that walkway during Sculpture by the Sea. Our experiments showed that, although all styles of science communication improved people's knowledge about the restoration project, using storytelling (either through a film or a podcast) was a more effective way of transmitting the science behind crayweed restoration than simply listing the facts.

We also discovered many interesting perceptions about the local marine environment among our control group, with two in particular standing out.

First, we were surprised to discover that more than half the people surveyed believed water quality in Sydney's coastline had *worsened* in the past thirty years, when it has in fact significantly improved. This lack of awareness of positive environmental change among the general population, even when that change takes place right before our very eyes, is remarkable. It suggests that successful management actions are not being communicated effectively to the public.

Second, we learned that many Sydneysiders thought corals are the dominant habitat formers in our temperate waters, even though reef-building corals are found more than one thousand kilometers north of Sydney's coast. This signals a need to raise awareness about the importance of seaweed forests as the key foundation species supporting ecological communities in cooler water coastlines.

Finally, when we invited the public to offer their stories of connection to their coastal environment, we detected a pattern in the way people approached their responses. Through classifying the hundreds of stories people told us, we identified three broad themes:

1. Human Connections to the Coastline. People frequently made historical references to places, offered stories of growing up, listed favorite marine

pastimes, and spoke of knowledge gained from experience or emotions evoked. For example:

> We used to play on the rock platforms and catch jellyfish and paddle for all my child-hood . . . so I guess that's something that's more recreational rather than knowing what's actually under the water. (female, eighty-two years old)

2. Values of the Coastal Environment. Our participants expressed the importance of biodiversity, recreational use, and either functional values the coast provided (i.e., it is "life-sustaining") or they mentioned the inherent, intrinsic values of their coastal environment, such as beauty, or identity. For example:

> Well the marine environment probably affects every Sydneysider's image of them-selves and their city so even though you're from the west, the coastline is still iconic for us and we identify with it. (male, twenty-six years old)

3. Recommendations. The public offered suggestions, often emphatically, for how to protect, preserve, or improve conditions within their local marine "patch." This recommendation was often accompanied by variations of the reason "for future generations." Many also expressed strong support for the crayweed restoration project. For example:

> As a childcare teacher I think it's important to teach kids—as they are our future—how human beings are connected to the environment and how the environment connects to our life. (female, twenty-six years old)

LESSONS LEARNED FROM A THREE-WAY INTERACTION BETWEEN SCIENCE, ART, AND THE PUBLIC

By sharing and listening to stories from the public and engaging in a three-way exchange between the art, the science, and the public, we learned about how the public connects and responds to changes along an urban coast and the life below its waters.

In particular, our work shines a light on the need to better communicate the effectiveness of management actions and positive environmental stories more generally. Although our oceans face major unresolved problems (overfishing and climate change being prime examples), much has improved along our coastlines following decades of management efforts by local, regional, and national governments. The major improvement of water quality in Sydney is a leading example of successful management, and yet surprisingly little seems to be known about these effective actions. This may be related to the way environmental stories are reported by fear-based media, which aims to attract audiences by focusing on negative consequences (Serani 2011).

However, there is also a growing appetite for positive environmental news, as illustrated by the #oceanoptimism social media movement, which brings together conservation scientists; policy makers; artists; engineers; and representatives of business, media, and philanthropy to celebrate conservation success stories (Knowlton 2019). Even the media seems to have become increasingly balanced and now reports on stories of environmental successes (Johns and Jacquet 2018). Further, there is growing evidence that optimism inspires and promotes conservation engagement (McAfee et al. 2019).

In this human-dominated era of the Anthropocene during which there is much environmental damage to fix, we argue that celebrating positive conservation stories and making science more accessible through art not only broadens the reach and influence of the information, but can also help reconnect people to the natural environment.

REFERENCES

Benayas, José M. Rey, Adrian C. Newton, Anita Diaz, and James M. Bullock. 2009. "Enhancement of Biodiversity and Ecosystem Services by Ecological Restoration: A Meta-analysis." *Science* 325 (5944): 1121–24.

Bennett, Scott, Thomas Wernberg, Sean D. Connell, Alistair J. Hobday, Craig R. Johnson, and Elvira S. Poloczanska. 2016. "The 'Great Southern Reef': Social, Ecological and Economic Value of Australia's Neglected Kelp Forests." *Marine and Freshwater Research* 67 (1): 47–56.

Campbell, Alexandra H., Ezequiel M. Marzinelli, Jon Gelber, and Peter D. Steinberg. 2015. "Spatial Variability of Microbial Assemblages Associated with a Dominant Habitat-forming Seaweed." *Frontiers in Microbiology* 6: 230.

Campbell, Alexandra H., Ezequiel M. Marzinelli, Adriana Vergés, Melinda A. Coleman, and Peter D. Steinberg. 2014. "Towards Restoration of Missing Underwater Forests." *PLOS One* 9 (1): e84106. doi: 10.1371/journal.pone.0084106.

Case, R. J., S. R. Longford, A. H. Campbell, A. Low, N. Tujula, P. D. Steinberg, and S. Kjelleberg. 2011. "Temperature Induced Bacterial Virulence and Bleaching Disease in a Chemically Defended Marine Macroalga." *Environmental Microbiology* 13 (2): 529–37. doi: 10.1111/j.1462-2920.2010.02356.x.

Coleman, M. A., B. P. Kelaher, P. D. Steinberg, and A. J. K. Millar. 2008. "Absence of a Large Brown Macroalga on Urbanized Rocky Reefs around Sydney, Australia, and Evidence for Historical Decline." *Journal of Phycology* 44 (4): 897–901. doi: 10.1111/j.1529-8817.2008.00541.x.

Dobson, Andy P., A. D. Bradshaw, and A. J. Baker. 1997. "Hopes for the Future: Restoration Ecology and Conservation Biology." *Science* 277 (5325): 515–22.

Groffman, Peter M., Cathlyn Stylinski, Matthew C. Nisbet, Carlos M. Duarte, Rebecca Jordan, Amy Burgin, M. Andrea Previtali, and James Coloso. 2010. "Restarting the Conversation: Challenges at the Interface between Ecology and Society." *Frontiers in Ecology and the Environment* 8 (6): 284–91.

Johns, Lisa N., and Jennifer Jacquet. 2018. "Doom and Gloom Versus Optimism: An Assessment of Ocean-related U.S. Science Journalism (2001–2015)." *Global Environmental Change* 50: 142–48. doi: https://doi.org/10.1016/j.gloenvcha.2018.04.002.

Johnson, C. R., S. C. Banks, N. S. Barrett, F. Cazassus, P. K. Dunstan, G. J. Edgar, S. D. Frusher, et al. 2011. "Climate Change Cascades: Shifts in Oceanography, Species' Ranges and Subtidal Marine Community Dynamics in Eastern Tasmania." *Journal of Experimental Marine Biology and Ecology* 400 (1–2): 17–32. doi: 10.1016/j.jembe.2011.02.032.

Knowlton, Nancy. 2019. "Earth Optimism—Recapturing the Positive." *Oryx* 53 (1): 1–2.

Ling, S. D., C. R. Johnson, S. D. Frusher, and K. R. Ridgway. 2009. "Overfishing Reduces Resilience of Kelp Beds to Climate-Driven Catastrophic Phase Shift." *Proceedings of the National Academy of Sciences of the United States of America* 106 (52): 22341–45. doi: 10.1073/pnas.0907529106.

Marzinelli, E. M., A. H. Campbell, A. Vergés, M. A. Coleman, B. P. Kelaher, and P. D. Steinberg. 2014. "Restoring Seaweeds: Does the Declining Fucoid *Phyllospora comosa* Support Different Biodiversity Than Other Habitats?" *Journal of Applied Phycology* 26 (2): 1089–96. doi: 10.1007/s10811-013-0158-5.

Marzinelli, Ezequiel M., Martin R. Leong, Alexandra H. Campbell, Peter D. Steinberg, and Adriana Vergés. 2016. "Does Restoration of a Habitat-Forming Seaweed Restore Associated Faunal Diversity?" *Restoration Ecology* 24 (1): 81–90.

Marzinelli, Ezequiel M., Zhiguang Qiu, Katherine A. Dafforn, Emma L. Johnston, Peter D. Steinberg, and Mariana Mayer-Pinto. 2018. "Coastal Urbanisation Affects Microbial Communities on a Dominant Marine Holobiont." *NPJ Biofilms and Microbiomes* 4 (1): 1.

McAfee, Dominic, Zoë A. Doubleday, Nathaniel Geiger, and Sean D. Connell. 2019. "Everyone Loves a Success Story: Optimism Inspires Conservation Engagement." *BioScience*.

Moser, Susanne C. 2010. "Communicating Climate Change: History, Challenges, Process and Future Directions." *Wiley Interdisciplinary Reviews: Climate Change* 1 (1): 31–53.

Serani, Deborah. 2011. "If It Bleeds, It Leads: Understanding Fear-Based Media." *Psychology Today* 7.

Smale, Dan A., and Thomas Wernberg. 2013. "Extreme Climatic Event Drives Range Contraction of a Habitat-Forming Species." *Proceedings of the Royal Society B: Biological Sciences* 280 (1754): 2012–829. doi: 10.1098/rspb.2012.2829.

Steneck, R. S., and C. R. Johnson. 2014. "Kelp Forests. Dynamic Patterns, Processes, and Feedbacks." In *Marine Community Ecology and Conservation*, edited by M. D. Bertness, J. F. Bruno, B. R. Silliman, and J. J. Stachowicz, 315–36. Sunderland, MA: Sinauer Associates.

Vergés, A., C. Doropoulos, H. A. Malcolm, M. Skye, M. Garcia-Piza, E. M. Marzinelli, A. H. Campbell, E. Ballesteros, A. S. Hoey, A. Vila-Concejo, Y. M. Bozec, and P. D. Steinberg. 2016. "Long-term Empirical Evidence of Ocean Warming Leading to Tropicalization of Fish Communities, Increased Herbivory and Loss of Kelp." *Proceedings of the National Academy of Sciences* 113 (48): 13792–96.

Voyer, Michelle, Kate Barclay, Alistair McIlgorm, and Nicole Mazur. 2017. "Using a Well-Being Approach to Develop a Framework for an Integrated Socio-economic Evaluation of Professional Fishing." *Fish and Fisheries* 18 (6): 1134–49.

Wernberg, T., S. Bennett, R. C. Babcock, T. de Bettignies, K. Cure, M. Depczynski, F. Dufois, J. Fromont, C. J. Fulton, R. K. Hovey, E. S. Harvey, T. H. Holmes, G. A. Kendrick, B. T. Radford, J. Santana-Garcon, B. J. Saunders, D. A. Smale, M. A. Thomsen, C. A. Tuckett, F. Tuya, M. A. Vanderklift, and S. K. Wilson. 2016. "Climate-Driven Regime Shift of a Temperate Marine Ecosystem." *Science* 353 (6295): 169–72.

Wernberg, Thomas, Kira Krumhansl, Karen Filbee-Dexter, and Morten F. Pedersen. 2019. "Status and Trends for the World's Kelp Forests." In *World Seas: An Environmental Evaluation*, 2nd edition, edited by Charles Sheppard, 57–78. New York: Academic Press.

Chapter Nineteen

Buoyant Ecologies

Interspecies Cooperation
for Sea Level Rise Adaptation

Adam Marcus

An environment is not a neutral, empty box, but an ocean filled with currents and
surges.

—Timothy Morton, *Humankind: Solidarity with
Nonhuman People* (2017, 189)

Conversations about coastal resilience, responses to climate change, and adaptation to sea level rise often focus on "macro" infrastructural strategies at the systemic scale, such as regional stormwater management plans, massive seawalls, or expansive wetland restoration projects. Informed by principles of coastal engineering, this ethos reflects an anthropocentric mindset, motivated by a defensive impulse to protect property and capital that humans have developed along the shoreline. The Buoyant Ecologies project proposes an alternative response to these challenges, instead emphasizing the "micro" scale of a material substrate and its capacity for nonhuman habitation as a catalyst for positive vectors of change that can reverberate at an ecosystemic scale.

The project is a collaborative, transdisciplinary research initiative that synthesizes architectural design, marine ecology, and advanced composite manufacturing to develop new approaches to constructing resilient waterfront structures. Its focus is on biofouling—the unchecked growth of barnacles and other invertebrates on hard surfaces like boat bottoms, docks, and other waterfront structures—and how this behavior can be encouraged, optimized, and ultimately leveraged for the benefit of the surrounding ecosystem. It imagines a coastal ecology of humans in cohabitation with nonhuman marine animals, in which the two are linked in mutual contingency. Engineered floating structures provide optimal habitats for the animals, while the resultant accumulation of marine life provides wave-attenuation capacity that in turn helps protect the vulnerable shoreline.

Biofouling is typically seen as a nuisance, as the growth of marine life can compromise the integrity of waterfront structures and create resistance for vessels in motion. Indeed, there is an entire subset of the marine industry devoted to developing antifouling coatings (which are often toxic) and material textures that inhibit the growth of marine life underwater. This project inverts such logic, instead positing fouling communities as an asset and proposing that controlled upside-down settlements can become an ecological resource across multiple scales. The vehicle for this research is a customized, digitally fabricated substrate made of fiber-reinforced polymer composite material (commonly known as *fiberglass*, which is often used to make sailboats and other water craft). Computational models employ statistical models and empirical data to optimize the surface geometry such that its upside-down hills and valleys vary in size to create pockets of space that protect smaller invertebrates from larger predators, thereby encouraging biodiversity in the immediate environs.

The parameters driving the surface geometry have been tested and refined through a series of nearly two dozen full-scale substrate prototypes deployed underwater at various sites in San Francisco Bay and Monterey Bay, where they are monitored for their performance as animal habitats (Figure 19.1). The success of these prototypes in producing diverse communities of marine invertebrates—tunicates, bryozoans, worms, mussels, sea urchins, oysters, nudibranchs, and more—confirms the hypothesis that variable geometries can yield variable settlement patterns along an upside-down surface. Furthermore, the extraordinary accumulation of invertebrates, particularly on surfaces that take on more vertical dimension, suggest the possibility that such "sponges" of thriving animal communities could help mitigate wave action.

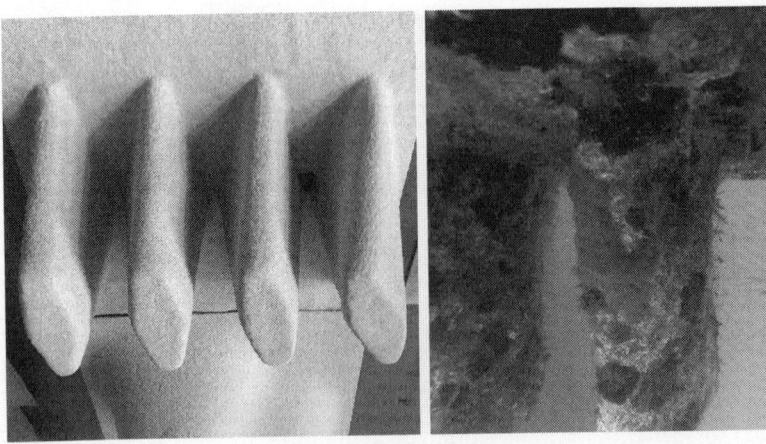

Figure 19.1. Views of ecological substrate prototype before deployment (left) and after twelve weeks underwater, colonized by invertebrate animals (right).
Image credit: Architectural Ecologies Lab, California College of the Arts.

It is this capacity for wave attenuation that presents a potential alignment between nonhuman and human interests. Faced with the prospects of increasing sea levels and flooding events caused by more frequent extreme weather events, coastal areas across the globe have embraced the notion of "resilience," which melds climate adaptation with a decidedly defensive posture driven by the desire to protect and preserve capital.[1] Cities are investing heavily in fortifications like seawalls and barrier islands in an attempt to keep the ocean at bay. But these permanent structures are an enormously expensive form of technology, and it is unlikely that such efforts will materialize for cities in marginalized and developing countries, which are expected to disproportionately experience the effects of sea level rise.[2] The notion of a floating breakwater that derives its wave attenuation capacity from the surrounding ecosystem of marine animals offers a softer, more agile alternative to this paradigm of costly, permanent structures that have become the default apparatus of resilience. It suggests a model for interspecies collaboration that is rooted in mutual cooperation and benefit rather than purely human self-interest.

This work builds on several important precedents. Visionary architect Buckminster Fuller patented two designs for floating breakwaters in the 1970s, which proposed flexible membrane structures to absorb wave energy (Fuller 1975, 1979). Fuller was motivated by a desire to develop a more flexible and economical alternative to the heavy, permanent masonry breakwaters commonly constructed to protect coastal settlements from erosion. But these designs did not take into account marine life; although they were never tested, it is likely that Fuller's structures would be compromised by the effects of biofouling, as the designs do not anticipate this inevitability. The Buoyant Ecologies research, conversely, *embraces* fouling communities as a resource that can both contribute to ecological diversity and build up wave attenuation capacity. In this regard, the project is comparable to many examples of artificial reefs that seek to foster restored habitats for marine life. A recent notable example is Living Breakwaters, a project by SCAPE currently in development in New York, which proposes a large reef of optimized oyster habitats to be constructed off the coast of Staten Island (Orff 2016, 237–68). Although the SCAPE project integrates marine life as a critical component of its ecological performance, it nonetheless remains a fixed, permanent breakwater. Buoyant structures offer advantages in their modularity, portability, and expandability over time—perhaps offering more flexibility for coastal areas adapting to the effects of sea level rise.

The most recent phase of the Buoyant Ecologies project builds on the proof-of-concept success of the early experiments by testing the optimized ecological substrate and wave attenuation capacity at a larger scale. The Float Lab, a prototype for a floating breakwater, incorporates a large underwater landscape with gradated topography that will allow further refinement of the optimization logics. It also serves as a floating platform from which more vertically proportioned prototypes can be suspended, to test the wave attenuation behavior of the large sponges of invertebrates that accumulate on these structures. The top side of the prototype provides an additional function, channeling and collecting water into a simulated tidal wetland that will attract birds and other terrestrial species (Figure 19.2).

The Float Lab is bean-shaped in plan and roughly the size of an automobile (Figure 19.3). The prototype consists of two identical parts that form the top and bottom hulls, joined together like a clamshell. Its geometry is designed to anticipate future deployment of multiple modules that tile together into networked archipelagos of floating breakwaters to form a protective barrier for vulnerable coastlines (Figure 19.4). Although the interior of the Float Lab prototype is not sized for human occupation, its scalability suggests a floating architecture in which humans could in fact inhabit the interior of the ecologically optimized breakwater. Larger-scale structures could take the

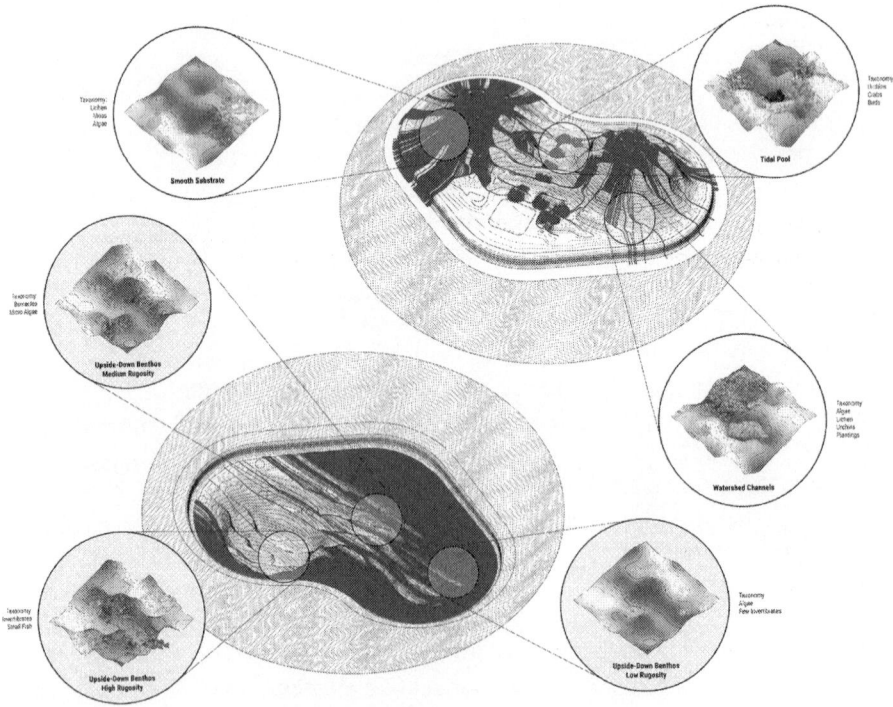

Figure 19.2.   The Float Lab is designed to perform as an ecological habitat both above and below the water.
Image credit: Architectural Ecologies Lab, California College of the Arts.

Figure 19.3. View of the Buoyant Ecologies Float Lab prototype.
Image credit: Architectural Ecologies Lab, California College of the Arts.

form of floating housing, a typology not uncommon in regions of the world where people are accustomed to living on the water, and an alternative that is perhaps increasingly relevant in an era of sea level rise. The functionality of the top surface as a tidal habitat even suggests the potential for aquaculture and other forms of food cultivation that could support human habitation in a

Figure 19.4.   These drawings speculate on the capacity of the Float Lab to scale up to a chain or necklace of floating breakwaters that achieve wave-attenuation capacity through underwater invertebrate habitats.
Image credit: Architectural Ecologies Lab, California College of the Arts.

similar manner to how the lower surface channels food and nutrients to the invertebrate communities. This kind of floating urbanism of self-sufficient buildings offers a countermodel to conventional notions of resilience: one that provides enhanced adaptability to climate change while also contributing productively to the broader ecosystem (Figure 19.5).

The prototype's primary function in the immediate short term upon its deployment in San Francisco Bay is to advance the substrate research, but its latent implications—networks of invertebrate habitats that perform as breakwaters, floating human habitats, artificial buoyant wetlands, aquacultural cooperation between species—imagine a kind of waterborne courbanism of people and animals. Architectural historian and theorist Todd Gannon suggests that one way for architects to grapple with the implications and challenges of the Anthropocene is to develop an understanding of the "complex feedback loops through which environments and their inhabitants co-construct one another" (2017, 142).[3] This project's ambition is to cultivate such an awareness by leveraging advanced workflows of design optimization and digital fabrication that are increasingly accessible and common in archi-

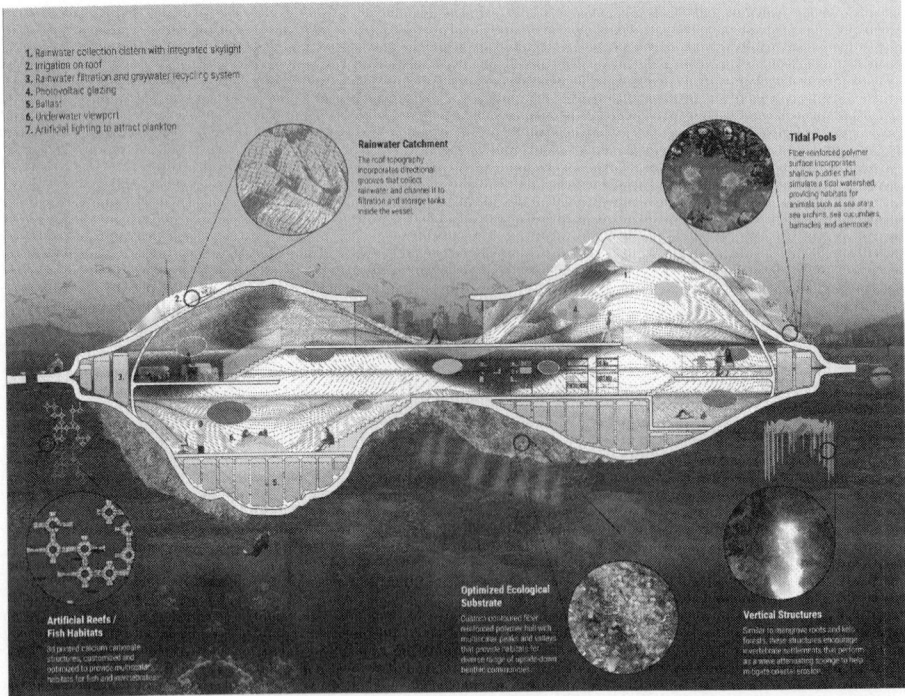

Figure 19.5. This drawing imagines a floating architecture of the future, in which human and non-human species cohabit in mutual benefit and cooperation.
Image credit: Architectural Ecologies Lab, California College of the Arts.

tectural practice. These tools have the capacity to link material performance with ecological performance, but they are often employed at the scale of the single building rather than a broader ecosystem. The project seeks to redirect such expertise and to channel it toward broader ecological agency.

By advocating a multispecies approach to ecological adaptation, the Buoyant Ecologies research invokes principles of mutualism, a biological phenomenon in which two species engage in a mutually beneficial relationship. Landscape architect and SCAPE founder Kate Orff notes, however, that mutualism does not go far enough in mitigating humanity's effect on the environment, and that more ambitious and proactive strategies are necessary (2016, 83–84). Orff's concept of "cohabit" pushes beyond mutual benefit and advocates for mutual regeneration: "in order to begin reversing the trends of extinction and creating an obverse trend, we need to start designing in a way that actively benefits a diversity of species and fosters a regenerative context that makes room for nonhuman animals, shaping urban spaces to support more biodiversity" (2016, 84).

This notion of cohabitation—humans living in concert with nonhumans, contingent on one another for ecological survival—constitutes the very technology for ecological regeneration across scales and species. In contrast to anthropocentric technologies of "resilience," which itself is an entirely human-centered concept of self-preservation, the Buoyant Ecologies project offers an alternative strategy for human adaptation to ecological change: one that both depends on and supports the health and diversity of nonhuman marine species. It leverages human innovation in design, computation, and manufacturing, but it does so in a way that is motivated and informed by the needs and desires of marine invertebrates. In a broader sense, the project recognizes and embraces what Donna Haraway calls the "obligatory confederacy" (2004, 146) between species, advocating models of coexistence that link the well-being of humans, nonhumans, and the broader oceanic realm.

ACKNOWLEDGMENTS

Buoyant Ecologies Project Credits:

Project Leaders: Adam Marcus, Margaret Ikeda, Evan Jones
Design Assistants: Taylor Metcalf, Georine Pierre, Jared Clifton
Marine Ecology: Benthic Lab, Moss Landing Marine Laboratories—John Oliver, Kamille Hammerstrom, Daniel Gossard
Fabrication: Kreysler and Associates
Naval Architecture and Engineering: Tri-Coastal Marine—Andrew Davis

NOTES

1. For a thorough overview of the architectural and urban design discourse on resilience, see Ross Exo Adams's article "Notes from the Resilient City" (2014).

2. There are numerous scholarly studies of how climate change will have disproportionate consequences in the developing world. For a comprehensive study of climate inequity, see the report *Climate Change 2014: Impacts, Adaptation, and Vulnerability* (Field et al. 2014).

3. In his book, Gannon channels Timothy Morton's notion of the "strange loop" in advocating ecological awareness (Morton 2016).

REFERENCES

Adams, Ross Exo. 2014. "Notes from the Resilient City." *Log* 32 (Fall 2014): 126–39.

Field, C. B., V. R. Barros, D. J. Dokken, K. J. Mach, M. D. Mastrandrea, T. E. Bilir, M. Chatterjee, et al. editors. 2014. *Climate Change 2014: Impacts, Adaptation, and Vulnerability. Part A: Global and Sectoral Aspects. Contribution of Working Group II to the Fifth Assessment Report of the Intergovernmental Panel on Climate Change.* Cambridge: Cambridge University Press.

Fuller, Richard Buckminster. 1975. Floatable Breakwater. U.S. Patent 3,863,455, filed December 10, 1973, and issued February 4, 1975.

———. 1979. Floating Breakwater. U.S. Patent 4,136,994, filed September 19, 1977, and issued January 30, 1979.

Gannon, Todd. 2017. "Strange Loops: Toward an Aesthetics for the Anthropocene." *Journal of Architectural Education* 71:2 (2017): 142–45. https://doi.org/10.1080/1 0464883.2017.1340787.

Haraway, Donna. 2004. "Otherworldly Conversations; Terran Topics; Local Terms." In *The Haraway Reader*, edited by Donna Haraway, 125–50. New York: Routledge.

Morton, Timothy. 2016. *Dark Ecology: For a Logic of Future Coexistence.* New York: Columbia University Press.

———. 2017. *Humankind: Solidarity with Nonhuman People.* London: Verso.

Orff, Kate. 2016. *Toward an Urban Ecology.* New York: Monacelli Press.

Chapter Twenty

South Korean Reef Metropolis

Amaia Sánchez-Velasco,
Jorge Valiente Oriol, and Gonzalo Valiente

Since the end of the Korean War in 1953, South Korea has faced permanent diplomatic pressure because of intricate relations with its neighbors, China and North Korea. In the aftermath of the war, a rural exodus prompted the exponential growth of its capital city. Nowadays, Seoul is a global metropolis, but because of the country's geographical and geopolitical confinement, it has struggled to expand both its urban fabric and its productive hinterland. As a consequence, South Korea has shifted its geopolitical agenda to focus on the oceans. This has not been in vain—the country has rapidly and strategically spread a myriad of maritime productive facilities such as artificial reefs, sea ranches, and marine forests, thus giving form to the nation's obsession of becoming a twenty-first century maritime pioneer.

An archipelago of artificial reefs extends the city into the seabed and expands the country's urban condition beyond the coastline in the form of a nature-culture continuum. These invisible structures host marine life and are designed to progressively degrade. Beyond their infrastructural role, they also perform as cultural, geopolitical, and architectural artifacts, organizing both human and nonhuman bodies under selective conservation, mass production, and defense premises across multiple scales.

GEOGRAPHICAL AND LEGISLATIVE CONSTRAINTS

South Korea's dramatic topography restricts the arable land to a thin littoral stripe, while the elongated peninsula gives access to the seas, which provide 45 percent of the proteins consumed in the country. Over the past decades, the effects of global warming, industrial pollution, agricultural runoff, population growth, domestic sewage, and overfishing practices have dramatically

261

reduced fish and invertebrate populations, thus affecting national fisheries and positioning the Yellow Sea on the World Wildlife Fund for Nature's list (2017) of the most environmentally critical regions on Earth.

The legislative changes defined by the 1982 United Nations Law of the Sea Convention expanded most countries' jurisdictions to cover much larger maritime areas because the exclusive economic zones (EEZs) were extended from twelve nautical miles offshore to two hundred nautical miles. However, South Korea's EEZ partially overlaps with other EEZs. Because of the proximity of other countries such as Japan, China, and North Korea, the EEZs are in this case demarcated by the equidistant line between the countries' coastlines, which is often located closer than two hundred nautical miles. The amalgam of national boundaries means that the closest international waters to South Korea are further away from their shores, resulting in its seas effectively being enclosed. The difficulty South Korean fisheries face trying to reach the distant high seas, together with the depletion of marine species and the demands of an increasing population, has led to the intensification of maritime productivity within its EEZ, as well as the country's investment in the extensive implementation of artificial reefs, sea ranches, and marine forests.

SOUTH KOREAN METROPOLITAN PROJECT

Mirroring the evolution of its inland urban growth, South Korea has rapidly increased its underwater construction over the last few decades. Seoul was devastated after the Korean War, but its rapid urban growth in the following years raised the country's total urban population rates from 27.1 percent in 1960 to 82.6 percent in 2016 (World Bank Group 2018). Today it hosts one of the five largest metropolitan populations on Earth, and is home to half of the South Korean people. In line with pace of this growth, a constellation of artificial reefs was progressively built around the peninsula, eventually occupying an area 3.5 times greater than that of Seoul—across 210,000 hectares of seabed. The cost of this growth—turning South Korea's seabed into a reef metropolis of intensive maritime production—required investment from both government and private corporations.

In the past decades, South Korea has applied a developmental approach to its economic growth (Nae-Hui 2018) and aligned its urban policy to neoliberal agendas. Terms such as *financialization, zoning,* and *gentrification* are common in the glossary of the city's urban transformation. Do the same terms remain valid terminology when describing the urban economies performing underwater?

UNDERWATER URBAN GROWTH

Similar to regulated urban developments, the South Korean Reef Metropolis passes through sequential phases such as planning, zoning, and implementation under the regulation of construction guidelines. During the preliminary stages, the projects are negotiated among government and private stakeholders. The first phase entails the analysis of existing seabed conditions, as well as the assessment of risks and the profitability of the investments. The central government develops the planning and budgets for artificial reefs and local governments are in charge of their construction and installation. The National Fisheries Research Development Institute assesses appropriate sites and evaluates the potential productivity of each location. The national government contributes 80 percent of the costs and the provincial government the remaining 20 percent. Competitive tender processes allow private corporations to win contracts for further deployments. The South Korean government's process of approval lasts two to three years, during which time artificial reef designs are deployed and further monitored. Finally, the artificial reefs are evaluated on the basis of construction cost, economic efficiency, and quality.

Artificial reefs do not operate as isolated artifacts. Over the past few hundred years, they have evolved from being mere piles of debris to purpose-driven, designed structures. They consist of an extensive infrastructural network able to organize inhabitation and circulation patterns. They create a multilayered circulation system driven by efficiency, investment, production, and conservation across multiple scales. From the master plan implementation and location of reefs to the definition of construction details and material regulations, the underwater reef metropolis involves a multiscalar and sophisticated design and execution process.

CUSTOMIZED REEFS

The idea of a surrounding or building as Third Natures explores the possibility of artificially modifying our environments to form complex assemblies—ecologies—in which living and inert materials, different social groups and technologically and culturally imbued objects come together in a state of constant interaction and friction. (Moreno and Grinda 2016, 37)

This productive underwater seascape is designed as a highly tailored environment. The spatial and material characteristics of reef module designs include surface texture, reef profile and orientation, shelter and shading, reef sizes, internal surface area, reef configuration, hydrological factors, interstitial spaces between singular modules, and the possibility of human inhabitation or social usage (e.g., space for fishers or divers). Body sizes, preferences, and habits of the targeted marine species drive the construction of particular reef designs with specific shapes, voids, surfaces, and profiles.

The Guidelines for the placement of artificial reefs defined by the United Nations Environment Program indicate that "the higher the surface area available for the settlement of algae and invertebrates, the greater source of food for other levels of the reef community and, therefore the greater productive capacity" (London Convention 2009, 27). The complexity of the reef modules is key to ensuring greater biodiversity. Some species prefer blind-ended holes, and others prefer open-ended ones. The higher the variety of cavities, crypts, refuges, and shapes, and the bigger the surface area of the module, the more diverse and abundant the ecosystem that inhabits them will be. However, the more complex the design, the higher the cost. Often, cost-effective, short-term implementation of reef structures will apply a reductionist construction of the custom-made ecosystems. If architectural typologies respond to the demands of diverse population types, artificial reefs also engage with the ethologic behavior of species. Like the segregation of dissimilar sociocultural groups in cities, artificial reefs use urban and architectural forms to attract and suit different subjects. Indeed, the spatial qualities of each artificial reef module fosters the growth and survival of certain targeted species over others. In both cases, more "profitable" inhabitants are more likely to be targeted. Distinct types of urban settlements and different forms of architectural typologies attract different demographics—after all, a loft downtown does not appeal to the same inhabitants as those of a suburban family house. Accordingly, in each case different demographics are targeted. Similar to neoliberal cities, this underwater metropolis perpetuates the violence of economically driven urban developments, in which the "nontargeted" and "undesirable" subjects have more chances to be expulsed.

REEFS AS LIVING RUINS

The South Korean Reef Metropolis is conceived as a living ruin, as it is built to progressively degrade. The Ministry for Maritime Affairs and Fisheries defines guidelines for reef construction that regulate material characteristics, as well as structural and degradation requirements (Kim 2011, 15–18). In more than 90 percent of the cases, steel and concrete reef modules configure this underwater-scape, as these materials can partially decompose while increasing their porosity to host marine inhabitants. However, the use of concrete is becoming increasingly controversial as its decomposition liberates chemical components and its production involves high emissions of carbon dioxide. Polyvinyl chloride and plastics are excluded from the material palette as they last too long; timber is also excluded—its rapid degradation would not benefit the marine ecosystems either.

Paradoxically, the progressive decay of the reef modules often reflects the exuberant takeover of the reef structures' surfaces, cavities, and profiles by all kinds of marine creatures. Degradation is intrinsic to the life cycle of the reef modules and the creatures that inhabit them, which turns the reefs into a form of techno-mollusk living assemblage in perpetual becoming.

THE SEA AS THE NEW CONTINENT: A POLITICAL AND EDUCATIONAL PROJECT

As with every urban and architectural project, South Korea's metropolitan expansion is also a political venture with ambitions to position South Korea as an oceanic pioneer. Educational institutions such as high schools and universities that specialize in maritime affairs contribute to the oceanic project through the formation of experts on marine-related disciplines, but also in the reinforcement of the collective imaginaries and ideologies. For instance, the Maritime and Ocean University's online website displays a welcoming message from the president of the institution, Dr. Park Han-il. He addresses the young generations of South Korea and highlights the importance of "competing against other advanced countries for ocean resources" such as logistic routes, minerals, food, and energy. He urges students to turn their "focus from the continent to the ocean and prepare themselves for competition in the ocean age." More importantly, he describes the ocean as South Korea's "new continent," which will host the "future of the human race" (2016). This patriotic discourse frames South Korea as a "seafaring nation"; both the national identity and the future urban expansion are invariably oriented toward the oceans.

Likewise, the Minister of Maritime Affairs and Fisheries Kim Young-Choon presented his ambition to defend South Korea's maritime-territorial integrity and enhance its status as global marine leader on the ministry's official website (2017a). He distributed the Inverted World Map at the National Assembly, the central government, and local governments as an instrument to raise public interest in the immense extractive possibilities of the sea. In his words, "This innovative idea encourages my country to infinitely extend its capability towards waters, even uncharted areas" (2017b). However, the inverted map also illustrates the ambition to liberate the country's psyche from the weight of the continental mass of its neighboring countries, thus allowing

South Korea to present itself as an ideological island that turns its back to the mainland and projects its geopolitical strategy toward the seas.

WARFARE REEF-URBANISM

Against the geography of stable, static places, and the balance across linear and fixed sovereign borders, frontiers are deep, shifting, fragmented, and elastic territories. Temporary lines of engagement, marked by makeshift boundaries, are not limited to the edges of political space but exist throughout its depth. Distinctions between the "inside" and the "outside" cannot be clearly marked. . . . The architecture of the frontier could not be said to be simply "political" but rather "politics in matter." (Weizman 2007, 4–5)

The maritime ambitions of the country are also fueled by South Korea's political disagreements with its continental neighbors. China and North Korea have been its political adversaries ever since the end of the Korean War. The demilitarized zone, and the contested maritime boundary at the Northern Limit Line (NLL), is a form of geopolitical theatre: the stage for tensions, mutual provocations and disputes. Additionally, these local tensions have global reach, because of their respective alliances with world powers (Figure 20.1).

On August 30, 1953, the US-led United Nations Command unilaterally delineated the position of the NLL. Just one month after the Korean Armistice Agreement on July 27, 1953 (Young-Koo 2009), this maritime boundary extended the Korean demilitarized zone between North Korea and South Korea into the Yellow Sea. The demarcation did not trace an equidistant line between the continental coastlines of the two countries; instead, five islands belonging to South Korea pushed the NLL toward the North Korean coastline—it was only fifteen kilometers away from shore in some areas. North Korea never accepted this maritime boundary, as it would drastically reduce their access to maritime resources.

Since 1953, North Korea has put forward several proposals for alternative boundaries, but an agreement over the NLL has not yet been reached. However, on April 27, 2018, the two countries adopted the Panmunjom Declaration of Peace, Prosperity and Unification of the Korean Peninsula, which led to the agreement that the NLL would be converted into a maritime peace zone, with the aim of avoiding military clashes and ensuring the safety of fishing and conservation activities.

From the armistice to the recently implemented declaration of peace, the disagreement over the maritime border has led to armed conflicts in relation to

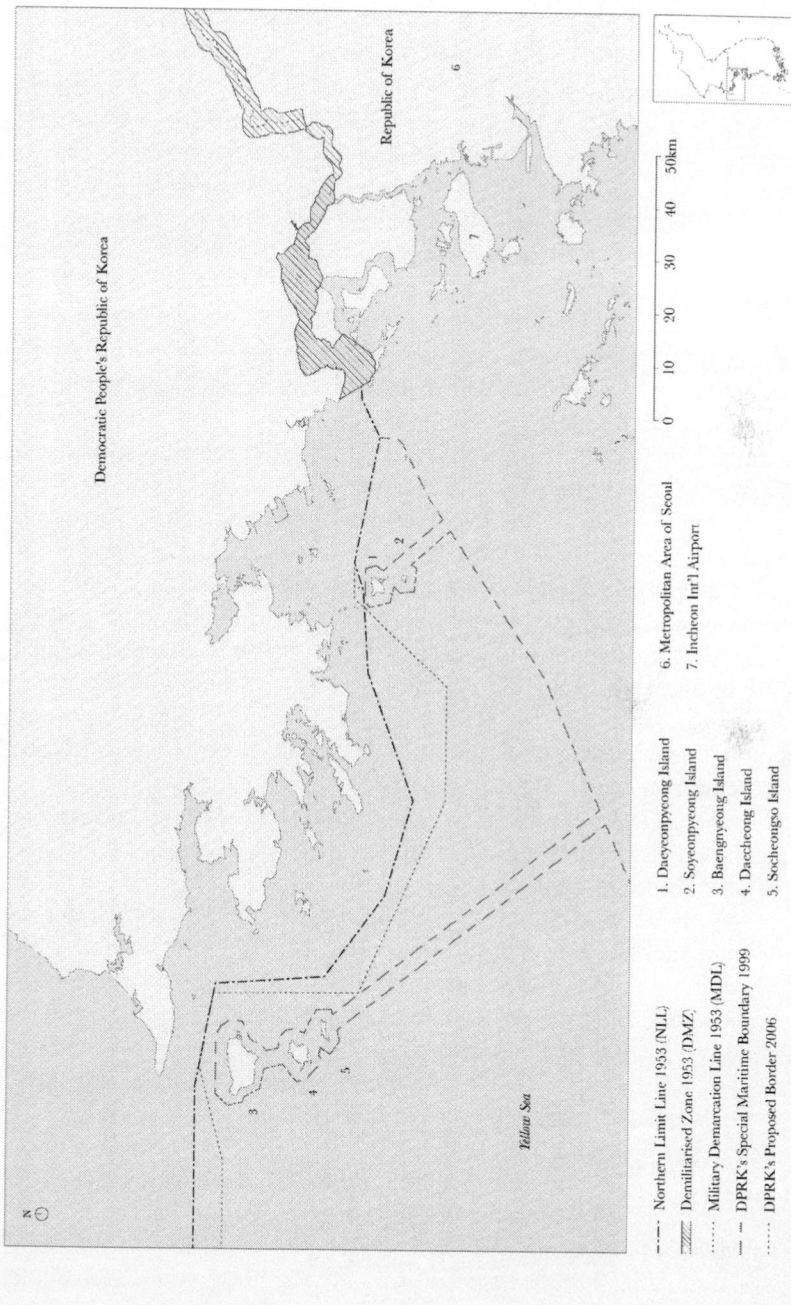

Figure 20.1. Northern Limit Line demarcation lines.
Map elaborated by the authors Amaia Sánchez-Velasco, Jorge Valiente Oriol, Gonzalo Valiente, and Isaac Harrisson.

fishing resources and turned the region into a hotly contested maritime zone. As the two Koreas never agreed to a common maritime boundary for the NLL, Chinese fishing vessels have used this ambiguity to maneuver within the disputed waters. The highly valued blue crab, *Portunus trituberculatus*, is a principal cause of tension: it is one of the main fishing resources in this region, valued both by North and South Korea (Ho 2016). In recent years, an increasing number of Chinese fishermen have ventured into this zone to catch these crabs because of their rising prices. Two to three tons of crabs can cost up to US$70,000. The tensions and the harvesting are synchronized. Conflicts are linked to the crabs' life cycles: they increase during the peak blue crab catchment season, which runs only between May and June (fishing is prohibited from July 15 to September 30). As North Korea has not officially recognized the border, fishing vessels that operate in the proximity or over the border are escorted by North Korean naval boats. Targeted marine species, military patrols, and fishing vessels from North Korea, South Korea, China, and the United Nations Command all perform as geopolitical artifacts in the context of this conflict-ridden seascape.

In 2016, South Korea shifted its strategy: the metropolitan expansion in this area turned into a form of a warfare urban development. The government planned, constructed, and immersed approximately eighty artificial reefs that would turn the ambiguous maritime frontier into a constructed environment. Between June and November, the structures were strategically sited around Yeonpyeong and Daecheong Islands, two of the five controversial islands along the NLL.

Bigger and heavier than other artificial reef typologies (Figures 20.2 and 20.3), these monumental structures weigh dozens of tons. The dimensions of their base range from ten to fifteen meters and their height oscillates between five and eight meters. The structural frame is made of steel and can include concrete or stone elements. The upper part of these "razor reefs" includes hooks purposely designed to cut the fishing nets of Chinese and North Korean commercial vessels operating in the area. Camouflaged as productive and conservation devices, theses reefs are passive-defensive artifacts that render visible the South Korean government's political position in regard to the contested NLL. The dimension of the reefs' modules requires the use of cranes and turns the installation into a mediatic opportunity to broadcast South Korea's national maritime project globally.

South Korean artificial reefs are cultural, political, infrastructural, and architectural artifacts. They organize both human and nonhuman bodies and are capable of reconfiguring hybrid ecological, political, and ecosystemic relationships. Artificial reefs are often invisible structures that dissolve into

Figure 20.2. Axonometric: Architectural drawing of the artificial reef deployed at the Northern Limit Line.
Drawing elaborated by the authors Amaia Sánchez-Velasco, Jorge Valiente Oriol, Gonzalo Valiente, and Joumana Elomar.

processes of conservation, production, and defense across multiple scales. They are structures in perpetual becoming, some form of object-animal that belongs to a much broader physical, conceptual, cultural, and infrastructural network. They expand the city of Seoul and serve as a nature-culture continuum that blurs the boundaries between land and sea.

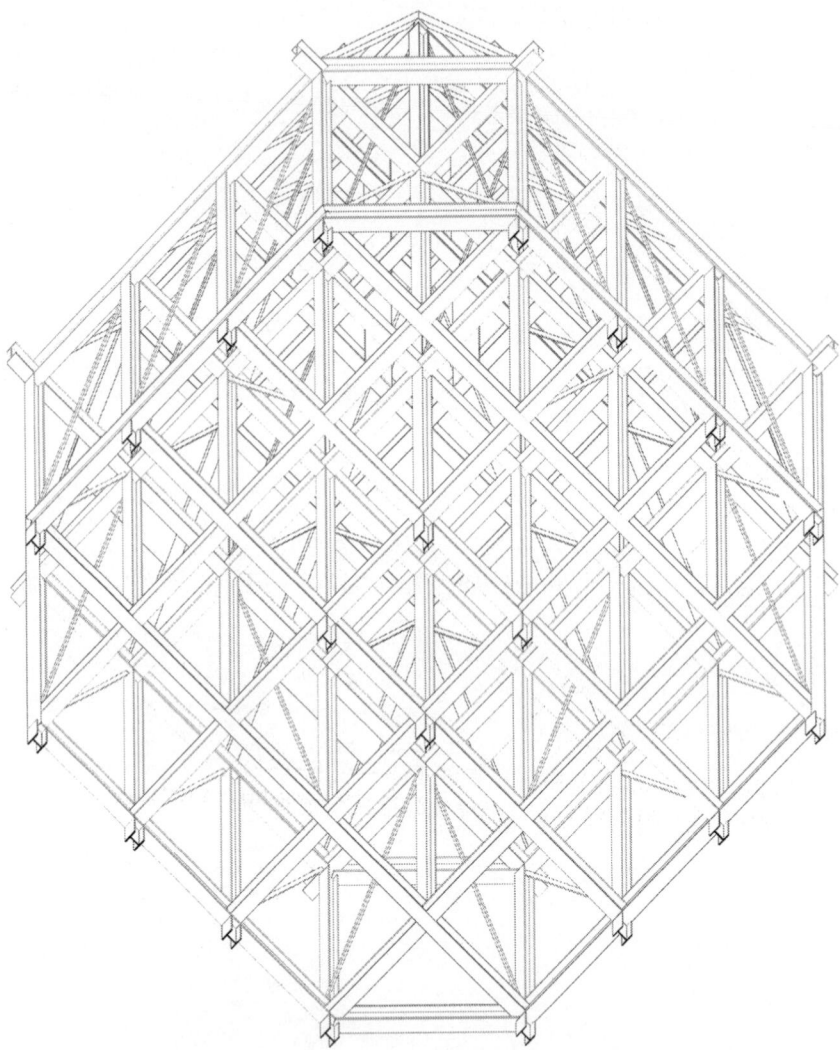

Figure 20.3.   Worms-view axonometric: Architectural drawing of the artificial reef deployed at the Northern Limit Line.
Drawing elaborated by the authors Amaia Sánchez-Velasco, Jorge Valiente Oriol, Gonzalo Valiente, and Joumana Elomar.

South Korea plays a pioneering role in the competitive development of artificial reefs in a global context. The extensive deployment of these structures turns the country's seabed into a reef metropolis and provides a critical scenario for investigating the current maritime urban planning strategies, architectural typologies, construction policies, and oceanic geopolitical boundaries. Under the effects of global warming, pollution, and increasing populations, oceans are, and will be, particularly under pressure. Ultimately, it is crucial to develop alternative political imaginaries and critical trans-disciplinary epistemologies that move beyond the extractivist, profit-driven growth, and real-state logics of the inland urban developments.

ACKNOWLEDGMENTS

We would like to acknowledge Urtzi Grau for his support during the development of this body of work, as well as Joumana Elomar and Isaac Harrisson for their contribution in the development of the images that this article contains.

REFERENCES

Han-il, Park. 2016. "President's Message." Korea Maritime and Ocean University. Accessed July 10, 2019. http://www.kmou.ac.kr/english/cm/cntnts/cntntsView.do?mi=459&cntntsId=455.

Ho, Shawn. 2016. "Tensions in the Yellow Sea: Crabs and the Inter-Korea Border Dispute." *RSIS Commentary*. CO16172. https://www.rsis.edu.sg/wp-content/uploads/2016/07/CO16172.pdf.

Kim, Chan Gil. 2011. "Artificial Reefs in Korea." *Fisheries* 26, no. 12: 15–18.

London Convention and Protocol/United Nations Environment Program. 2009. London Convention and Protocol/UNEP Guidelines for the Placement of Artificial Reefs. London: United Nations.

Moreno, Cristina Diaz, and Efren Garcia Grinda. 2016. "Third Natures: Ten Years of Mundane, Carnal Assemblies." *El Croquis* 184, no. 37: 23–39.

Nae-hui, Kang. 2018. "The Financialization and Gentrification of Space in Seoul: A Cultural Political Economic Analysis." Accessed July 10, 2019. http://insurgentnotes.com/2018/08/the-financialization-and-gentrification-of-space-in-seoul-a-cultural-political-economic-analysis/.

Weizman, Eyal. 2007. *Hollow Land: Israel's Architecture of Occupation*. New York: Verso.

World Bank Group. 2018. "Urban Population (% of total)." The World Bank Group. Accessed June 25, 2018. https://data.worldbank.org/indicator/SP.URB.TOTL.IN.Z S?end=2016&locations=KR&start=1960.

World Wide Fund for Nature. 2017. "Priority Places: Marine." Accessed July 10, 2019. http://wwf.panda.org/knowledge_hub/where_we_work/yellow_sea/.

Young-Choon, Kim. 2017. "Message from the Minister." Minister of Oceans and Fisheries. Accessed July 10, 2019. http://www.mof.go.kr/eng/content/view.do?menuKey=474&contentKey=160.

Young-Koo, Kim. 2009. "A Maritime Demarcation Dispute on the Yellow Sea: Republic of Korea." *Journal of East Asia and International Law 2*, no. 2: 1–12.

Chapter Twenty-One

Living Breakwaters

SCAPE Landscape Architecture

SCAPE Landscape Architecture

Scenarios of increased coastal inundation, sea-level rise, energy supply, water quality, and ecological resiliency were once relatively abstract threats, but today, neighborhoods, cities, and nations are awakening to realize the implications of how we inhabit the world and how our behaviors directly affect our shared built-natural environment. At this moment, when nations are struggling to develop economic and financial models that redirect the most destructive aspects of our carbon-heavy lifestyles, the vision of a new world remains unclear and without consensus. What is clear is that our past approach to problem solving, with one-dimensional, super-infrastructural solutions, has largely failed. Many design firms are working to develop multifunctional approaches that integrate ecological and social systems into infrastructural design. Our work at SCAPE, a landscape architecture and urban design studio in New York City and New Orleans, aims to develop projects and design methodologies that go beyond static, monofunctional, and purely physical solutions to embrace the interrelationships between physical structures, their social and ecological contexts, and our individual and collective actions. Living Breakwaters, a proposal developed for the Rebuild by Design competition off the Staten Island coastline (Figure 21.1), is used in this chapter as an example of this approach.

New York City's historic coastline was crucial to the city's early survival and growth. Its cliff-like geography at the "Narrows" in the outer harbor was enhanced for military fortification, and coastal reefs and marshes provided an abundant source of food and fuel. The geography of this coastline also served to protect inland areas from storms—the outer harbor absorbed wave energy and shielded the inner harbor, where settlement concentrated in Lower Manhattan. Archipelagos of small islands, abundant oyster beds, sandy shoals, protective dunes, and inland swamps formed a complex watery

273

Figure 21.1.   Living Breakwaters off the South Shore of Staten Island.
© SCAPE Landscape Architecture.

edge, while human settlement tended to occur along roads located on ridge lines on high ground. Over time, this three-dimensional bathymetry—or underwater topography—was dredged to make way for shipping vessels and oil tankers to support commerce. At the same time, streams were filled in and canalized, and coastal edges straightened. Eventually, the intertidal zone was transformed into a flat, low-lying area inhabited by the City's burgeoning industries. Although this large-scale industrial transformation of the harbor supported robust economic and cultural development, it erased the "ecological infrastructure" of marshes, maritime forests, dunes, oyster reefs, and other intertidal ecosystems that once buffered and protected our shore.

The vulnerabilities of a hardened and extended shoreline became clear on October 29, 2012, when Superstorm Sandy made landfall in the northeast United States. Sandy affected thirteen states and cost more than $65 billion in damages and economic loss. In the wake of the storm, President Barack Obama launched the Hurricane Sandy Rebuilding Task Force under the direction of the Department of Housing and Urban Development, setting the seeds for a design competition to develop innovative design solutions to address the Sandy-affected region: Rebuild by Design. Rebuild by Design was a year-long, collaborative, interdisciplinary, and community-driven design process that resulted in innovative resilience proposals for some of the most affected communities in the region. SCAPE's project for the initiative, Living Breakwaters, focused on restoring the function of the shallows while creating contemporary living infrastructure that would protect vulnerable neighborhoods close to the water's edge. The project started with a conceptual design presented as part of the Rebuild by Design competition, but became a funded project for the Governor's Office of Storm Recovery with $60 million of Community Development Block Grant Disaster Recovery (CDBG-DR) funding allocated by the US Department of Housing and Urban Development. The project is in the implementation phase.

LIVING BREAKWATERS PROJECT

Living Breakwaters was conceived to connect physical, social, and ecological resilience (Figure 21.2). The proposal is a necklace of offshore breakwaters that reduce risk, revive aquatic ecosystems, and connect residents and educators to Staten Island's southeast shoreline. The Living Breakwaters are strategically sited relative to the South Shore of Staten Island to reduce risk for coastal communities, build reef habitat for juvenile fish and shellfish, and enhance waterfront recreation and stewardship. In areas of high land loss and risk to infrastructure, they are located to encourage sedimentation and help build back the beach. In more ecologically or economically sensitive zones, such as hard clam habitat areas, their footprint is minimized or realigned. The physical structures provide habitat to the Raritan Bay's rich ecosystem of marine life, and an on-land Water Hub fosters social and stewardship programs that enhance the awareness and care of the resource. The project is in collaboration with many partners, including the New York Harbor School and the Billion Oyster Project, who are working to restore oysters in Raritan Bay and teach the next generation of ecological stewards about protecting Staten

Figure 21.2. Physical and Ecological Resilience.
© SCAPE Landscape Architecture.

Island's fragile coastline. The project strengthens ongoing shoreline improvements and creates opportunities for rethinking the coastal edge, engaging people with the water rather than building barriers and walls.

THE PHYSICAL BREAKWATERS

Breakwaters are a common technique used by engineers worldwide to reduce wave action and build shorelines. The Living Breakwaters project has similar engineering goals, and is designed to also increase underwater habitat types through physical microcomplexity that is beneficial for a diversity of species, including juvenile finfish and shellfish (Figure 21.3). The breakwaters provide habitat space throughout the water column, from a subtidal structure to above-water islands. Unlike typical breakwaters, Living Breakwaters incorporates underwater, small-scale pockets—or reef streets—into its edges to provide foraging and shelter for juvenile fish. Reef streets are made of stone boulders and ecological concrete units, for diversity and habitat generation. Ecological concrete has been used in multiple applications in Israel, New York City, and Europe for seawall construction, tide pool creation, and ar-

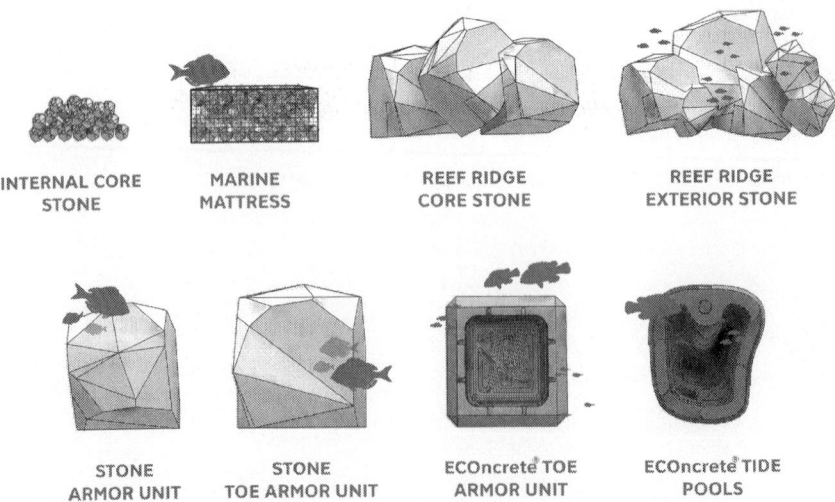

Figure 21.3. Breakwater Construction and Habitat Techniques.
© SCAPE Landscape Architecture.

mor unit construction, and has been shown to foster marine life and biogenic buildup at a level higher than typical concrete. Above water, the breakwaters will potentially host harbor seals and nesting birds, providing a habitat free from predators. Juvenile fish will hide from predators in the matrix of ECOncrete and stone and feed in the narrow waters of the reef streets. This constructed habitat will attract fish of all ages, and in turn, fishermen looking for a perfect fishing hole.

OYSTER RESTORATION

Reefs and leased oyster beds once extended across the shallow water flats of Raritan Bay, reducing storm effects and filtering water. The Living Breakwaters project incorporates oyster restoration as one of many critical habitats of the breakwater design. The team worked with project partners at the New York Harbor School and the Billion Oyster Project (Figure 21.4) to propose multiple techniques—from oyster gardens to spat on ECOncrete units—to restore oysters to the Raritan Bay in a measured and monitored way. Along with filtering the water, oysters will biogenically build along with the threat of climate change to increase the strength of the protective breakwater system, adding habitat while enhancing physical integrity over time.

Figure 21.4. Community Engagement for Living Breakwaters.
© SCAPE Landscape Architecture.

BUILDING SOCIAL RESILIENCY

The competition design proposed that Staten Islanders reconnect to the coast through a network of highly visible and programmed Water Hubs to become places that foster social cohesion and provide orientation, information, storage space, and group gathering spaces, as shown in Figure 21.5. Water Hubs are to be designed through community design charrettes where programs and needs are identified and incorporated into the structures. Each site condition and community need create a different type of hub—embedded, floating, cantilevered, or elevated. At Tottenville, potential Water Hub programs include community kayak storage suggested by Kayak Staten Island, wet-lab science tools for use by Tottenville High School, and bird- and seal-watching platforms replacing those lost at Conference House Park during the storm. These programs, combined with amenities such as bathrooms and water fountains, vastly expand the age range and experience level for water-based activities available to residents. The Water Hub acts as a beacon for orientation, a place of gathering, and bridges the layered dune system below.

Conventional techniques that promise protection by erecting a barrier between people and water ultimately sever our visual and physical relationship to the water. The Living Breakwaters project is representative of a larger layered, ecological approach that makes communities safer while increasing our perception of risk and landscape process by building a landscape scale intervention that integrates aquatic habitat and community access. The project is well on its way to becoming a constructed reality, as the proposal

Figure 21.5.   Resilient Water Hub and Shoreline.
© SCAPE Landscape Architecture.

was awarded to New York State and is currently being implemented by the Governor's Office of Storm Recovery with $60 million of CDBG-DR funding allocated for this project. Pending permit approvals, construction will take place over two years, and is anticipated to be complete by the beginning of 2021. The physical construction marks just the beginning of a cycle of growth, stewardship, and adaptation that can set a new path for resilient infrastructure design in a changing climate.

Chapter Twenty-Two

Sustaining the Seas through Interdisciplinary Songwriting

Kim Williams, Sarah M. Hamylton, Lucas Ihlein, and Leah Gibbs

Swim, swim
We swim in a rising sea
If parrotfish had political teeth
They'd bring the colour back to the Barrier Reef
Rock the boat, rock the boat

The health of Australia's Great Barrier Reef is in decline, largely because of the increased frequency of coral bleaching events (Hughes et al. 2017). Global warming, driven by carbon emissions from fossil fuels, elevates sea surface temperatures. This is the primary cause of coral bleaching, a phenomenon whereby the thermal threshold of a coral is exceeded, and the photosynthetic algae that live in its skin (from which the coral derives its color) is lost, leaving the coral without its primary source of food and susceptible to mortality (Hoegh-Guldberg 1999). Several global mass coral bleaching events have occurred since 1983, during which record high temperatures triggered pantropical coral mortality. On the northern Great Barrier Reef, a recent extreme mass coral bleaching event in 2016 killed up to 80 percent of live coral on reef crests (Terry Hughes et al. 2018).

The first serious environmental challenges to the Reef arose in the 1960s, when the Queensland State Government supported plans to mine the Reef for fertilizer, cement, gas, and oil (McCalman 2017, 77). From 1965 to 1975, Australian poet Judith Wright, artist John Busst, and forester Len Webb worked together in a decade-long campaign to protect the Reef. Combining their talents and resources, Wright, Busst, and Webb used the power of poetic language, the persuasiveness of scientific evidence, and the good fortune of political connections to battle the Queensland Government's support of commercial exploitation of the Reef. Ultimately, they were successful and their advocacy resulted in the Reef's designation as a Marine Park in 1975, with special protections. In 1981, the Great Barrier Reef was made a World Heritage Site. Yet Wright was under no illusions. Later, she wrote of the ever-present threats to the Reef, "The Reef's fate is a microcosm of the fate of the planet. The battle to save it is itself a microcosm of the new battle within ourselves" (2014, 186).

Today the challenges to the Reef are local and global: among the key threats, chemical and sediment run-off from land-based agriculture, coal mining, and overfishing are the responsibility of state and federal governments, and climate change represents a far more complex set of global problems. Marine scientists aim to convey to decision-makers the seriousness of the Reef's predicament. The data are unequivocal. Many in the scientific community are frustrated that after decades of research, effective large-scale responses to coral reef decline are still not forthcoming. Scientists have made it clear that, although land-based practices play a role in water quality, the biggest threat to coral reefs around the world is climate change (Lesley et al. 2018). In Australia, government leadership fails to develop appropriate emissions reduction policy; as of April 2019, it is not on track to reach its targets under the Paris Agreement. Given that science research and communication are not generating policy transformation, alternative approaches are needed.

INTERDISCIPLINARY COLLABORATION IN *MAPPING THE ISLANDS*

In recent years, marine scientists have acknowledged the limited capacity of their work to bring about decisive action on climate change (Hamylton 2018). Some have turned to collaborating across disciplines to experiment with new approaches, combining methods and exchanging knowledges to encourage stronger responses to reef decline. Working across disciplinary boundaries unsettles established patterns of thinking and behavior. It has the potential to shift paradigms in both the academy and environmental management.

It is this current academic trend to foster interdisciplinary collaboration to solve "real-world" problems that sparked the project "Mapping the Islands: How Art and Science Can Save the Great Barrier Reef." Funded by the University of Wollongong's Global Challenges Program, a strategic research initiative to encourage transformative interdisciplinary research under the theme "Sustaining Coastal and Marine Zones," the project's initial brief was for coastal geographer Sarah Hamylton to remap a number of islands on the Reef that had been mapped forty-five years previously, with a view to observing changes to the islands. Hamylton was interested in how artists might approach mapping and how the involvement of other academic disciplines could influence her work. Human geographer Leah Gibbs and artists Lucas Ihlein and Kim Williams were invited to collaborate. Hamylton organized a fieldtrip: seven days on a boat in the central and northern Great Barrier Reef with marine scientists, a human geographer (Gibbs), an artist (Williams), and a host of others, visiting four islands.

Hamylton led the task of mapping the islands. The other stated objective was to create space for an unspecified collaborative process between Hamylton, Gibbs, and Williams. What emerged is the subject of this chapter: collaboration through music. The journey moves from mapping to music and beyond, through open-ended play and experimentation. Songwriting and singing are examined as potential ways to forge collegial relationships, bring our research to nonacademic audiences, and open new possibilities for academic research and impact. We explore how, in bringing a new voice to pressing environmental issues, collaboration across disciplines might offer useful ways of doing environmental research.

MUSIC IN ENVIRONMENTAL AND SOCIAL JUSTICE CAMPAIGNS

The arts have played a key role in environmental and social justice campaigns over many decades. The growing field of science communication recognizes the value of pictures, theater, music, and poetry to prompt deeper engagement with environmental issues, particularly climate change (Evans 2014). This builds on a considerable historical national and global foundation. At the same time as Judith Wright was making passionate pleas to protect the Great Barrier Reef through writing and poetry (Wright 1977), the turbulent politics of the 1960s and 1970s saw the rise of antiwar, feminist, and civil rights movements. Within these movements, songs of protest and social commentary have galvanized widespread campaigning and political action, notably by American songwriters such as Joan Baez, Nina Simone, Pete Seeger, and Woody Guthrie, who symbolized these movements.

In the 1960s, Ted Egan wrote "Gurindji Blues" (1969) with Vincent Lingiari, at the height of the Gurindji land rights struggle and in the wake of the Wave Hill walk off in the Northern Territory. Elsewhere, Chilean songwriter and musician Violetta Parra pioneered the Nueva Cancion (New Song) movement, which aligned itself with the democratic presidential campaign of Salvador Allende. Into the 1970s, Parra inspired a generation through songs of social commentary, preserving folk traditions through compositions that honored regional Chilean musical styles. This political song movement "[gave] a voice to the people, bringing out what actually matters to people that are not quite empowered and giving them more power" (Ford and Rojas 2017). In Australia, rock band Midnight Oil confronted political issues over several decades through anthemic songs such as "Blue Sky Mine" (1990), and Yothu Yindi's "Treaty" (1991) sang of Indigenous civil rights. Kev Carmody and Paul Kelly's "From Little Things Big Things Grow" (1993) built on stories previously told about the Gurindji land rights struggle, and spoke of the growing reconciliation movement. The song continues to be recorded and performed by artists today.

Poetry and songwriting continue to have a role in contemporary environmental campaigns, as can be seen in public demand for stronger action on climate change. In 2017, the journal *Plumwood Mountain* published a volume of collected poems "Poets Speaking up to Adani" that emerged from a day of action, staged online, in which forty-three poems were posted as text, audio, or video, carrying messages about the environmental impact of the proposed Adani coal mine (Elvey 2017). In December 2018, thousands of protesters attended the Stop Adani march and rally in Sydney. Chants rippled through the crowd, bolstered by percussion, a brass band, a bagpipe player. Chanting with voices, rhythm, and repetition brought cohesion and a shared sense of purpose. It was a powerful and simple technique that "gave voice" to people's concerns about the exploitation of coal in the Galilee Basin and the impact of coal on both the Great Barrier Reef and global climate change.

These traditions of political and social commentary across the arts offer rich context from which our explorations of music proceed. We examine the capacity of popular protest music to act as a conduit for shared ideas, expressing environmental issues simply and anthemically. Central to this creative strategy is collaboration; concurrently we explore the mechanisms and impacts of interdisciplinary collaboration. We ask the question: how can music—and particularly songwriting—provide an avenue for interdisciplinary collaboration?

WRITING A SONG ON THE GREAT BARRIER REEF

On the Great Barrier Reef, we spent the week moving between four low, wooded islands on a sea in constant motion. The research vessel rocked back and forth, causing seasickness in some of the passengers. Days were spent conducting aerial drone surveys on the islands, underwater surveys from the support vessel, and scuba dive surveys on the fringing reefs. Land-based surveys were carried out on foot by circumnavigating each island. Each of these islands, though unremarkable from a tourist's perspective, offered a rich array of material for the scientist, the geographer, and the artist. Underwater, small recruits of baby living corals were scattered thinly among a landscape largely composed of bleached or algae-covered dead coral. On the shore, the natural flotsam thrown up by waves, currents, and winds was entangled with a vast array of human detritus: thongs, fridge doors, fluorescent tubes, and, most starkly, plastics of all kinds.

Back on the boat, evenings were spent in the saloon. In this relaxed atmosphere, spontaneous singing and playing led to the collective writing of a song about the Great Barrier Reef, inspired by the rocking boat. Songwriting provided a light-hearted activity that was a timely and welcome antidote to the physical and mental strains of days in the field. Collaborative writing of the lyrics to *Rock the Boat* (Williams et al. 2018) wove together marine science, environmental politics, and rhythm and meter. The process revealed the capacity of music and song to bring people together to share a joyful experience and transmit complex ideas through lyric and melody. It revealed our knowledge of and passionate views about the Reef, evident through our lyrics, a blend of science, pop poetry, and activism. Later, we continued the songwriting process by developing a set of extensive footnotes, embedded in the musical notation of the song, to capture some of the cross-disciplinary discussion that took place on the boat. The footnotes allow diverse knowledges to expand the compact text of the lyrics into a broader conversation. For brevity, we have included only a few footnotes; a full version is available alongside the audio files of the recorded track (https://bluespottedrays.bandcamp.com/releases) and also with the published musical notation (Gibbs et al. 2019).

Rock the Boat

Verse 1: *Say you wanna see anemones*
Float around the reef awhile[1]
Take a little time to cool it down
Gather all your friends and make a sound
Chorus:
Rock the boat . . . rock the boat
Rock the boat . . . rock the boat
Verse 2:
Coral ain't made for a boilin' pot
Nemo likes the water cool not hot
Makin' way for the blue spotted ray
Wave your little fins and have your say
Chorus

Bridge:
Swim . . . swim
We swim in a rising sea
If parrotfish had political teeth[2]
They'd bring the colour back to the Barrier Reef
Chorus
Verse 3:
On and on the turtles hatch
Damselfish, keep tending their patch[3]
If you want your reefs to stick around
Better keep the carbon in the ground[4]
Chorus to finish

SINGING AND SONGWRITING
FOR INTERDISCIPLINARY COLLABORATION

Alongside the pragmatics of mapping the reef islands and writing a song on the boat, the possibilities opened up by the process of interdisciplinary collaboration have emerged as a further focus of our work together. Philosopher Margrit Shildrick regards meaningful interdisciplinary collaboration as a form of Deleuzian assemblage (Shildrick et al. 2018, 5). The concept of assemblage asserts that sense or meaning is provided by the *connection between* specific concepts and the *arrangement of* those concepts (Deleuze and Guattari 1988). It is through this specificity that songwriting offers insight for our collaborative research.

Shildrick's research explicitly brings together the arts, biosciences, and humanities "to explore the complexities of heart transplantation without privileging any one discourse." Adopting the idea of a research assemblage, Shildrick found that working alongside her collaborators "breaks through disciplinary silos to enable a fuller comprehension of the significance and experience of heart transplantation in both theory and practice." Through her experience of collaborative work Shildrick poses a series of questions, which we adopt to help us examine our own project:

- What happens when people collaborate across disciplines?
- How can we do it well?
- What new insights might emerge?
- How might collaboration change the way we think?

WHAT HAPPENS WHEN PEOPLE
COLLABORATE ACROSS DISCIPLINES?

The published literature on interdisciplinary research methods foregrounds the challenge of bringing people from different disciplines together, including the tension of reconciling quantitative and qualitative methodologies (Lach 2014) and the need to define a new focal scope in the absence of disciplinary boundaries (Lyall et al. 2015). That no individual researcher in our team works with a full knowledge of the project's ultimate end-point has sustained these tensions. These tensions were productive in terms of building relationships that truly incorporate different disciplinary viewpoints. This requires each collaborator to experience a degree of disciplinary discomfort, faced with unfamiliar expertise, and accept that the emphasis and focus of the project may continue to shift through time.

Creating the song *Rock the Boat* gave us a collective confidence that none of us had alone. Some of us could play guitar or ukulele, others were happy to sing or shake maracas. None of us are professional musicians. As the group puzzled together over lyrics, the conversation turned from damselfish grazing habits to rates of sea level rise to the number of syllables we could squeeze into a line of the song to the hopelessness of Australian energy policy and political corruption. Most of us had never written a song before. The process took us beyond the scope of our individual experience, skills, and knowledge, but more importantly, it granted us a freedom to reach beyond our customary practices and disciplinary conventions for conveying our understanding of, and opinions about, the environment.

HOW CAN WE DO IT WELL?

Interdisciplinary work takes time. It entails building relationships, including respect for, and trust in, each other's judgment, as well as learning about the objectives, underlying philosophies and practices of other disciplines, and accepting the limits of one's own knowledge domain (Bridle et al. 2013). An interdisciplinary collaboration is therefore well supported by participants who have patience with the research process. Drawing on the idea of a "knowledge democracy," in which all actors have access to and ability to put their knowledge forward in the process of solving societal problems, Bunders and colleagues (2010) distinguish two dimensions of interdisciplinary collaborative research as the *degree of knowledge input* in the project, and the *degree to which nondominant actors are explicitly involved in the decision-making that shapes the research agenda*. The act of songwriting emerged as an inclusive experience in the sense that all participants were present and actively engaged in writing the song, with adequate consideration of their contributions, viewpoints, and expertise. The later inclusion of footnotes to the song was a useful tool for representing broader dialogue that occurred as the lyrics were put together, and formed an additional layer to our interdisciplinary work. This allowed the song to go beyond an assemblage of human voices and musical notation, providing a novel basis for connecting the ideas within the song lyrics to scientific data and political commentary.

Identifying and selecting a single initial objective—creating a song—provided a focus for working together. It demanded choosing a disciplinary approach, which in turn required all of us to sit with the unknown, to work toward unfamiliar objectives that demanded a leap of faith. Our emergent relationship of respect, and practice of listening and collaborative decision-making, helped each of us trust the process.

WHAT NEW INSIGHTS MIGHT EMERGE?

Returning to the conceptualization of interdisciplinary collaboration as assemblage, the process of songwriting offers a new way to *connect* and *arrange* concepts. As the scientist's familiar delivery of content moved from the prescriptive template of introduction, methods, results, and discussion, so decisions about what ideas were of greatest interest, and how to connect and present them, evolved. When each of us writes outside of our discipline, we do so in an emerging interdisciplinary field that is not so thoroughly mapped as the established disciplines to which we belong. As the format of what is being written shifts from journal article to song, the objective of the writing changes. This opens up intellectual space where researchers have a wide set of ideas from which to draw, and a broad range of methods and practices for arranging ideas. It's not just about what is communicated, but *how*. For example, in the lyric "Coral ain't made for a boilin' pot / Nemo likes the water cool not hot," the songwriting process allows us to strategically adjust scientific findings for literary and affective ends. For example, the popular character from the 2003 Pixar film *Finding Nemo* is used, metonymically, to represent the entire Great Barrier Reef ecosystem. Also, poetic license is taken with two scientific facts to dramatize the threat to the reef by rising sea temperatures: strictly speaking, sea surface temperatures are currently well under that of a "boilin' pot," and the optimal aerobic temperature for a reef fish is 29°C, not exactly cool. Thus, the literary conventions of songwriting compress a set of complex ecological facts into a catchy couplet that goes beyond the standard manner of science communication, potentially allowing research insights to be shared *affectively* with broader and different communities.

HOW MIGHT COLLABORATION
CHANGE THE WAY WE THINK?

In the context of the ecological sciences, Scheffer and colleagues (2015) outline a duality in the way scientists typically think, based on both intuition and reasoning. *Intuitive* thinking can help us see novel solutions and associations instantly, but is prone to error; *reasoning* checks and modifies results and represents the generally accepted ways of doing science. They emphasize the value of unstructured socializing time, shared exploratory experiences, and cooperation between artists and scientists, for stimulating scientific progress by encouraging intuitive thinking. In this regard, songwriting in our project emerged as a productive approach to interdisciplinary collaboration as it created the conditions for shared structured and unstructured dialogue from which generative discussions emerged. These went beyond educating each other about our respective disciplines to engaging in each other's disciplinary practices and also to entering into a *new shared space not fully belonging to any of us.* The production of a song contrasted with the standard outputs valued in many disciplines; it demanded a reconceptualization of the validity of different output forms, including creative outputs.

An authentic collaborative project that involves researchers working together to design a research project from the outset demands that varying disciplinary objectives are encountered early. Whereas a scientist may seek to reduce the world to a measurable unit that can be manipulated with the intention of posing and testing a hypothesis, an artist may seek to open out the thinking of their audience or provoke a discussion, and a human geographer might aim to critique and reveal the sociocultural processes underlying an issue. These three approaches adopt fundamentally different approaches to the research task. When taken together, they constitute an entirely new research question, shifting, for example, away from the cliché of using art to communicate science.

CONCLUSION

Through our interdisciplinary collaboration, a sense of collective confidence that would have eluded each of us alone has emerged from the process of writing a song about the environmental plight of the Great Barrier Reef. This confidence allows us not only to move beyond our individual skills and knowledge, but to extend beyond our customary practices and methods, and reach new audiences with our work. Songwriting has emerged as a democratic form of knowledge production, in the sense that it allows all researchers to be actively engaged in writing lyrics, arranging and performing, and to have their knowledge represented. The work is broadened through the addition of footnotes to the lyrics, placing the research in a hybrid space between essay, in which footnotes are conventional, and song lyrics, which reflect the traditions of popular music. Adopting an unfamiliar objective from the outset takes each of us outside our customary way of doing research. In bringing together artists, scientists, and a social scientist to engage in a shared research exploration, the act of writing a song combines intuitive thinking with reasoning to provide a new avenue for connecting and arranging concepts in generative dialogue.

ACKNOWLEDGMENTS

This work was funded by a University of Wollongong Global Challenges Program grant: "Mapping the Islands: How Art and Science Can Save the Great Barrier Reef" and supported by a residency at the Bundanon Trust Artist-In-Residence program.

NOTES

1. The Great Barrier Reef (GBR) attracts more than 1.6 million visitors per year (Australian Government, n.d.). It is the world's most extensive coral reef ecosystem. In 1981, it was listed as a World Heritage Site, based on four of the World Heritage Convention's ten criteria: (vii) "The GBR is of superlative natural beauty above and below the water"; (viii) it "is a globally outstanding example of an ecosystem that has evolved over millennia"; (ix) "the globally significant diversity of reef and island morphologies reflects ongoing geomorphic, oceanographic and environmental processes"; and (x) "it is one of the richest and most complex natural ecosystems on earth, and one of the most significant for biodiversity conservation" (UNESCO, n.d.). As well as protecting tremendous ecological and physical diversity, the GBR contributes more than $5 billion to the Australian economy, and generates about sixty-three thousand jobs (UNESCO, n.d.).

2. Who does have political teeth? The Grattan Institute report *Who's in the Room? Access and Influence in Australian Politics* (Wood and Griffiths 2018) highlighted the vulnerability of Australian politics to well-resourced interests (big businesses and industry coalitions) that use money, resources, and relationships to influence policy to serve their interests, often at the expense of the public interest. In *The Coal Truth*, David Ritter (2018) exposed multiple interweaving sources of political power in the Australian coal industry, stemming from excessive spending on political lobbying and donations and exaggerated claims of the national economic significance of coal. Another strong influential force within Australia's political systems is the continual shifting of individuals in and out of roles in politics, fossil fuel industry groups, and energy and mining companies in a phenomenon dubbed the "revolving door" (Krien 2017). Such influence operates through promises made, factions formed, donations offered, and royalties and taxes agreed, and is thought to underpin remarkable changes in the attitudes of elected prime ministers toward coal (noteworthy transitions include Tony Abbott, Julia Gillard, Malcolm Turnbull, and Scott Morrison). Against such powerful lobbying, how can the political teeth of parrotfish—and other nonhumans invested in the survival of the Great Barrier Reef—be sharpened?

3. Herbivorous damselfish are notoriously territorial and adopt "farming" behaviors to maintain their algal "lawns" (Lassuy 1980).

4. A report by the Climate Council (Steffen 2015) estimates that if the coal from the 250,000-km^2 Galilee Basin in western Queensland was burned, it would release

705 million tons of carbon dioxide each year, more than 1.3 times Australia's current annual emissions. Consequences for the reef have been reported by scientists for decades. They include increased ocean acidification, which reduces the capacity of corals to build their calcified skeletons (Kleypas et al. 1999) and coral bleaching (Hoegh-Guldberg 1999).

REFERENCES

Australian Government Department of the Environment and Energy. n.d. *World Heritage List: The Great Barrier Reef, Queensland.* http://www.environment.gov. au/heritage/places/world/gbr.

Bridle, Helen, Anton Vrieling, Monica Cardillo, Yoseph Araya, and Leonith Hinojosa. 2013. "Preparing for an Interdisciplinary Future: A Perspective from Early-career Researchers." *Futures* 53: 22–32.

Bunders, Joske F. G., Jacqueline E. W. Broerse, Florian Keil, Christian Pohl, Roland W. Scholz, and Marjolein B. M. Zweekhorst. 2010. "How Can Transdisciplinary Research Contribute to Knowledge Democracy?" In *Knowledge Democracy*, edited by Roeland J. in't Veld, 125–52. Heidelberg: Springer.

Deleuze, Gilles, and Félix Guattari. 1988. *A Thousand Plateaus: Capitalism and Schizophrenia.* New York: Bloomsbury.

Elvey, A. 2017. "Poets Speaking Up to Adani." *Plumwood Mountain* 5, no. 2: 1–322.

Evans, Eleri. 2014. "How Green Is My Valley? The Art of Getting People in Wales to Care about Climate Change." *Journal of Critical Realism* 13, no. 3: 304–25.

Ford, Andrew, and Daniel Rojas. 2017. "Violetta Parra: 100 Years of a Nueva Canción Pioneer," *The Music Show*, ABC Radio National, October 8, 2017. Available at https://www.abc.net.au/radionational/programs/musicshow/violeta-parra-100-years/9025072.

Gibbs, Leah, Kim Williams, Sarah M. Hamylton, and Lucas Ihlein. 2019. "'Rock the Boat': Song-Writing as Geographical Practice." *Cultural Geographies.* DOI: 10.1177/1474474019886836.

Hamylton, Sarah. 2018. "Make Your Science Count." *Nature* 554, no. 7690: 137.

Hoegh-Guldberg, Ove. 1999. "Climate Change, Coral Bleaching and the Future of the World's Coral Reefs." *Marine and Freshwater Research* 50, no. 8: 839–66.

Hughes, Lesley, Annika Dean, Will Steffen, and Martin Rice. 2018. "Lethal Consequences: Climate Change Impacts on the Great Barrier Reef." Climate Council of Australia Limited. Available at https://www.climatecouncil.org.au/resources/climate-change-great-barrier-reef/.

Hughes, Terry P., Michele L. Barnes, David R. Bellwood, Joshua E. Cinner, Graeme S. Cumming, Jeremy B. C. Jackson, Joanie Kleypas, et al. 2017. "Coral Reefs in the Anthropocene." *Nature* 546, no. 7656: 82.

Hughes, Terry P., James T. Kerry, Andrew H. Baird, Sean R. Connolly, Andreas Dietzel, C. Mark Eakin, Scott F. Heron, et al. 2018. "Global Warming Transforms Coral Reef Assemblages." *Nature* 556, no. 7702: 492.

Kleypas, Joan A., John W. McManus, and Lambert A. B. Meñez. 1999. "Environmental Limits to Coral Reef Development: Where Do We Draw the Line?" *American Zoologist* 39, no. 1: 146–59.

Krien, Anna. 2017. *Quarterly Essay 66 The Long Goodbye: Coal, Coral and Australia's Climate Deadlock*, volume 66. Carlton, Victoria, Australia: Quarterly Essay.

Lach, Denise. 2014. "Challenges of Interdisciplinary Research: Reconciling Qualitative and Quantitative Methods for Understanding Human–Landscape Systems." *Environmental Management* 53, no. 1: 88–93.

Lassuy, Dennis R. 1980. "Effects of 'Farming' Behavior by *Eupomacentrus lividus* and *Hemiglyphidodon plagiometopon* on Algal Community Structure." *Bulletin of Marine Science* 30, no. 1: 304–12.

Lyall, Catherine, Ann Bruce, Joyce Tait, and Laura Meagher. 2015. *Interdisciplinary Research Journeys: Practical Strategies for Capturing Creativity*. New York: Bloomsbury.

McCalman, Iain. 2017. "Linking the Local and the Global. What Today's Environmental Humanities Movement Can Learn from Their Predecessor's Successful Leadership of the 1965–1975 War to Save the Great Barrier Reef." *Humanities* 6, no. 4: 77.

Ritter, David. 2018. *The Coal Truth: The Fight to Stop Adani, Defeat the Big Polluters and Reclaim Our Democracy*. Crawley: University of Western Australia Press.

Scheffer, Marten, Jordi Bascompte, Tone Bjordam, Stephen Carpenter, Laurie Clarke, Carl Folke, Pablo Marquet, et al. 2015. "Dual Thinking for Scientists." *Ecology and Society* 20, no. 2.

Shildrick, Margrit, Andrew Carnie, Alexa Wright, Patricia McKeever, Emily Huan-Ching Jan, Enza De Luca, Ingrid Bachmann, et al. 2018. "Messy Entanglements: Research Assemblages in Heart Transplantation Discourses and Practices." *Medical Humanities* 44, no. 1: 46–54.

Steffen, W. 2015. *Unburnable Carbon: Why We Need to Leave Fossil Fuels in the Ground*. Climate Council of Australia. Available at https://www.climatecouncil.org.au/resources/galilee-basin-unburnable-coal/.

United Nations Educational Scientific and Cultural Organization (UNESCO). n.d. *Great Barrier Reef*. https://whc.unesco.org/en/list/154.

Williams, Kim, Sarah Hamylton, Leah Gibbs, Lucas Ihlein, Rafael Carvalho, and Mystery Carnage. 2018. *Rock the Boat*. Recorded by Silver Sound Recording Studio, Wollongong. Available at Bandcamp: https://bluespottedrays.bandcamp.com/releases.

Wood, Danielle, and Kate Griffiths. 2018. *Who's in the Room? Access and Influence in Australian Politics*. Melbourne: Grattan Institute.

Wright, Judith. 1977. *The Coral Battleground*. West Melbourne: Thomas Nelson Australia.

Wright, J. 2014. *The Coral Battleground*, 3rd edition. Melbourne: Spinifex Press.

Section 6

THINKING WITH SEAS

Chapter Twenty-Three

The Sea Is Time

Contestations of Temporality in
J. P. Clark-Bekederemo's The Raft

Henry Obi Ajumeze

As my body enters the sea, I enter an indigenous thought-world stimulated by cultural memory, imagination, perception, and understanding. Time becomes a condition of action in the sea. . . . Entering the time of the sea alters the locus of power.

Karin Amimoto Ingersoll, *Waves of Knowing*

In September 2002, UCLA's Fowler's Museum of Cultural History hosted an exhibition titled *Ways of the Rivers: Arts and Environment of the Niger Delta*, which invokes the term "watery ethos" (Anderson and Peek 2002, 12) to illustrate how the myriad rivers and creeks of Nigeria's Niger Delta not only define a geographic region but also form a cultural confluence. The region is the vast basin consisting of several tributaries where the River Niger empties itself in the Atlantic Ocean, and extends across the southern region of Nigeria that borders along the Gulf of Guinea, covering Cameroon, Equatorial Guinea, São Tomé and Príncipe, Gabon, and Angola. This cultural understanding triangulates the connection between the coastal geography, the water spirits, and the communities of riverine people whose lives are connected to the ways of the river. It is in this sense that one can think about how the waters of the Delta frame a specific culture of involvement—one in which the waters are implicated in the exercise of ecological relationship that underlines human entanglement with the sea.

Tracking a drift in the representation of the rivers in J. P. Clark-Bekederemo's *The Raft* (1964), this chapter finds that a sustained sense of involvement with the waters reveals multiple temporal systems of the coastal people of the Niger Delta. This aspect of knowledge ecology in the context of time suffices as a position of unification of nature and culture that characterizes the epis-temic practices in Africa and needs clarification from the much scalar regime of time in the Anthropocene discourse.

The Raft does not resist the Anthropocene time in its vast narrative trope precisely because it imagines time in its version of transition and navigational power, but figures time more extensively in the form in which it aligns with the ecological practices of everyday coastal life. Published in 1963, the play tells the story of four lumbermen—Olotu, Ibobo, Kingebe, and Ogoro—who are adrift in a boat conveying timber across the River Niger. They have been hired by some middlemen in Warri, a fast growing commercial township in the Delta region, on account of their experiences in ferrying merchandise across the river. Notwithstanding their knowledge of the sea, it is unclear what cut the raft loose as it was firmly fastened to the mooring with reeds in seven places. Yet, "all the knottings are unfastened at once," setting the raft adrift in what figures in the play as an exploration of the Delta's geography of waters. The navigational aesthetics of the play do not seem to emphasize the mobility of the boat as it roams aimlessly in the river; rather, it is revealing of the transformations in the River Niger that are enacted in the process of the oceanic exploration. In other words, *The Raft* is not a play that particularly tropes on the meaning of drifts and flows, but more reflectively about changes performed through the theatricality of exploration in real time. In several in-stances, the conversation of the lumbermen turns to the dangers in the River Niger, the fear that the raft may run into "some sand bank" where baskets and stakes for catching shrimps are accidentally deposited in the river floor, caus-ing a kind of treachery to the shrimps and fishes. The playwright is ostensibly referring to the precarious condition of the sea that results from pollution by local fishermen, if in a less pernicious level than industrial modernity, in the toxicities that happen in the River Niger.

From this perspective, the dangers of the river expand to cover the mul-tispecies that populate the coastal plains, especially highlighted by Ibobo's fears about the predatory presence of the sea cows: "I rather fear it's sea cows. We may/Have, without knowing, planting ourselves/Right in the field of their grazing, and now/They have come and eaten us out of our roots" (95). When the raft sinks into the "great Osikorobo whirlpool" described as a zone of oceanic danger from which the boatmen "may never find release from its vice," the story turns into a struggle to death against forces that climax in resistance against those who journey through the River Niger. In other words, located in the middle of the river, Osikorobo whirlpool represents unspeak-able fear of nature among the Niger Delta people:

Ogro: And we are water-logged here

In Osikorobo—the confluence of all

The creeks!

Olotu: The drain pit of all the earth,

Or are you too caught by fear to say it?

Ibobo: Hold your tongue. (104)

Following the disastrous plunge, the raft breaks into two parts; the part bearing the log takes Olotu away into the tide, prompting Ogro's death as he dives into the river for Olotu's rescue—at this point the play focuses on a clear and distant view of an approaching Royal Niger vessel, *Naragula*, that causes Ogro's end, "caught in the mortal arms/Of that stern-wheeling machine" (119)—leaving Kengide and Ibobo in the domineering spread of fog over the entire seascape. The recourse to early capitalist modernity in its threnodic implication is equally significant in thinking about the transformation that the play announces and in establishing these changes as setting for the condition that Bruno Latour—in thinking about the Anthropocene—calls "agitated and sensitive Earth" (Latour 2014, 4). If time imposes precarity, *The Raft* appears to argue for the cultural invention of time to negotiate and contest the tragic consequences of the anthropogenic backlashes. Enacting several days of the raft's aimless drift in the River Niger to portray how the riverine people survive the perturbations of the sea, the play mobilizes everyday factors of experience to mediate the entanglement of the sea's presence, and to come to terms with the fatalistic incursion of modernity through the ocean's gateway.

In what follows, I track the inherent praxis of temporality in the coastal culture of Nigeria's Niger Delta—rather than the timescale of the Anthropocene that prompted it—to articulate what Stephanie LeMenager describes as "everyday coastal aesthetic" (LeMenager 2014, 43), which implies strategies of ordinary coastal people that are deployed as modes of survival. Although the Anthropocene time is instrumental in considering the ecological and environmental changes in the region's universe of wetlands, this chapter takes its departure in thinking about how cultural mechanisms of temporality have always existed in people's quotidian involvement with the sea. The idea of the Anthropocene—which in itself is a category of temporal conception proposed to mark "the central role of mankind in geology and ecology" (Crutzen and Stoermer 2000, 17)—has implicit transformative significance that is evident in the river's material turbulence. The sea's capacity to produce time appears to mirror the possibility of evolutionary time to alter the status of the waters, a certain sense of the ubiquity of time in the becoming and status of nature. In other words, in thinking about the geologic time, Clark-Bederemo's *The Raft* reminds us that the seas and rivers always already embody the apparatus of time—if as strategy of surviving the tragic apocalypse. Early plays about

Nigeria's Delta illustrate this temporal schema to highlight the complex involvement of the ecology of the region, which James Tsaaior affectively describes as "a congregation of ageless swamps, calm creeks, and primeval estuaries" (Tsaaior 2005, 72).

To be sure, the four lumbermen represent an assembly of Deltans with whom Clark-Bekederemo literally navigates the different nuances of everyday experience of the River Niger to which they are geographically bound up. Through the quartet, Clark-Bekederemo gives us a valuable glimpse into the seascape of the Niger Delta rivers that is a mixtape of ecological agencies, portraying their individualism in the economic and political setting of the region, yet bringing them together in a common space of watery origin—what is fondly described as *Deltaness*—as people united in a "democracy of rivers" (Green 2013, 53). In fact, the region is inhabited by people of different ethnicities who migrated in the early sixteenth century—which is noticeable in the play's characterization that underpins ethnic pluralities. So the play tasks itself with making kin—and culture—bound by the flows of waters. This fact is elaborated when the raft drifts to the pit of Osikorobo whirlpool—an event that presents ocean entanglements. At this point, the lumbermen are shown deploying knowledge of the waves and waters to negotiate the mysteries of Osikorobo, knowledge systems that reflect diverse ethnic differences, yet signifying shades of fluvial ontologies. Here we see Olotu, whose exposure to life in urban centers located outside the Delta region, inscribes him in the frame of what Anna Tsing describes as "possibility of life in capitalist ruins" (Tsing 2014, 1)—and this is instantiated in the dramatization of his wristwatch as well as his death. Clark-Bekederemo is not content with merely showing "the chaotic but rhythmic turbulence of the sea" or demonstrating the vitality of the human effort at maneuvering the forces of the waters, but to register a protocol of interconnection as an ongoing way of living in the region. In emphasizing the nature of this entanglement as the composition of everyday life in the Niger Delta region, I seek to understand how various temporal regimes are culturally invented as modalities of living and surviving the wetlands.

Theater is often, and for very clear reasons, easily inscribed in the spatial turn, but its intersection with temporality is gaining momentum. Helen Gilbert and Joanne Tompkins speak of the contingency of timescape in the unfolding of history in postcolonial drama, describing it as "mythical time that occupies the spatial field, that prefigures the landscape as a temporal category of timelessness" (Gilbert and Tompkins 1996, 137). Engaging with the question of temporality in the early Niger delta drama is diagnostic of significant cultural paradigms, especially considering the suggestion of ways in which the waters reproduce multiple versions of time. Drawing on the season of tides that characterize the Atlantic Ocean, Niger Delta dramatists imagine narratives that articulate the dynamics of the tides in a manner that concatenates with the quotidian life of the characters. Through the tides, nature is endowed with the capacity to exercise agency in which the human subject is drawn in the web of seasonal rounds. Clark-Bekederemo particularly makes the tide the pivot of his dramatic plot, rendering temporality within the context of aesthetic and social agencies, thereby constituting not only the dramatic plot around which the narrative is structured, but also the socioenvironmental lives of the characters in the play in ways that evoke and mirror the experience of living in the timelessness of the Delta. To put it perceptively, this can be explained as the nuance of the dramatic structure with which the play captures the idea of living in the Delta, one that is defined in the manner that time-space is unified, and contested, in the geography of the region.

In *The Raft*, time is largely cast in the mold of the sea and measured in the rhythm and flow of the sea's tidal rise and ebb, producing a broad temporal tradition of seasonality. This way it offers multiple temporal categories, especially for unexpected events that cannot be accommodated within the seasonal structures of time. The scholar Ato Quayson describes this category of time as "[T]he implicit déjà vu of rememory [that] is meant to define a momentous event in the past that produces an excess beyond temporality, thus coexisting with the present as thought-picture with a peculiar experiential force" (Quayson 2003, 134). In Clark-Bekederemo's earlier play *The Masquerade* (1964a), for instance, villagers in the delta community gossip in the opening scene about the rising tide holding out for too long to suspend the market economy and compare it with a similar event that happened "several floods ago." Perhaps *The Masquerade* might be read as a companion to *The Raft* in its figuration of the tide as praxis of time in river thinking. Both speak to the littoral force of flooding in relation with the principality of the moon to reflect on tragic futurity. This spectacularly tragic future is heralded in *The Masquerade* where the villagers—First Neighbor and Second Neighbor—consider the fate of the community in what appears the elemental interplay that climaxes in a disaster of planetary significance:

First Neighbor: Anyone might think it was

The annual flood already

Second Neighbor: Look up, and see! the moon's fresh

bowl

Is quite upturned. It is clearly

Spilling over towards my left

First Neighbor: Yes, that's right; the tilt is prominent

It is never so but there is disaster

General down the whole delta (51)

In this light, the flood appears to produce a structure of time in the historical and cultural imagination of the Delta people, offering a thought-picture in the experience of the community such that frames the referentiality of time. This temporalized memory seems to underwrite a distinct linguistic and metaphoric category in *The Raft* with instances that are registered and heightened by ways in which the nature of the sea supplies the language of time. A good example is the scene in which Ibobo calls Ogro a somnambulist who slips into the sea: "Surely you lose track of events as swiftly/As the tide washes off all floating/well, weren't you up/In your sleep that night, pottering about/ With pole erect in your hands, like a girl/Doing the lost fertility dance?"

The account of Ogro's sleep-walking is described in terms of slipperiness that is characteristic of the sea, making the event assume meaning in the body of the sea's nature. In Quayson's view, this praxis of temporality is framed around the everyday unfolding of experiences. He informs, "The passage of time is measured not just by the hands of the clock . . . , but by registering impressions of seemingly mundane events" (Quayson 2003, 125). This everyday passage of events that dissolve and rememorialize through time's alignment with the flows of the river is key in thinking about the travails of the lumbermen. In focusing on the characters' quotidian habitus and memory of the rivers, the play invokes direct ecological consequences of the dystopic realities of the seascape. It presents a human community struggling within and against the tide of a volatile river. Indeed, Clark-Bekederemo shows that it is at the level of the everyday that one apprehends nature's dystopia, that one captures it in terms that transcend the veneers of metaphor. As Michel de Certeau reminds us, every place is inhered with a constraining order that requires everyday routine to negotiate—a negotiation that compels a space and situation in which one must produce unexpected results while deploying inevitable choices of plurality and creativity (de Certeau 1984, 168). Clark-Bekederemo appears to emphasize this point in the play, privileging it as a kind of eco-drama of everyday coastal experience. In a bid to control the turbulent river, to read the tidal waves, or to cultivate the river's fishes—as the play narrates in different scenes—the playwright demonstrates a plurality of creativity and improvisation. "Habit," writes Stephanie LeMenager, "frays in this unique moment of global ecology, and such fraying indicates a potential shift in human understanding of the everyday" (LeMenager 2014, 527). In *The Raft*, Clark-Bekederemo shows that habit culminates at the autochthonic base of the communities' mode of negotiating their littoral experience, emphasizing a connection between lived experience and the sociotechnological demands of the sea. Throughout the aimless drift of the raft, every material is called to question in their relation with the river, echoing Shewry's view that "an oceanic ecopoetics will have to start with the recognition that our relation to the sea is always already technological" (Shewry 2011, 247). Ogro's use of cigarette cup to collect excreta as bait for fishing vividly illustrates the prevalence of everyday technology in the Delta. As Kengide puts it, "Ogro's is not/ The only fruitful, if somewhat unfamiliar/Manner of fishing"(108), pointing to ways in which the play recognizes the "unfamiliar" and unorthodox techniques with which the Delta communities mobilize everyday experience in surviving the precarious moments of the waters. This everyday awareness of the modalities of living and survival in the littoral universe calls attention to what Stephanie LeMenager describes as "the everyday Anthropocene," which complicates the "time monument" that underwrites the category of the epoch. Arguing that the logic of the epochs is unsuitable in considering the shifts implicit in the everyday lived experiences through ecological and environmental crises, LeMenager recommends that we pay attention to the routines of the everyday in thinking about the Anthropocene. In her words, "epochs are not attentive to the wearing away of the bodies, their slow depletion" (LeMenager 2017, 225). *The Raft* depicts this everyday Anthropocene in ways that reveal how technicalities of everyday experiences forge temporal signatures. To devise multiple temporal mechanisms is a coping strategy in an increasingly precarious sea—but also a caution about the temporality of the sea, an indicator of how systems of time reside in the bodies of the sea.

The result of this is that the tragic experience of the lumbermen is as much seaborne as it is temporal. The play suggests that to lose track of the flow of the tide is symptomatic of loss of time. Given the constraints of human ability to control the navigation of the waves, particularly at the pitch of the night, knowledge of time becomes necessary in tracking the drift of the raft. To that extent, the raft appears not only entangled with the sea's hostile turbulence, but also with the fissures of time. Knowing the time offers clue to the necessity of escape from the inevitable depth of Osikorobo whirlpool. Described as "the great Osikorobo whirlpool"—a pernicious pit from which "one may never find release"—the site offers an interesting paradox in the epistemology of the sea. It is a flow-less mass of water in the middle of the Niger River that captures the boats and ships with a centripetal force. This force contests and resists the flowscape of the Niger, thereby complicating time by the manner in which it stalls navigation. The decision to test the direction of the tide as means of restoring the boat from sinking in the whirlwind points to, and is an expression of, how time functions in configuring the status of natural hazards. Placing a bowl in the river—a traditional oceanographic practice in the Delta that determines the movement of the tide—Kengide observes a rightward swing to suggest a sea-bound drift: "It means ebb tide/And that we are heading to the sea." However, Olotu doubts Kengide's test because he believes that the sea and time are complicit in the tragic derailment of the boat. He asks: "What if the raft/Had swung completely round while we slept? It then would mean we are moving inland—in the direction of Odi" (97). This doubt prompts the quartet to request alternative modalities for telling time that is not implicated in the sea's order of temporal knowledge. Using Olotu's wristwatch—an object acquired through his exposure and travels to the different cities in Nigeria—Clark-Bekederemo introduces the modernist temporal paradigm:

Olutu: Why can't you rely on my watch? I bought

In Lagos—at the Kingsway stores, in fact.

Kengide: As if it is not common knowledge

Hausa men hawk the best of them in the streets

And will barter them for a pair

Of tattered trousers.

Ibobo: Now, who was it that wanted some sign

In the sky?

Ogro: I said I want to hear the crowing

Of a cockerel, welcoming in the dawn

Of another day. (98)

The play therefore provides some kind of revolt against Western scientism and modernity, offering a dramatic enactment that sought to demystify what Walter Mignolo describes as "colonial matrix of power" (Mignolo 2011, xvi). Mignolo locates the origin of western modernity in the European Enlightenment period and argues that it "serves not all humanity, but only a small portion of it that benefits from the belief that in terms of epistemology there is only one game in town" (Mignolo 2011, xii). He argues, for instance, that the notion of time is "caught and woven into the imaginary of the modern/colonial world-system." In a related argument on the evolution of material objects in the post-Enlightenment modernity, Alf Hornborg critiques the "cultural assumption that material objects are politically innocent" (Hornborg 2014, 133). Such arguments implicate the instrumentalities of time as resources of ideological alignments, making them assume positions that function as signs of cultural endorsement. This draws attention to the use of Olotu's wristwatch as a representation of modernist time, in which case it highlights how objects of colonial modernity become politically and culturally useful in the project of subjugating indigenous temporal systems. It asserts the trope of multiple temporalities and, in so doing, offers time as front for decolonial knowledge contestation. The subversion of modernity's universal logic of time is recognized in Kengide's mockery of Olotu's watch when he asks, "Has paralysis/ Ever left that brat of a machine?" and thereafter describes it as "that brat/Of a machine has probably been at work without/Stop, and so outstepped the old woman in the moon" (98). If the reference to an old woman in the moon inscribes time in the body of Delta fables, affirming time as "a category belonging to culture" as Mignolo has argued, it also elaborates on aspects of cultural interactions that undermine modernity's "denial of coevalness."

Environmentalist Rachel Carson similarly informs that "the story of how the young Earth acquired an ocean . . . is founded on the testimony of the earth's most ancient rocks, on other evidence written on the face of the earth's satellite, the moon" (Carson 1989, 3). Carson's description of the moon as a "great tidal wave" resonates with Clark-Bekederemo's enactment of the seascape in his plays. Whereas the Niger Delta lays claim to the involvement of water, the moon represents an ethereal archetype in the imagination of Delta writers and, as *The Raft* shows, inspires fables that infuse perspectives of time and water.

Read in this light, time might prove particularly useful in negotiating not only new ways of surviving the sea's turbulence, but also ways of contesting the hegemonies of colonial modernity. As Harry Garuba and other decolonial scholars remind us, the success of the modern-colonial world system is premised on how subjects, socially located on the colonial divide, are conscripted to "think epistemically like the ones in dominant positions" (Garuba 2013, 44). This assessment captures the character of Olotu, whose representation underwrites the play's engagement with early industrial modernity and its enabling epistemic systems. Once the raft sinks into the pit, the story shifts to demonstrate Olotu's estrangement from the ways of the sea by his persistence in punting the raft off the whirlpool—against Ibobo's caution: "Ten such poles, tied end to end, will not plumb the floor of Osikorobo" (102). The playwright is perhaps employing Osikorobo to open paradoxical questions of a thoroughly arcane knowledge of River Niger that is fixated and bound in mystery, highlighting the nonnegotiability of ecological relations of power and knowledge in the Delta region as we find in Kengide's assertion:

I see why we of the Delta
Never will make good. You believe all
The tales tampering with the stars
That are told you abroad, but never anyone

At home about your own rivers. Truly
We are a castaway people. (102)

In other words, the play prompts the question: with what knowledge
of time might we approach ecological perturbations in the Niger Delta
wetlands? Olotu appears to represent what Alp Hornborg describes as an
"ecologically alienated individual" (Hornborg 1998, 4), highlighting a de-
tachment from the cultural involvement with the seascape epistemology. In
representing the River Niger as a form of flowing waterbody of knowledge,
Clark-Bekederemo suggests the question of sustained experience in the occasion of rupture and alienation. Thus, Olotu's alienation from the Delta, which
the play figures in his sojourn in foreign lands—culminating in his modernist orientation, appears to offer insight into how the project of modernity
aspires to change the natural order of things. The idea of alienation is further
elaborated when the raft breaks into half, revealing that Olotu cannot swim
as he is borne away by the tide. As Kengide informs, "Mark me, the truth/Is
that he can't swim. Which of you has seen/Him bathe without his clinging
to the raft/Like a snail" (111). In this sense, Olotu contrasts with the other
boatmen, especially Kengide, who is described as "wall-gecko gone grey
at home." Perhaps Clark-Bekederemo is stating, as Mignolo has done, that,
"[T]he point . . . is not where you reside but where you dwell. Césaire and
Fanon . . . dwelled in the history of the Middle Passage, of the plantations, of
slavery and of the runaway slaves" (Mignolo 2011, xiii). Tim Ingold makes a
similar point with his "dwelling perspective," suggesting that human genes do
not account for how knowledge of place is transmitted, but rather the process
of dwelling (Ingold 1995, 77). Olotu may be an indigene of the Delta—and
may in fact have the genes of Deltaness—yet he dwells in the many cities of
his sojourn across Nigeria. His indwelling in the city appears to have stripped
him of involvement in, and connection with, the sea. He appears cut off, not
only from the spatial link with the seas, but also emptied epistemologically.
For instance, to insist on his wristwatch as the veritable way of reading the
rhythms of the tide is to lay emphasis on the universal rhetoric of modernity.
His claims of the superiority of the wristwatch—which he bought at the
Kingsway chain of stores—over indigenous temporal orientations, resonates
with what Catherine Walsh terms "subalternizing and invisibilizing other
epistemes" (Walsh 2007, 224). In this light, Olotu represents the consequence
of dislocation from place and time, completely incapable of surviving the sea
from his location in the position of "capitalism time." As Karin Amimoto
Ingersoll writes, "time is created through place in a way that place, in fact,
creates time, and neither place nor time is void" (Ingersoll 2016, 115). In *The
Raft*, time is apprehended through systems of knowledge that is enabled by
the sea's connection with the community, and thus offers a future that will

resist the tactics of colonial temporal domination. This fact aligns with what Jacques Rancière describes as the "distribution of the sensible" in which time is constituted in politics in the context of community. As he writes:

> This apportionment of parts and positions is based on a distribution of spaces, times and forms of activity that determines the very manner in which something in common lends itself to participation and in which individuals have a part in this distribution. . . . It is a delimitation of spaces and times, of the visible and the invisible, of speech and noise, that simultaneously determines the place and the stakes of politics as form of experience. Politics resolves around what is seen and what can be said about it, around who has the ability to see and the talent to speak, around the properties of spaces and the possibilities of time. (Rancière 1999, 12–13)

The way in which epochs are articulated, Imre Szeman notes, is significant in understanding the direction of the future (Szeman 2007, 805). Conversely, capturing the everyday helps to establish the formation of the present, which designates a significant outline of survivalism that can be understood in its economic and ecological ramifications; and that can be retooled to forge a front of resistance against the hegemony of colonial modernity. Niger Delta dramatists imagine the universe of waters in its quotidian involvement with the communities, and in *The Raft* Clark-Bekederemo depicts time as a cultural force that frames, and is framed, by the sea in its multiple representations of history and experience. The way the boat drifts ashore the River Niger draws attention to what Madhu Benoit describes as "temporal signifiers float[ing] loosely on a . . . stream of sliding sign fields" (Benoit 1998, 98). Hence, the travails of the boatmen drifting in the sea's turbulence—itself a sign of time in performance—remind us to capture the discourse of disaster with the immediacy of the everyday.

REFERENCES

Anderson, Martha, and Philip Peek. 2002. "Ways of the Rivers: Arts and Environment of the Niger Delta." *African Arts* 35, no. 1: 12–25.

Benoit, Madhu. 1998. "Circular Time: A Study of Narrative Technique in Arundhati Roy's *The God of Small Things.*" *World Literature Written in English*. 38.1: 98–106.

Carson, Rachel. 1989. *The Sea Around Us*. New York: Oxford University Press.

de Certeau, Michel. 1984. *The Practice of Everyday Life*. Berkeley: University of California Press.

Clark-Bekederemo, John Pepper. 1964a. "The Masquerade." In *Three Plays*. London: Oxford University Press.

———. 1964b. "The Raft." In *Three Plays*. London: Oxford University Press.

Crutzen, Paul, and Eugene F. Stoermer. 2000. "The Anthropocene," *IGBP [International Geosphere-Biosphere Programme] Newsletter* 41: 1–20.

Garuba, Harry. 2013. "On Animism, Modernity/Colonialism, and the African Order of Knowledge: Provisional Reflections." In *Contested Ecologies: Dialogues in the South on Nature and Knowledge*, edited by Lesley Green, 42–51. Cape Town: HSRC Press.

Gilbert, Helen, and Joanne Tompkins. 1996. *Post-colonial Drama: Theory, Practice, Politics*. London: Routledge.

Green, Lesley. 2013. "A Second Intervention—Space, Time, Life." In *Contested Ecologies: Dialogues in the South on Nature and Knowledge*, edited by Lesley Green. Cape Town: HSRC Press.

Hornborg, Alf. 1998. "Ecological Embeddedness and Personhood: Have We Always Been Capitalists?"*Anthropology Today* 14, no. 2: 3–5.

———. 2014. "Technology as Fetish: Marx, Latour, and the Cultural Foundations of Capitalism." *Theory, Culture and Society* 31, no. 4: 119–40.

Ingersoll, Karin Amimoto. 2016. *Waves of Knowing: A Seascape Epistemology*. Durham, NC: Duke University Press.

Ingold, Tim. 1995. "Building, Dwelling, Living." In *Shifting Contexts: Transformations in Anthropological Knowledge*, edited by Marilyn Strathern. London: Routledge.

Latour, Bruno. 2014. "Agency at the Time of the Anthropocene." *New Literary History* 45, no. 1: 1–18.

LeMenager, Stephanie. 2014. *Living Oil: Petroleum Culture in the America*. Oxford: Oxford University Press.

———. 2017. "Climate Change and the Struggle for Genre." In *Anthropocene Reading: Literary History in Geologic Times*, edited by Tobias Menely and Jesse Oak Taylor, 220–38. University Park: Penn State University Press.

Mignolo, Walter. 2011. *The Darker Side of Western Modernity: Global Futures, Decolonial Options*. Durham, NC: Duke University Press.

Quayson, Ato. 2003. *Calibrations: Reading for the Social*. Minneapolis: University of Minnesota Press.

Rancière, Jacques. 1999. *The Politics of Aesthetics*. Translated by Gabriel Rockhill. New York: Continuum.

Shewry, Teresa. 2011. "Pathways to the Sea: Involvement and Commons in Works by Ralph Hotere, Cilla McQueen, Hone Tuwhare, and Ian Wedde." In *Environmental Criticism for the Twenty-First Century*, edited by Stephanie LeMenager, Teresa Shewry, and Ken Hiltner, 247–60. New York: Routledge.

Szeman, Imre. 2007. "System Failure: Oil, Futurity, and the Anticipation of Disaster." *South Atlantic Quarterly* 106, no. 4: 805–23.

Tsaaior, James. 2005. "Poetics, Politics and the Paradox of Oil in Nigeria's Niger Delta Region." *African Renaissance* 2, no. 6: 72–80.

Tsing, Anna Lowenhaupt. 2014. *The Mushroom at the End of the World: On the Possibility of Life in Capitalist Ruins*. Princeton, NJ: Princeton University Press.

Walsh, Catherine. 2007. "Shifting the Geopolitics of Critical Knowledge: Decolonial Thought and Cultural Studies 'Other' in the Andes."*Cultural Studies* 21, no. 2–3: 224–39.

Chapter Twenty-Four

Thinking from the Southern Ocean

Charne Lavery

HOW A PENGUIN LOOKS AT THE WORLD

In early 2017, the cartographer Frans Blok published a world map centered on the South Pole. Rather than the familiar Mercator projection, which centers on Europe, this one places Antarctica at the center, surrounded by the vast expanse of the Southern Ocean. Blok explains that all projections distort whatever lies at their edges (the "Law of the Conservation of Trouble"), which is usually the Pacific and the poles. The Mercator projection, which effectively magnifies the landmasses of the Northern Hemisphere, places Europe "nicely in the middle." But what of a world map that places instead the South at the center, Antarctica in the middle? If every world map reveals a lot about whoever made it, Blok wonders if this might be "how a penguin looks at the world" (Blok 2017). For Blok the proposition that climate change may eventually make the frozen Antarctic continent more inhabitable prompts this anticipatory tilting of the globe. The resulting map (Figure 24.1), not particularly useful for Europeans, is more of a "penguin's world map" (Blok 2017).

Interestingly, a projection centered on the North Pole results in very little visible distortion, because "there simply is not that much land in the Southern Hemisphere" (Blok 2017). The Antarctic Projection, however, stretches to absurd proportions the lands of the north because they are situated all the way toward its edges, where the distortion is greatest. In positions of centrality, instead, are the countries of southern Africa, South America, and Australia and New Zealand, which retain their recognizable shape. This map of the world is also overwhelmingly blue, as a result of the vastly greater relative size of the Southern Ocean in relation to the polar seas (Blok 2017). It therefore places at the center of the world simultaneously the poorer countries of

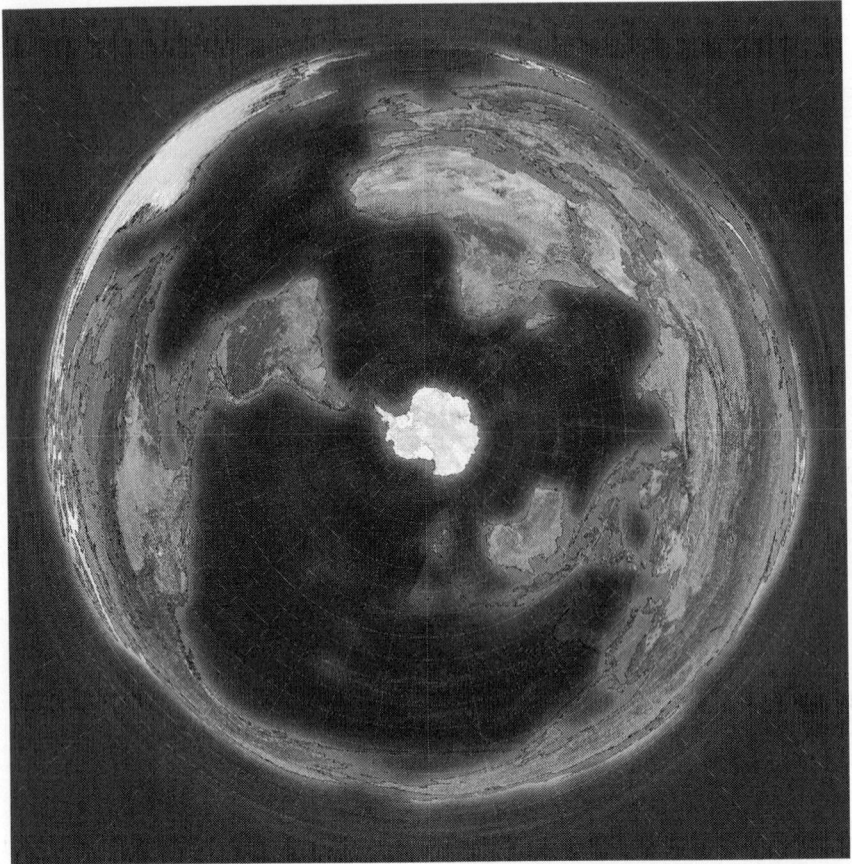

Figure 24.1. **"The Penguin Projection," from "The Antarctic Projection: a Penguin's World Map," published 11 April 2017.**
Image by designer Frans Blok.

what is known in political-economic terms as the "global South," and the geographical Southern Hemisphere with its predominating Southern Ocean.

This chapter asks if something similar might be done in conceptual rather than cartographical terms: how to bring together questions that pertain to the Global South—from the still-decolonizing countries of the world, centering social and economic justice—while also registering the interrelationship between changing global climates and the currents of the Southern Ocean—from the oceanic Southern Hemisphere, centering environmental justice? This chapter explores ways of thinking from the Southern Ocean, turning to literary and cultural texts for their imaginatively elastic approaches to this multiscalar-multispecies conundrum. Although largely uninhabited (and therefore largely unnarrated), points of contact with the Southern Ocean appear in works of fiction. Here I explore the ways in which perspectives from drifting ships in southern seas are imagined in both the most Euro-canonical of Southern Ocean literature, by Samuel Taylor Coleridge and Joseph Conrad, and also in narratives from and about the Global South, starting here with Kanishk Tharoor and looking ahead to others by Yvette Christiansë and Abdulrazak Gurnah.

THINKING FROM . . .

The economically unequal relationship between the Global North and Global South has historically been accompanied by a corresponding dominance of scholarship and thought, an unreflectively parochial Northern perspective on what should be global questions. As a corrective, anthropologists Jean and John Comaroff issued a call for the urgent need for "theory from the South," not only to retain a focus on redressing inequality, but because the South could in fact serve as a laboratory for northern futures (Comaroff and Comaroff 2015). This has become more widely, if critically, embraced as the project of "thinking from the South" (see, for instance, Hofmeyr 2018). One of its guises is the "Thinking from the South" distinguished lecture series held at the University of the Witwatersrand in Johannesburg in the first half of 2019; related projects are "writing the world from an African metropolis" (Mbembe and Nuttall 2004) and "thinking from the Cape" (Samuelson and Musila 2011).

Alongside these calls for southern and subaltern perspectives has emerged another trajectory, summed up by oceanic studies scholar Hester Blum as the challenge to "take the oceans' nonhuman scale and depth as a first critical position and principle," a call to think from the ocean (2012, 24). This is part of an oceanic turn motivated partly by rising sea levels as the most visible marker of climate change in the Anthropocene era (DeLoughrey 2017). Initially a move beyond nation- and area studies, oceanic studies is animated by an urgent awareness of environmental destruction that centers on the climate-crucial oceans and is influenced by ecocritical, new materialist, and posthumanist thought (Bystrom and Hofmeyr 2017).

One way of bringing together these two positions, the ocean and the south, is via the term *oceanicity*—a term I have borrowed from meteorology, indicating the degree to which a place is overall subject to the influence of the sea (Antonello 2017, 296). As the opposite of *continentality*, *oceanicity* becomes one way of describing a key difference between the Northern and Southern Hemispheres (as comparing the two polar projections makes clear). The

Southern Hemisphere has 20 percent more ocean than the Northern, as Meg Samuelson pointed out in "Southern Hemisphere Blues" (and elaborated in our "The Oceanic South"), and can therefore be described as the part of the world with much higher oceanicity (Samuelson 2017; Samuelson and Lavery 2019). Whereas *oceanicity* is usually employed in relation to the climate of a seaside city, this expanded definition points to the decisive significance of the Southern Ocean for regulating climate on a planetary scale (Riffenburgh 2006, 467).

. . . THE SOUTHERN OCEAN

Although oceans in general comprise "unboundedness, drift and solvency" (Blum 2013, 152), in the Southern Ocean each of these oceanic characteristics is exaggerated. Because its winds and currents can circulate endlessly without encountering a continental barrier, the Southern Ocean is characterized by perpetual movement: waves can "lap" themselves, in theory, drawing on infinitely circulating fetch (Antonello 2017, 296). Oceanographic data is drawn overwhelmingly from the Northern Hemisphere, because of the proximity of oceanographic centers and the density of shipping lines, and because the harshness of the Southern Ocean environment remains a barrier to research (Riffenburgh 2006, 467). Given that it represents a "gap in the knowledge," the Southern Ocean also then exemplifies the alienness of the sea (Helmreich 2009).[1]

   This is not only a matter of physics but meaning, as the question of nomenclature bears out. Historian Alessandro Antonello notes that the Southern Ocean only achieved a stable name as late as the 1970s, having been since its first European exploration two hundred years earlier "variously named the 'Antarctic seas,' the 'Antarctic ocean,' the 'southern seas,' the 'southern oceans,' 'The Icy Sea,' as the southerly extent of the Atlantic, Indian or Pacific Oceans" (2017, 296). Perhaps the most common confusion is between the Southern Ocean and the South Seas, cognate terms that designate quite different oceans—the tropical Pacific versus the freezing Antarctic seas (although they share properties like immense, landless expanse and frontier symbolism) (Cohen 2010, 50). As Antonello goes on to say, "the shifting names of the ocean articulate more than definitional debates, but speak to its history" (2017, 297). Although Antarctica and the Southern Ocean have been so little inhabited and traveled, their emptiness has in fact been inversely proportional to their exploitation, as histories of whaling—and prior to that, sealing—to the verge of extinction demonstrate (Antonello 2017, 315). In other words, Southern Ocean exploration has resulted in rampant overexploitation that has remained largely unnamed, and hardly narrated.

The Southern Ocean is characterized by both physical and semiotic drift, in some of its most extreme forms, which makes it a very difficult place to think about, or to think from (on which more later). Yet, as its currents and creatures are being irrevocably changed in ways that are crucial to global futures, it becomes important to try to understand the attempts. The next sections explore the nature of Southern Ocean experiences in literary works that undertake these imaginative voyages south.

LITERARY VOYAGES SOUTH

The most iconic literary treatment of Antarctic seas is Coleridge's *Rime of an Ancient Mariner*, in which an aged sailor tells a passerby about a disastrous southern voyage. The ship sets off heading south, but soon encounters a storm so intense and long-lasting that the ship is blown across the equator and clear into the Southern Ocean. There they encounter a realm of ice—"The ice was here, the ice was there, / the ice was all around" (Coleridge 1798, 9)—and remain entirely at the mercy of a relentless wind driving them farther and farther south, until an albatross appears. The bird, a symbol of good luck, brings with it a change in the wind, which, blowing now from the south back northward, allows them to finally escape the "wond'rous cauld" (Coleridge 1798, 8). Of course, the speaker kills the albatross for no given reason and the ship is becalmed in sluggish seas, no longer surrounded by endless ice but endless water: "water, water every where, / nor any drop to drink" (Coleridge 1798, 13). The ship drifts on this "wide, wide sea" (now the Pacific, the poetic version of the semantic slippage registered previously) for much of the second part of the poem, registering the horror of dead shipmates, the "slimy things" that fill the ocean, and a series of increasingly strange visions (Coleridge 1798, 13, 22).

In the "Rime," the experience of drifting in the ocean is so disorienting, in its landlessness and isolation, that it produces a kind of madness; in the end, however, it also delivers an urgent moral, which the Mariner must go from place to place repeating. His message is that the ship's curse, brought on by shooting the albatross, is finally broken by a recognition of the beauty of the "water-snakes" that swim beneath the ship, creatures of wonder and strangeness just like the albatross (Coleridge 1798, 24). The moral is a naturalist one—"he prayeth well, that loveth well, / both man and bird and beast" (Coleridge 1798, 50). It may even be described as a posthuman ethic that prefigures more contemporary environmental theories.

However, as Paul Gilroy, author of the influential *The Black Atlantic* (1993), which led theorizing about oceans and race, argues in a more recent work, posthumanist thought—linked closely to invocations of the Anthropocene—risks falling into much older habits of antihumanist thought. These have long been critiqued by the black radical tradition that, required by slave-driven capitalism, addressed the boundaries between human and nonhuman, strenuously resisting the construction of black humanity as an object of property, or indeed as "beast" (Gilroy 2018, 2–3). This uneasy relationship between environmentalism and racism is far more evident in the other most defining text of the Southern Ocean, Edgar Allan Poe's *The Narrative of Arthur Gordon Pym of Nantucket*. Coleridge's plot of the southern voyage and the drifting ship appear again in Poe but with a more overt racism, as the all-black islanders who populate the southern continent are portrayed as vicious and uncivilized, while the novel ends on a gigantic white figure that combines the whiteness of race with the whiteness of ice, a pernicious and persistent association.[2]

A more complicated, although certainly not innocent, relationship between Southern Ocean drifting and imperial race relations is addressed in a lesser-known short story by the ambivalently colonial Joseph Conrad in "Falk" (1903). The story is set, like most of Conrad's sea fiction, in an upriver harbor in the Indian Ocean. Most of the narrative is a foiled imperial romance, involving the narrating Captain, a pilot, Falk, and a Dutch girl with whom he has fallen in love. The love story is thwarted by a secret from Falk's past, which is revealed to the Captain at the end of the frame narrative, a vivid story-within-a-story of Falk's disastrous voyage on a "southern-going ship" that goes much too far south (Conrad 1946, 3).

Falk had signed up as an officer on the *Borgmester Dahl*, a steamer headed for Wellington. But, "in latitude 44° south and somewhere halfway between Good Hope and New Zealand, the tail shaft broke and the propeller dropped off" (Conrad 1903, 136). The crew falls quickly into lassitude, passing several sailing ships and whalers but waiting for a steamer to tow them to safety. Unfortunately for them, "there were very few steamers in those latitudes then" (Conrad 1903, 137). They "drove south" until they "drifted south out of men's knowledge" and "the edge of the polar ice-cap rose from the sea and closed the southern horizon like a wall" (Conrad 1903, 137). They continue to drift aimlessly on "a grey sea ruled by iron necessity and with a heart of ice" for so long that one day a man is killed in "the cold, cruel dawn of Antarctic regions," and the remaining crew turn to cannibalism (Conrad 1903, 141).

Falk is the only one who survives, and the story overall is more concerned with the long-term consequences for his life in an outpost of empire than with shipwreck narrative, a traditionally popular form (Cohen 2010). The domesticity of the ships in harbor and the stifling local politics of the port contrast starkly with, and overwhelm, the tale of gruesome survival on the wide, wide sea. The two are linked, however, by the question of racial belonging. Falk's experiences on a disabled steamer in the Southern Ocean have left him, among other things, a vegetarian. This makes him the enemy of the hotelkeeper Schomberg: "'A white man should eat like a white man, dash it all,' he burst out impetuously. 'Ought to eat meat, must eat meat. I manage to get meat for my patrons all the year round. Don't I? I am not catering for a dam' lot of coolies'" (Conrad 1903, 98). Falk betrays his race by adopting the eating habits of the racial Other, employing a cook from Madras and allowing his lascar sailors relatively free rein (Conrad 1903, 97). Conrad's treatment of the relationship between southern voyages and shifted perspectives of interspecies as well as interracial relations here is a troubled, ironic treatment, of course, but one that begins to imagine the potentially transformative, tilting effect of a journey to the south.

ALL AT SEA, AT THE BOTTOM OF THE WORLD

It is perhaps significant that all these voyages to the Southern Ocean are associated with shipwreck or breakdown, the ships driven south by the wind and left to drift on the open, icy sea. In his better-known novel, *Lord Jim*, Joseph Conrad describes the existential effects of being in a lifeboat on the open ocean:

> There is something peculiar in a small boat upon the wide sea. Over the lives borne from under the shadow of death there seems to fall the shadow of madness. When your ship fails you, your whole world seems to fail you; the world that made you, restrained you, took care of you. It is as if the souls of men floating on an abyss and in touch with immensity had been set free for any excess of heroism, absurdity, or abomination. Of course, as with belief, thought, love, hate, conviction, or even the visual aspect of material things, there are as many shipwrecks as there are men. (Conrad 1900, 88)

Despite Conrad's characteristic descriptive excess, there is a claim here about position and perspective—that bobbing in a boat, out of sight of land and over an abyss of watery depth, has the power to alter one's view of the world.

It is a position that comes close to embodying what Gilroy has called a "sea-level" perspective, his proposal for addressing the tendency in Anthropocene thought to forget "slavery's pelagic theatre of power" (Gilroy 2018, 4). One way to develop a properly "planetary humanism" that proceeds from the history of antiracist thought rather than serves as racism's mask, is to adopt a "lowly watery orientation," allowing for "glimpses of offshore humanity that might, in turn, yield an *offshore* humanism" (Gilroy 2018, 14, 8, 16, emphasis in original).

Jennifer Wenzel, too, describes the incommensurabilities of social and environmental questions by invoking a language of globe, world, and planet, asking in particular "how might one calibrate the *globe* in globalization with the *planet* at risk in environmental crisis" (2014, 19). She suggests that one way to approach the problem is to analyze moments in which "marginalized literary characters or documentary subjects situate their own precarious local condition within a broader, transnational context," a method that might "offer glimpses of a subaltern planetary subjectivity" (Wenzel 2014, 19). These are "scenes of world-imagining from below," where "belowness" involves "not only *class position*, in the familiar idiom of subalternity, but also *spatial position*: altitude and perspective in a more literal, scopic sense" (Wenzel 2014, 20). Whereas she glosses this as "imagining from the earth, from the ground," in Gilroy's version the same impetus includes a shift off-shore to describe a "lowly, watery orientation" (Gilroy 2018, 6).

But Wenzel also notes that for the Northern Hemisphere in the age of European imperialism, being high up on the map was conflated with being metaphorically and morally "above": "Europe hovers over the lands below, from a position of scopic and epistemological privilege" (Wenzel 2014, 20). In addition to class and spatial position, the attempt here is to include global position too, in the lowly, watery hemisphere—tilting to a planetary perspective from below.

OFFSHORE HUMANISM ON ICE

How then to begin to imagine a position in the Southern Ocean but also metaphorically from the South? I turn now to a collection of stories by Kanishk Tharoor, *Swimmer among the Stars*, reviewed by writer Amitav Ghosh as one example of the kind of writing that may find a way to incorporate the improbable probabilities of climate change into the limits of realist narration (Ghosh 2015). The final short story of the collection, "Icebreakers," is set entirely in the Southern Ocean. Here, the ships are disabled in a different way, not becalmed as in Coleridge or suffering from mechanical failure as in Conrad, but stuck in the ice. The story begins with a Russian research vessel, the first icebreaker, whose asthmatic Captain, "misled by weather forecasts and satellite imagery," ventured too deep down a channel of ice-free water known as a *polynya*, only to find it close around him (Tharoor 2017, 234). They call the nearest icebreaker, a Chinese vessel that comes to their rescue.

However, it gets stuck, too, followed by each of the subsequent ships: Australian, Japanese, Canadian, Swedish, Chilean, and South African (Tharoor 2017, 238). By the end of the story, all but one of the world's icebreakers form a "shrinking chain to the horizon, all tilting to one side like sleeping tramps" (Tharoor 2017, 239).

What perspective does their lowly, icy orientation, drifting with the sea ice in the far south, produce? It is tinged with environmental apocalypticism, for starters, as the icebound position speaks paradoxically of climate change—despite the more well-known effect of melting polar ice. As the onboard researchers tell the Captain, while the glaciers on the continent are precipitously melting, the sea ice around Antarctica is paradoxically growing, for unknown reasons. The story is focalized by the Russian ship's Indian photographer, who quietly records the slowly worsening situation, the ship's crews, and the aurora. The line of ships appears to her "disjointed, composed from random blocks, rising from the ice desert like the ruins of lost civilizations"—a postapocalyptic, Ozymandian vision (Tharoor 2017, 237). At the end, as the world's last icebreaker nuzzles the edge of the ice, her camera battery dies before she can record whether it succeeds or fails (Tharoor 2017, 239).

But the line of ships also forms a kind of humanist international cooperation at sea. As the line of trapped ships grows longer, "tenuous lines of exchange form on the sea ice, sashimi traded for schnapps, red wine for dried sausage" (Tharoor 2017, 238). It is a cooperation practiced on skis and snowshoes, and facilitated by the language of the sea. The Captain of the Russian and Chinese ships are able to give each other counsel because "they speak in English, which is the language of the sea, of the air, and of space, even if it will never fully conquer the land" (the language of ice, of course, is Russian) (Tharoor 2017, 232). It is a fragile but tangible set of connections, an "offshore humanism," that is reminiscent of older forms of maritime cosmopolitanism that the narrator invokes (for instance, describing the Chinese ship's crew who are mostly Filipinos as "the Ishmaels of our time") (Tharoor 2017, 232).[3] The cast of ships also reflects the mixed southern and northern make-up of the Antarctic Treaty nations, which, even though these memberships depend in turn on colonial and settler histories, signal a not entirely Eurocentric view (Antonello 2017, 310).

At various points, the trapped travelers from their different ships descend to the ice to exchange food and information with the other ships, but also to watch the aurora, take measurements, dance. In doing so, they experience directly the thickening crust of sea ice on the Southern Ocean that the "dizzying labor of NASA satellites" has not helped to predict or explain (Tharoor 2017, 230). With a lowly scopic, spatial position, they are better placed to appreciate the improbable probabilities of climate change than the high-altitude satellites (Wenzel 2014, 19). The focalizing photographer who quietly records the slowly worsening situation directs the reader's gaze increasingly northward from the southern ice. The story then goes some way to centralizing the Global South, at sea (ice) level.

AN UPSIDE-DOWN WORLD

Another story in Tharoor's collection, "Letters Home," addresses the famous medieval map of the cartographer-monk Fra Mauro that, drawn partly from Arab traditions of cosmography, places the south at the top and the north at the bottom. In this particularly gnomic story, the upside-down map highlights the image of an Indian caravel at the top of the map, which sailed from India to southern Africa and then continued west for thousands of miles. It encounters only more and more sea, missing the ice, the South Atlantic islands, and the Americas. Although, as the story ends humorously, "In this way, Indians missed the best chance they had to discover Indians," they discover the Southern Ocean instead (Tharoor 2017, 163).

The Mappa Mundi, in this story, represents another tilting of the imagined world that, like the penguin projection, highlights suppressed Southern perspectives. This chapter places this kind of upside-down, south-centered map of the world in conversation with the oceanicity of the Southern Hemisphere, posing the Southern Ocean as a problem (following Bystrom and Slaughter [2017] who do so for the South Atlantic). What region does it convene—connecting the settler states of Australia, southern Africa and Argentina, for instance as Samuelson and I suggest in "The Oceanic South" (2019)—and, alternately, what vision might it offer? Representing an opening foray in this direction, this chapter has explored literary perspectives floating or icebound in the Southern Ocean, which may serve as "scenes of world-imagining from below" and off-shore, that assume a different position from which to think more equal futures for an imperiled planet.

ACKNOWLEDGMENTS

The research was funded by the South African National Research Foundation (South African National Antarctic Programme) and the A. W. Mellon Foundation. Funding for the trip to the *Sustaining the Seas* conference in Sydney was provided by the Environmental Humanities South Centre at the University of Cape Town.

NOTES

1. For further discussion of the understudied south Atlantic, see Bystrom and Slaughter (2017); for the southern Indian Ocean, see Bremner (2015).

2. See discussion of, for instance, J. M. Coetzee's *Elizabeth Costello* in "The Oceanic South" in which an African novelist on a cruise in the Southern Ocean describes himself as "the one black face in this sea of white" (Samuelson and Lavery 2019).

3. This offshore humanism is taken off-planet in another story in the collection, "A United Nations in Space," in which each of the characters is identified only by their country name (Tharoor 2017, 59).

REFERENCES

Antonello, Alessandro. 2017. "The Southern Ocean." In *Oceanic Histories*, edited by Alison Bashford, David Armitage, and Sujit Sivasundaram, 296–318. Cambridge Oceanic Histories. Cambridge: Cambridge University Press. https://www.cambridge.org/core/books/oceanic-histories/southern-ocean/EC3E27F511E37258BC-6037D2631A387E.

Blok, Frans. 2017. "The Antarctic Projection: A Penguin's World Map." *3Develop Image Blog* (blog). April 11, 2017. http://www.3develop.nl/blog/antarctic-projection-penguins-world-map/.

Blum, Hester. 2012. "Melville and Oceanic Studies." In *The New Cambridge Companion to Herman Melville*, 22–36. Cambridge: Cambridge University Press.

———. 2013. "Introduction: Oceanic Studies." *Atlantic Studies* 10, no. 2: 151–55.

Bremner, Lindsay. 2015. "Fluid Ontologies in the Search for MH370." *Journal of the Indian Ocean Region* 11, no. 1: 8–29.

Bystrom, Kerry, and Isabel Hofmeyr. 2017. "Oceanic Routes: (Post-it) Notes on Hydro-Colonialism." *Comparative Literature* 69, no. 1: 1–6.

Bystrom, Kerry, and Joseph R. Slaughter. 2017. *The Global South Atlantic*. New York: Fordham University Press.

Cohen, Margaret. 2010. *The Novel and the Sea*. Translation/transnation. Princeton, NJ: Princeton University Press.

Coleridge, Samuel Taylor. 1798. *Lyrical Ballads: With a Few Other Poems*. J. & A. Arch.

Comaroff, Jean, and John L. Comaroff. 2015. *Theory from the South: Or, How Euro-America Is Evolving toward Africa*. New York: Routledge.

Conrad, Joseph. 1900. *Lord Jim*. Edited by Jacques Berthoud. New Edition. Oxford World's Classics. Oxford: Oxford University Press.

———. 1903. *Typhoon and Other Tales*. Edited by Cedric Watts. Revised Edition. Oxford World's Classics. Oxford: Oxford University Press.

———. 1946. *The Mirror of the Sea: Memories and Impressions; A Personal Record: Some Reminiscences*. Collected Edition of the Works of Joseph Conrad. London: Dent.

DeLoughrey, Elizabeth. 2017. "Submarine Futures of the Anthropocene." *Comparative Literature* 69, no. 1: 32–44.

Ghosh, Amitav. 2015. "A Voice for the Anthropocene." *Chrestomather* (blog). October 15, 2015. http://amitavghosh.com/blog/?p=7313.

Gilroy, Paul. 1993. *The Black Atlantic: Modernity and Double Consciousness*. London: Verso.

———. 2018. "'Where Every Breeze Speaks of Courage and Liberty': Offshore Humanism and Marine Xenology, Or, Racism and the Problem of Critique at Sea Level: The 2015 Antipode RGS-IBG Lecture." *Antipode* 50, no. 1: 3–22.

Helmreich, Stefan. 2009. *Alien Ocean: Anthropological Voyages in Microbial Seas*. Berkeley: University of California Press.

Hofmeyr, Isabel. 2018. "Against the Global South." In *The Global South and Literature*, edited by Russell West-Pavlov, 307. Cambridge: Cambridge University Press.

Mbembe, Achille, and Sarah Nuttall. 2004. "Writing the World from an African Metropolis." *Public Culture* 16, no. 3.

Riffenburgh, Beau. 2006. *Encyclopedia of the Antarctic*. New York: Routledge.

Samuelson, Meg. 2017. "Southern Hemisphere Blues." Presented at Sustaining the Seas: Fish, Oceanic Space and the Politics of Caring, Sydney, Australia, December 11–13.

Samuelson, Meg, and Charne Lavery. 2019. "The Oceanic South." *English Language Notes* 57, no. 1: 37–50.

Samuelson, Meg, and Grace Musila. 2011. "Locations and Locutions: Which Africa, Whose Africa?" *Social Dynamics* 37, no. 3: 424–26.

Tharoor, Kanishk. 2017. *Swimmer among the Stars*. New York: Picador.

Wenzel, Jennifer. 2014. "Planet vs. Globe." *English Language Notes* 52, no. 1: 19–30.

Index

abalone: extinction pressures, 12, 14,
20; as food, 100, 103; habitat, 239,
245; poaching, 174, 178; red tide,
11, 23, 239
Aboriginal people, 43–44, 213;
connections to landscape, 46, 51–52,
54; dispossession of, 215
acidification. *See* climate change
activism, 18, 21, 31, 46–48, 285. *See
also* protests
advocacy: campaigns, 127, 169n2, 197,
281, 283–84; groups, 129, 165
Africa, 34; Lake Chad, 190–91;
southeastern coast, 204; and slavery,
33–34, 304, 312–13. *See also*
Madagascar; Mauritius; Niger Delta;
Seychelles; South Africa
Agent Orange, 219, 223
Alaska, 132
albatross, 135n1, 142, 311
algae, 20, 188, 264, 281, 285
algal bloom, 14, 19–20. *See also*
harmful algal blooms
Antarctica, 307, 310; in literature,
311–12, 315
anthropocene, 17, 20, 22–24, 30–31,
208; architectural responses, 258;
and the everyday, 301; and fish
stocks, 142; Gilroy, 312–13; and

temporality, 297; and seafood, 127–
28; and water, 216
Aotearoa. *See* Māori; New Zealand
apartheid, 11, 20, 183n4
aquaculture, 115–16, 120, 158, 188,
257; Australian, 81; Chinese, 29, 38;
and climate change, 190; Indonesian,
114
aquaecology, 186, 191, 196–97. *See
also* Mills, Elyse
architecture, 255, 258, 260n1, 264. *See
also* floating structures the Arctic,
93, 145
army. *See* military
Arnold, Chester L. Jr., and Gibbons, C.
James, 213–14
artificial reefs, 255; building materials,
265; South Korean, 261–64, 268–70
arts and environmental campaigns,
283–84
athwart, 15–16
Atlantic Fishery Commission, 145,
194
Atlantic Ocean: overfishing, 188; social
history, 33–34; tides, 299
Australia, 53; biosecurity, 78;
Blackwattle Bay, 44–45, 47,
51–52; border defense policy,
208; carbon emissions, 290n4;

About the Authors and Artists

Henry Obi Ajumeze holds a PhD in African studies from the University of Cape Town, specializing in the interdisciplinary connection between drama, ecology, and petroculture. He is currently working on a book project titled *Material Ensembles: African Theatre and Drama in the Posthuman Times*, which rethinks the anthropocentric culture of theater assemblages by exploring the life worlds of the nonhuman materials in the meaning-making process of theater production. Born in the Delta State of Nigeria, Henry is the author of the poetry collection *Dimples in the Sand*.

Will Allen is an independent systems scientist and action researcher. He has more than twenty-five years of experience in sustainable development and natural resource management. Through his work, he seeks to bridge local, indigenous, and organizational perspectives, and help diverse groups develop a shared understanding around goals, actions, and indicators. He also manages the Learning for Sustainability website—https://learningforsustainability.net/—as an international clearinghouse pointing to online resources around collaboration and adaptation.

Mialy Andriamahefazafy is a PhD candidate in human geography within the Institute of Geography and Sustainability at the University of Lausanne, Switzerland. Her research interests focus on the political ecology of tuna fisheries in the Western Indian Ocean, especially in Madagascar, Mauritius, and Seychelles. She explores the politics of access to tuna resources, narratives around resources use, materiality, and regional cooperation.

João Afonso Baptista is an anthropologist based at the Institute of Social Sciences in the University of Lisbon. He is the author of the book *The Good Holiday: Development, Tourism, and the Politics of Benevolence in Mozambique* (2017), and his articles have been published *in Environmental Humanities, American Anthropologist,* and *Time and Society.*

Kate Barclay is professor of global studies at the University of Technology, Sydney. She researches the social aspects of fisheries, mainly in the Asia Pacific region. Since the late 1990s, she has researched the sustainable development of tuna resources in the Pacific Islands. Recent projects include multidisciplinary work to evaluate the social and economic contributions of fisheries and aquaculture to coastal communities, governance analyses of global supply chains for beche de mer and tuna, and improving consideration of gender equality and social inclusion in fisheries and aquaculture.

Paula Blackett is an environmental social scientist with National Institute of Water and Atmospheric Research. She specializes in mechanisms and processes to establish dialogue (e.g., engagement, collaboration, coproduction of knowledge, coinnovation, colearning, and serious games) between scientists, managers, and the community to understand challenges and achieve negotiated outcomes and solutions in environmental management and practice change more generally.

Kassandra Bossell is a multidisciplinary artist, working across sculpture and installation. She is concerned with nature and the human role within it, exploring ideas of transformation, interdependent connectivity, and collaborative systems. Working with a diverse range of materials, she draws on philosophy and science to offer new perspectives on interspecies relationships, reflected in fictional forms that float between the familiar and the uncanny.

Arum Budiastuti is lecturer in the Faculty of Humanities at Universitas Airlangga, Indonesia, and is currently completing her PhD in the Department of Gender and Cultural Studies at the University of Sydney. Sponsored by Indonesia Endowment Fund for Education, her PhD project looks at the transformation of Halal and Muslim identity through Halal food regulation, representation, and consumption in contemporary Indonesia. She is interested in research topics related to religion and culture, consumption studies, science and technology studies, and environmentalism.

Alexandra H. Campbell is a marine ecologist, science communicator, and lecturer in biology at the University of the Sunshine Coast. She is interested in solutions-focused research that solves environmental problems and addresses societal needs.

Nádia Carvalho Nunes has a master's degree in anthropology and is currently a PhD student in sociology in the Institute of Social Sciences at the University of Lisbon. Her PhD research, funded by Fundação para a Ciência e Tecnologia (FCT) [SFRH/BD/140401/2018], focuses on veganism and identity. Previously she was a research assistant in an Environmental and Social Research Council–funded project on food freshness, "Enacting Freshness in the UK and Portuguese Agri-food Sectors" (award number ES/N009649/1), led by Professor Peter Jackson.

Michaelie Crawford and **Jennifer Turpin**, of Studio TCS, are award-winning public artists. They create large-scale public, participatory, community, and environmental artworks and have a particular focus on developing artworks for environmental restoration projects. They aim to reconnect people with nature, in place and with community through the poetics of art. Their projects seek to reveal the invisible, remember the forgotten, and imagine the future. They aspire to facilitate inclusive, sustainable, and innovative change for communities in today's increasingly complex and urbanized world.

Kate Davies is a social scientist with eight years of experience working on collaborative and interdisciplinary research projects that inform coastal and marine governance and management at multiple scales. She is passionate about improving social and ecological outcomes, especially reducing vulnerabilities and risks for indigenous and local communities, through the coproduction of knowledge and practice. Much of her current work focuses on participatory efforts to address cumulative effects and implement ecosystem-based management.

Safina Echa studies geography at the University of Antsiranana in Madagascar. She has an interest in the socioeconomics of fisheries in the North of Madagascar. She has been undertaking surveys within fishing communities on their livelihoods and relations to the environment.

David M. Evans is professor of material culture at the University of Bristol. His research is located broadly within economic sociology and the sociology of consumption with specific interests in food and sustainability. His current work (funded by the Engineering and Physical Sciences Research Council) is focused on single-use plastics—packaging in particular—and their role in processes of economic organization. He is also researching and writing a book on supermarkets and social change.

Sonia Garcia Garcia is a PhD candidate at the University of Technology, Sydney. Her research explores regulatory pathways to bridge the disconnection in Australia between the regulation of sustainability in the harvest and in the postharvest space. With a background in literature and linguistics and a former career in international arts policy, she is particularly interested in how policy change may result from shifts in the definitions of problems and their solutions, from our collective negotiations over the meaning of things.

Leah Gibbs is a social scientist working on human-environment relations and has done extensive work on shark management in Australia.

Bruce Glavovic is Earthquake Commission Chair in Resilience and Natural Hazards Planning at Massey University. His research centers on the role of governance in building resilient and sustainable communities, with a focus on coastal communities and the roles of land-use planning, collaboration, and conflict resolution, and bridging the science-policy-practice interfaces. He is cochair of the Scientific Steering Committee of Future Earth Coasts (2016–2019). He is coordinating lead author of the chapter on sea-level rise in the Intergovernmental Panel on Climate Change's (IPCC) *Special Report on the Ocean and Cryosphere in a Changing Climate* (2017–2019); lead author of the chapter on Climate Resilient Development Pathways; and cross-chapter paper lead for the paper on Cities and Settlements by the Sea, in the IPCC's Working Group II contribution to the *Sixth Assessment Report* (2019–2021).

Lesley Green is professor of anthropology and deputy director of Environmental Humanities South at the University of Cape Town. A former Fulbright Scholar at the Science and Justice Research Center at the University of California at Santa Cruz and a Mandela Fellow at the W. E. B. Du Bois Research Institute at Harvard, her research focuses on understanding and strengthening justice-based environmental governance in Southern Africa.

   She is the editor of *Contested Ecologies: Dialogues in the South on Nature and Knowledge* (2013), coauthor of *Knowing the Day, Knowing the World* (2013), and author of *Rock | Water | Life: Ecology and Humanities for a Decolonising South Africa* (forthcoming 2020).

Alison Greenaway attends to coproduction of sustainable development knowledge and practice. She facilitates and evaluates processes that foster collective deliberation to address complex issues. Alison is currently exploring technologies for invasive species management, implementation of large-scale biodiversity restoration, pest control across public-private boundaries, and marine ecosystem-based management.

Jennifer Mae Hamilton is a teacher and researcher exploring the cultural politics of the climate crisis. Her chief interest is in the way stories about the weather are constructed in relation to human political ambitions. Her first book surveyed the changing significance of the storm in *King Lear*. *This Contentious Storm: An Ecocritical and Performance History of* King Lear (2017) is available via Open Access through Bloomsbury Collections. She is currently a lecturer in English literary studies at the University of New England on unceded Anaiwan land in Armidale, Australia. She has collaborated with Astrida Neimanis in developing concepts like "weathering" and "composting"; these concepts are tools to build and maintain an inclusive feminist ethos within contemporary environmental humanities scholarship.

Sarah M. Hamylton is a researcher in spatial analysis of coastal environments, with a specific focus on the Great Barrier Reef.

Daniel Hikuroa (Ngāti Maniapoto, Waikato-Tainui) is an earth system scientist and established world expert on weaving indigenous knowledge and science to realize the dreams and aspirations of the communities he works with. He works from flax roots community right through government, and is currently a member of the Environmental Protection Authority, statutory Māori advisory—Ngā Kaihautū Tikanga Taiao, Watercare and Pāmu Environmental Advisories, Pūniu River Care Board, Ngāti Whātua Ōrākei Kahui Rangahau (Research Committee), and various other hapū and iwi advisories. Dan was a member of the government Freshwater Reform, Science and Technical Advisory Group; reviewed the National Climate Change Risk Assessment framework; and assisted Te Papa with their recently launched Te Taiao—Nature exhibition.

Peter Jackson is professor of human geography at the University of Sheffield and codirector of the University Research Institute for Sustainable Food. His research focuses on the geographies of consumption, with a particular interest in food. Recent projects include an Economic and Social Research Council–funded project on the UK and Portuguese agrifood sectors and an European Research Area–Net project on food, convenience, and sustainability.

Kate Johnston is a postdoctoral research fellow at the Sydney Environment Institute working on the Australian Research Council—funded project Food-Lab Sydney (2018–2020) with partners including City of Sydney, Technical and Further Education New South Wales, and FoodLab Detroit. She previously worked in the Department of Gender and Cultural Studies (University of Sydney) on the Sustainable Fish Lab project. Her research interests include environmental and food justice, food systems, sustainability, blue humanities, and experimental and interdisciplinary methodologies.

Lana Kajlich is a socioecologist looking at the spread of stewardship actions for community members taking part in marine habitat restoration. Lana has a strong background in education and instructional design. She has a master's in science with research on mined land rehabilitation and is currently a PhD candidate at the University of New South Wales–Sydney. Her broader research interests include scientific storytelling for public audiences, education, and broader policy.

Christian Kull is a geographer within the Institute of Geography and Sustainability at the University of Lausanne, Switzerland. He has a particular interest in the social dimensions of environmental change in developing countries, islands, and highlands. His fieldwork has tended to concentrate around the Indian Ocean rim, particularly Madagascar but also eastern and southern Africa, India, Australia, and beyond.

Janet Laurence explores notions of art, science, imagination, memory, and loss examining our physical, cultural, and conflicting relationship to the natural world through both site-specific, gallery, and museum works. Working in varying mediums, Laurence creates immersive environments that navigate the interconnections between life world. Laurence has been a recipient of Rockefeller, Churchill, and Australia Council Fellowships, and the Alumni Award for Arts, University of New South Wales. Laurence was a trustee of the Art Gallery of New South Wales, a former board member of the Visual Arts Board (VAB) of the Australia Council, and was visiting fellow at the New South Wales University Art and Design. Laurence was the Australian representative for the COP21/FIAC, Artists 4 Paris Climate 2015 Exhibition, exhibiting a major work—Deep Breathing: Resuscitation for the Reef—at the Muséum National D'Historie Naturelle, in Paris, France. For her full biography, go to http://www.janetlaurence.com/biography/.

Charne Lavery is a researcher at the Wits Institute for Social and Economic Research at the University of the Witwatersrand in Johannesburg. She is co-investigator on the project "Oceanic Humanities for the Global South" (www.oceanichumanities.com) and her recent research publications include work on Indian Ocean fiction, Afrofuturist ecological imaginaries, and representations of Africa's relationship to Antarctica.

Nancy Lee wrote her PhD on the effects of media attention on our expectations of chefs and their work. She researches and writes about food and the ways that food media shape our understanding of what, how, and where we eat, and the people who cook and grow our food. She currently works in research strategy and translation at the University of Sydney.

Erena Le Heron's wide-ranging interests are connected through a focus on narrative, land, and ocean, and the relationships between them. She critically engages with practices of ethics-of-care, more-than-human agency, narrative power, and assemblage thinking. Theoretical contributions focus on reimagining, reassembling, and remediating economy-environment relations.

Richard Le Heron is professor of geography in the School of Environment at the University of Auckland, past chair of the International Geographical Union Research Commission on the Dynamics of Economic Spaces, and former vice president (humanities and social science) of the Royal Society of New Zealand. His early research drew on the geography of political economy literature to explore capitalist accumulation under globalizing conditions and investigate restructuring in pastoral, forestry, and horticultural industries and organizations in New Zealand. This research was reinvigorated and redirected as he engaged deeply with the provocations of the governmentality, Actor Network, assemblage, and other poststructural literatures. His recent research, as part of the Biological Economies project and the Participatory Processes project in the Sustainable Seas National Science Challenge is transdisciplinary, across science and social sciences. It includes coimagining and applying socioecological knowledge in conditions of the Anthropocene, the cogovernance potential of enactive knowledge under uncertainties, rural and marine economies as sites of progressive experimentation, transitioning challenges to a blue-economy, and new generation institutions and coagency capacities to embrace land-coast-marine interactions, interdependencies, and challenges.

Pamima Leste works in the agriculture sector in Mauritius. In 2018, she obtained her bachelor's degree with honors in agricultural science and technology at the University of Mauritius. Her dissertation was about "the opportunities and challenges regarding the adoption of biopesticides by small and medium scale vegetable producers in Mauritius." She has a special interest in marine ecology and agroecology.

June Logie, research associate for sustainable seas in the School of Environment at the University of Auckland, is a graduate of Auckland and Massey universities with a background in geographical education, business administration, and coastal and environmental management. Life fellow of the New Zealand Geographical Society, she is a past vice president and executive manager and is currently co–vice president and secretary of the Auckland branch.

Alana Mann is chair of the Department of Media and Communications, Faculty of Arts and Social Sciences, University of Sydney, Australia, and key researcher in the Sydney Environment Institute. Her research focuses on the communicative dimensions of citizen engagement, participation, and collective action in food systems planning and governance. She is chief investigator on the Australian Research Council–funded project FoodLab Sydney.

Adam Marcus is a licensed architect and educator based in Oakland, California. He directs Variable Projects, an independent architecture practice, and he is a partner in Futures North, a public art collaborative dedicated to exploring the aesthetics of data. Adam is currently associate professor of architecture at California College of the Arts (CCA), where he teaches design studios in design computation and digital fabrication, codirects CCA's Architectural Ecologies Lab, and collaborates with CCA's Digital Craft Lab.

Ezequiel M. Marzinelli is a marine ecologist and lecturer at the University of Sydney. Ziggy's research focuses on understanding the processes that generate, maintain, and affect marine ecosystems. He uses this information to develop sensible, practical solutions to environmental problems, for example, via restoration of degraded habitats. Ziggy has done research in temperate, tropical, and polar ecosystems around the world, and is passionate about teaching and communicating his research to the general public.

Astrida Neimanis is senior lecturer in the Department of Gender and Cultural Studies at the University of Sydney, on Gadigal land, in Australia. Her work focuses on water, weather, and bodies from intersectional feminist perspectives. Astrida's most recent book is *Bodies of Water: Posthuman Feminist Phenomenology* (2017).

Rob Nicholls is senior lecturer at the University of New South Wales Business School. His research interests encompass competition law and policy as well as the regulation of networked industries and the financial services sector. Rob has had a thirty-year career concentrating on competition, regulation, and governance, particularly in networked industries. He is the independent telecommunications adjudicator in a regime established to deal with wholesale disputes arising over both legacy services and migration to the national broadband network in Australia.

Clare Nicholson is a PhD candidate at the University of New South Wales Art and Design, Sydney, where she graduated with a master of visual arts in 2015 and was nominated for the Dean's Award for academic excellence. Nicholson has held solo shows, been involved in numerous group exhibitions, and presents at academic conferences. Her art is held in private collections, both within Australia and overseas.

Marieke Norton is a postdoctoral researcher at the University of Cape Town and master's program convenor for the African Climate Change and Development Initiative. Her research interests focus on the relational interaction between the environment and humanity, with a particular focus on social-ecological marine systems in times of change.

Susanne Pratt, a researcher, educator, artist and techno-scientific muser, explores how creative practice can foster social and environmental responsibility, with an emphasis on improving environmental health and collective flourishing. She is currently based in the Faculty of Transdisciplinary Innovation at the University of Technology Sydney where she cofounded the xFutures Lab.

Elspeth Probyn is professor of gender and cultural studies at the University of Sydney. She has published several groundbreaking monographs including *Sexing the Self* (1993), *Outside Belongings* (1996), *Carnal Appetites* (2000), *Blush: Faces of Shame* (2006), and *Eating the Ocean* (2016). She is currently building a new project on fish markets, gender, ethnicity, and labor in the Global South.

Rosemary Rayfuse is Scientia Professor in Law in the Faculty of Law at University of New South Wales, Sydney. She is a fellow of the Academy of the Social Sciences in Australia and conjoint professor in the Faculty of Law at Lund University, Sweden. She researches and teaches in the area of public international law in general and more specifically in the law of the sea, and has published widely on issues of oceans governance, high seas fisheries, protection of the marine environment in areas beyond national jurisdiction, and climate change and the oceans. She is on the editorial or advisory boards of a number of international law journals, is a member of the International Union for the Conservation of Nature Commission on Environmental Law, and Chair's Nominee on the International Law Association's Committee on International Law and Sea-Level Rise.

Grandeza is an art and architecture collective that operates between the fields of spatial practice, design, cultural production, and pedagogical exploration. Grandeza is also a research group formed by architects and academics (**Amaia Sanchez-Velasco, Jorge Valiente Oriol,** and **Gonzalo Valiente**) from the Faculty of Design Architecture and Building at the University of Technology Sydney. Their research and creative practice detects, denounces, and challenges the transformative violence that late-capitalism practices apply over subjects, spaces, and ecologies. In 2019, they curated the Golden Bee awarded Australian Pavilion at the XXII International Exhibition Triennale di Milano, in collaboration with Miguel Rodriguez Casellas. Their installation "Teatro Della Terra Alienata" addressed the urgency of current social and ecological threats at the Great Barrier Reef.

SCAPE is a design-driven landscape architecture and urban design studio based in New York. SCAPE believe landscape architecture can enable positive change in communities through the creation of regenerative living infrastructure and public landscapes. They work to integrate natural cycles and systems into environments across all scales, from the urban pocket-park to the regional ecological plan. They do this through diverse forms of landscape architecture—built landscapes, planning frameworks, research, books, and installations—with the ultimate goal of connecting people to their immediate environment and creating dynamic and adaptive landscapes of the future.

Emma L. Sharp is a social scientist in geography, focusing on food politics, gender, care, and alternative economies. She is based in the School of Environment at the University of Auckland. She is associate investigator at Te Pūnaha Matatini—The Centre for Complex Systems and Networks, a Centre of Research Excellence hosted at the University of Auckland.

Alexandra Söderlund is a digital communications specialist, with a passion for science and the environment. She completed an honors thesis examining the communication of climate change in the Australia media landscape and has worked at some of Australia's leading environmental nongovernmental organizations, including Solar Citizens, World Wildlife Fund–Australia, and the Climate Council.

Peter D. Steinberg is director and CEO of the Sydney Institute of Marine Science, professor of biology at University of New South Wales, Sydney, and visiting professor at the Singapore Centre for Environmental Life Sciences Engineering at Nanyang Technological University in Singapore. He has thirty years of experience in a diversity of biological and ecological fields and has been a Fulbright Scholar, a Queen Elizabeth II Fellow, and CEO of a publicly listed biotechnology company. His research interests include environmental change and coastal ecology, kelp and seaweed ecology, bacterial biofilm biology, ecoengineering, and restoration ecology.

Patsy Theresine is a senior research technician with the Seychelles National Parks Authority. She has more than five years of experience in the marine environmental service industry and marine research. Her team is responsible for the management of all the marine and national parks of the Seychelles islands. Patsy has a wealth of knowledge in conducting conservation monitoring programs, leading scientific research, and managing protected areas.

Monica Truninger is a sociologist and researcher at the Instituto de Ciências Sociais da Universidade de Lisboa (Portugal). Her research focuses on food, consumption, and sustainability. She led the Portuguese teams of international research projects funded by Horizon 2020, 7th Framework Programme for Research and Technological Development (FP7), Environmental Research Council, and Environmental and Social Research Council. Her most recent books include *Alimentação em Tempos de Crise: Consumo e Segurança Alimentar nas Famílias Portuguesas* (Food in Times of Hardship: Consumption and Food Security in Portuguese Families, written in Portuguese), coauthored with A. Horta, Sónia G. Cardoso, Fábio R. Augusto, José Teixeira, and Ana Fontes, published in 2019; and *The Routledge Handbook on Consumption*, coedited with Margit Keller, Bente Halkier, and Terhi-Anna Wilska and published in 2017.

Adriana Vergés is a marine ecologist based at University of New South Wales, Sydney, where she leads a marine conservation research group. She has worked in tropical coral reefs and temperate ecosystems from around the world and much of her research is experimental and takes place underwater, with a scuba tank strapped behind her back. She is passionate about communicating science to the wider public, especially through films, art, and new media.

Kim Williams and **Lucas Ihlein** are artists who have been collaborating with farmers and community members in Queensland, Australia, exploring the relationship between sugarcane farming and the health of the Great Barrier Reef.

Georgina Wood hails from Sydney's northern beaches and has been exploring the sea all of her life. She is an avid fan of marine life and the imaginative worlds that underwater places inspire. George holds a bachelor of arts and a bachelor of science from the University of Sydney and is currently completing her PhD in marine ecology at University of New South Wales, Sydney. Much of her time is spent underwater working on the reforestation of Sydney's lost crayweed forests and investigating how seaweed genetics influence restoration success. George is also a passionate science communicator—if she isn't off getting wet in the ocean you can often find her giving talks about it to the general public.